AF598085

Methods in Molecular Biology™

Series Editor
John M. Walker
School of Life Sciences
University of Hertfordshire
Hatfield, Hertfordshire, AL10 9AB, UK

For further volumes:
http://www.springer.com/series/7651

Structural Genomics

General Applications

Edited by

Yu Wai Chen

Randall Division of Cell & Molecular Biophysics, King's College London, London, UK

Editor
Yu Wai Chen
Randall Division of Cell
& Molecular Biophysics
King's College London
London, UK

ISSN 1064-3745 ISSN 1940-6029 (electronic)
ISBN 978-1-62703-690-0 ISBN 978-1-62703-691-7 (eBook)
DOI 10.1007/978-1-62703-691-7
Springer New York Heidelberg Dordrecht London

Library of Congress Control Number: 2013950895

Printed on acid-free paper

Humana Press is a brand of Springer
Springer is part of Springer Science+Business Media (www.springer.com)

Preface

More than a decade has lapsed since the ambitious Protein Structure Initiative (PSI) was launched by the U.S. National Institute of General Medical Sciences in 2000. With the initial enthusiasm curbed, the outcome of Structural Genomics (SG) can now be more proportionally assessed within an established context. To the researchers who are not directly involved in SG projects, they have been observing with sceptical eyes, pondering on the justification of these expensive endeavours. While the grand objective of populating the "protein structure universe" is yet far from completed, it is undeniable that alongside the course of pursuing this goal, the field of SG has produced many technological advances that transform and accelerate protein production, structural determination and analysis (refer to PSI-Nature StructuralBiology Knowledgebase Technology Portal; http://technology.lbl.gov). In this "yet another" SG-themed book, I steered clear of collecting interim reports of SG centres, as these are regularly updated in the literature. While staying close to the spirit of SG, this volume uniquely emphasises on the benefits to the wider structural research community. It is meant to strike a balance and fill some gaps—the target reader is an "average" structural biologist in a small or medium-sized laboratory.

The topics are grouped under three parts: (I) the cloning and production of proteins for structural studies, (II) experimental methods and (III) computational methods and data analysis. Half of this book is devoted to the first part, as recombinant protein production remains a major bottleneck in many structural projects. For experimental methods, I intentionally brought in a range of complementary technologies. As a result of high-throughput practices, structural data is generated at an ever-increasing rate. This calls for improved quality control and creative computational tools for data interpretation and visualisation—these topics are grouped into the third part. Overall, the spectrum of topics reflects the trend towards tackling more ambitious challenges of studying macromolecular machineries and complexes.

In compiling this volume, I witnessed the generosity of the SG community to share experiences and methods. Some authors were keen to make their work more readily accessible beyond this book. In the end, the outcome is most satisfactory: it represents a global effort with a shared vision. I would like to thank all the authors for their contributions. There are a few who were eager to contribute but were not able to; to those I express my gratitude all the same.

London, UK *Yu Wai Chen*

Contents

Contributors

THOMAS B. ACTON • *Center for Advanced Biotechnology and Medicine, Northeast Structural Genomics Consortium, Rutgers University, Piscataway, NJ, USA*
EMILY T. BEEBE • *Department of Biochemistry, University of Wisconsin-Madison, Madison, WI, USA*
ITXASO BELLON-ECHEVERRIA • *European Molecular Biology Laboratory, Grenoble Outstation, Grenoble, France; Unit of Virus Host Cell Interactions, UJF-EMBL-CNRS, Grenoble, France*
IMRE BERGER • *European Molecular Biology Laboratory, Grenoble Outstation, Grenoble, France; Unit of Virus Host Cell Interactions, UJF-EMBL-CNRS, Grenoble, France*
CHRISTOPH BIENIOSSEK • *European Molecular Biology Laboratory, Grenoble Outstation, Grenoble, France; Unit of Virus Host Cell Interactions, UJF-EMBL-CNRS, Grenoble, France*
JEROME BOISBOUVIER • *Institut de Biologie Structurale Jean-Pierre Ebel, CEA, Grenoble, France; Institut de Biologie Structurale Jean-Pierre Ebel, CNRS, Grenoble, France; Institut de Biologie Structurale Jean-Pierre Ebel, Université Joseph Fourier – Grenoble 1, Grenoble, France*
MASSIMILIANO BONOMI • *Department of Bioengineering and Therapeutic Sciences, Department of Pharmaceutical Chemistry and California Institute for Quantitative Biosciences (QB3), University of California San Francisco, San Francisco, CA, USA*
NICOLA A. BURGESS-BROWN • *Structural Genomics Consortium, University of Oxford, Oxford, UK*
SUSAN D. CHAPPLE • *Iontas Ltd., Cambridge, UK*
YU WAI CHEN • *Randall Division of Cell & Molecular Biophysics, King's College London, London, UK*
BEN CHISNALL • *King's College London, London, UK*
FLORENT CIPRIANI • *European Molecular Biology Laboratory, Grenoble Outstation, Grenoble, France; Unit of Virus Host Cell Interactions, UJF-EMBL-CNRS, Grenoble, France*
SCOTT CLASSEN • *Physcial Bioscience Division, Lawrence Berkeley National Laboratory, Berkeley, CA, USA*
ELODIE CRUBLET • *Institut de Biologie Structurale Jean-Pierre Ebel, CEA, Grenoble, France; Institut de Biologie Structurale Jean-Pierre Ebel, CNRS, Grenoble, France; Institut de Biologie Structurale Jean-Pierre Ebel, Université Joseph Fourier – Grenoble 1, Grenoble, France*
ZBIGNIEW DAUTER • *Synchrotron Radiation Research Section, Argonne National Laboratory, National Cancer Institute, Argonne, IL, USA*
MARCIN J. DOMAGALSKI • *Department of Molecular Physiology and Biological Physics, University of Virginia, Charlottesville, VA, USA*

IAN M. DONALDSON • *The Biotechnology Centre of Oslo, University of Oslo, Oslo, Norway; Department of Molecular Biosciences, University of Oslo, Oslo, Norway*
FLORINE DUPEUX • *European Molecular Biology Laboratory, Grenoble Outstation, Grenoble, France; Unit of Virus Host Cell Interactions, UJF-EMBL-CNRS, Grenoble, France*
KEVIN N. DYER • *Physcial Bioscience Division, Lawrence Berkeley National Laboratory, Berkeley, CA, USA*
MICHAEL R. DYSON • *Iontas Ltd., Cambridge, UK*
BRIAN G. FOX • *Department of Biochemistry, University of Wisconsin-Madison, Madison, WI, USA*
OPHER GILEADI • *Structural Genomics Consortium, University of Oxford, Oxford, UK*
SUSANNE GRÄSLUND • *Structural Genomics Consortium, University of Toronto, Toronto, Ontario, Canada*
CHARLES GREENBERG • *Department of Bioengineering and Therapeutic Sciences, Department of Pharmaceutical Chemistry, and California Institute for Quantitative Biosciences (QB3), University of California San Francisco, San Francisco, CA, USA*
MICHAL HAMMEL • *Physcial Bioscience Division, Lawrence Berkeley National Laboratory, Berkeley, CA, USA*
MARJORIE M. HARDING • *Centre for Translational and Chemical Biology, University of Edinburgh, Edinburgh, UK*
FRANZ HERZOG • *Department of Biochemistry and Gene Center, Ludwig-Maximilians-Universität München, Munich, Germany*
KUN-YI HSIN • *Open Biology Unit, Okinawa Institute of Science and Technology Graduate University, Okinawa, Japan*
YUANPENG JANET HUANG • *Center for Advanced Biotechnology and Medicine, Northeast Structural Genomics Consortium, Rutgers University, Piscataway, NJ, USA*
GREG L. HURA • *Physcial Bioscience Division, Lawrence Berkeley National Laboratory, Berkeley, CA, USA*
COURTNEY JOHNSON • *King's College London, London, UK*
MICHAEL A. KENNEDY • *Department of Chemistry and Biochemistry and the Northeast Structural Genomics Consortium, Miami University, Oxford, OH, USA*
RIME KERFAH • *Institut de Biologie Structurale Jean-Pierre Ebel, CEA, Grenoble, France; Institut de Biologie Structurale Jean-Pierre Ebel, CNRS, Grenoble, France; Institut de Biologie Structurale Jean-Pierre Ebel, Université Joseph Fourier – Grenoble 1, Grenoble, France*
YAVUZ KULABEROGLU • *King's College London, London, UK*
VIOLAINE LANTEZ • *Institut de Biologie Structurale Jean-Pierre Ebel, CEA, Grenoble, France; Institut de Biologie Structurale Jean-Pierre Ebel, CNRS, Grenoble, France; Institut de Biologie Structurale Jean-Pierre Ebel, Université Joseph Fourier – Grenoble 1, Grenoble, France*
KEREN LASKER • *Department of Bioengineering and Therapeutic Sciences, Department of Pharmaceutical Chemistry, and California Institute for Quantitative Biosciences (QB3), University of California San Francisco, San Francisco, CA, USA*
PRAVIN MAHAJAN • *Structural Genomics Consortium, University of Oxford, Oxford, UK*
SHIN-ICHI MAKINO • *Department of Biochemistry, University of Wisconsin-Madison, Madison, WI, USA*
VINCENT MARIAULE • *European Molecular Biology Laboratory, Grenoble Outstation, Grenoble, France; Unit of Virus Host Cell Interactions, UJF-EMBL-CNRS, Grenoble, France*

JOHN L. MARKLEY • *Department of Biochemistry, University of Wisconsin-Madison, Madison, WI, USA*
JOSÉ A. MÁRQUEZ • *European Molecular Biology Laboratory, Grenoble Outstation, Grenoble, France; UJF-EMBL-CNRS, Unit of Virus Host Cell Interactions, Grenoble, France*
GUILLAUME MAS • *Institut de Biologie Structurale Jean-Pierre Ebel, CEA, Grenoble, France; Institut de Biologie Structurale Jean-Pierre Ebel, CNRS, Grenoble, France; Institut de Biologie Structurale Jean-Pierre Ebel, Université Joseph Fourier – Grenoble 1, Grenoble, France*
ROBERT M. MCCARRICK • *Department of Chemistry and Biochemistry, Miami University, Oxford, OH, USA*
WLADEK MINOR • *Department of Molecular Physiology and Biological Physics, University of Virginia, Charlottesville, VA, USA*
GAETANO T. MONTELIONE • *Center for Advanced Biotechnology and Medicine, Northeast Structural Genomics Consortium, Rutgers University, Piscataway, NJ, USA*
TAKESHI MURATA • *RIKEN Systems and Structural Biology Center, Yokohama, Japan; JST, PRESTO, Chiba, Japan*
SHUISONG NI • *Department of Chemistry and Biochemistry, Miami University, Oxford, OH, USA*
YAN NIE • *European Molecular Biology Laboratory, Grenoble Outstation, Grenoble, France; Unit of Virus Host Cell Interactions, UJF-EMBL-CNRS, Grenoble, France*
MARJOLAINE NOIRCLERC-SAVOYE • *Institut de Biologie Structurale Jean-Pierre Ebel, CEA, Grenoble, France; Institut de Biologie Structurale Jean-Pierre Ebel, CNRS, Grenoble, France; Institut de Biologie Structurale Jean-Pierre Ebel, Université Joseph Fourier – Grenoble 1, Grenoble, France*
RICCARDO PELLARIN • *Department of Bioengineering and Therapeutic Sciences, Department of Pharmaceutical Chemistry, and California Institute for Quantitative Biosciences (QB3), University of California San Francisco, San Francisco, CA, USA*
ROBERT P. RAMBO • *Life Science Division, Lawrence Berkeley National Laboratory, Berkeley, CA, USA*
THERESA A. RAMELOT • *Department of Chemistry and Biochemistry and the Northeast Structural Genomics Consortium, Miami University, Oxford, OH, USA*
BARAK RAVEH • *Department of Bioengineering and Therapeutic Sciences, Department of Pharmaceutical Chemistry, and California Institute for Quantitative Biosciences (QB3), University of California San Francisco, San Francisco, CA, USA*
SABRY RAZICK • *The Biotechnology Centre of Oslo, University of Oslo, Oslo, Norway; Biomedical Research Group, Department of Informatics, University of Oslo, Oslo, Norway*
IVAN RODIC • *Physcial Bioscience Division, Lawrence Berkeley National Laboratory, Berkeley, CA, USA*
DANIEL RUSSEL • *Department of Bioengineering and Therapeutic Sciences, Department of Pharmaceutical Chemistry, and California Institute for Quantitative Biosciences (QB3), University of California San Francisco, San Francisco, CA, USA*
NATALIE J. SAEZ • *Architecture et Fonction des Macromolécules Biologiques, Aix Marseille Université, Marseille, France*
ANDREJ SALI • *Department of Bioengineering and Therapeutic Sciences, Department of Pharmaceutical Chemistry, and California Institute for Quantitative Biosciences (QB3), University of California San Francisco, San Francisco, CA, USA*

MICHAEL R. SAWAYA • *UCLA-DOE Institute for Genomics and Proteomics, University of California, Los Angeles, CA, USA*

DINA SCHNEIDMAN-DUHOVNY • *Department of Bioengineering and Therapeutic Sciences, Department of Pharmaceutical Chemistry, and California Institute for Quantitative Biosciences (QB3), University of California San Francisco, San Francisco, CA, USA*

MIKAKO SHIROUZU • *RIKEN Systems and Structural Biology Center, Yokohama, Japan; Division of Structural and Synthetic Biology, RIKEN Center for Life Science Technologies, Yokohama, Japan*

CLAIRE STRAIN-DAMERELL • *Structural Genomics Consortium, University of Oxford, Oxford, UK*

F. WILLIAM STUDIER • *Biosciences Department, Brookhaven National Laboratory, Upton, NY, USA*

JOHN A. TAINER • *Life Science Division, Lawrence Berkeley National Laboratory, Berkeley, CA, USA*

TAKAHO TERADA • *RIKEN Systems and Structural Biology Center, Yokohama, Japan; RIKEN Structural Biology Laboratory, Yokohama, Japan*

DEEPAK B. THIMIRI GOVINDA RAJ • *European Molecular Biology Laboratory, Grenoble Outstation, Grenoble, France; Unit of Virus Host Cell Interactions, UJF-EMBL-CNRS, Grenoble, France*

ELINA TJIOE • *Department of Bioengineering and Therapeutic Sciences, Department of Pharmaceutical Chemistry, and California Institute for Quantitative Biosciences (QB3), University of California San Francisco, San Francisco, CA, USA*

SIMON TROWITZSCH • *European Molecular Biology Laboratory, Grenoble Outstation, Grenoble, France; Unit of Virus Host Cell Interactions, UJF-EMBL-CNRS, Grenoble, France*

SUSAN E. TSUTAKAWA • *Life Science Division, Lawrence Berkeley National Laboratory, Berkeley, CA, USA*

ANDREI L. TURINSKY • *Molecular Structure and Function program, Hospital for Sick Children, Toronto, ON, Canada*

BRIAN TURNER • *Molecular Structure and Function program, Hospital for Sick Children, Toronto, ON, Canada*

JAVIER VELÁZQUEZ-MURIEL • *Department of Bioengineering and Therapeutic Sciences, Department of Pharmaceutical Chemistry, and California Institute for Quantitative Biosciences (QB3), University of California San Francisco, San Francisco, CA, USA*

THIERRY VERNET • *Institut de Biologie Structurale Jean-Pierre Ebel, CEA, Grenoble, France; Institut de Biologie Structurale Jean-Pierre Ebel, CNRS, Grenoble, France; Institut de Biologie Structurale Jean-Pierre Ebel, Université Joseph Fourier – Grenoble 1, Grenoble, France*

LAKSHMI S. VIJAYACHANDRAN • *European Molecular Biology Laboratory, Grenoble Outstation, Grenoble, France; Unit of Virus Host Cell Interactions, UJF-EMBL-CNRS, Grenoble, France; Amrita Centre for Nanosciences and Molecular Medicine, Amrita Institute of Medical Sciences and Research Centre, Amrita Vishwa Vidyapeetham University, Kochi, India*

RENAUD VINCENTELLI • *Architecture et Fonction des Macromolécules Biologiques, Aix Marseille Université, Marseille, France*

BENJAMIN WEBB • *Department of Bioengineering and Therapeutic Sciences, Department of Pharmaceutical Chemistry, and California Institute for Quanstitative Biosciences (QB3), University of California San Francisco, San Francisco, CA, USA*

ALEXANDER WLODAWER • *Protein Structure Section, Macromolecular Crystallography Laboratory, NCI at Frederick, Frederick, MD, USA*
SHOSHANA J. WODAK • *Molecular Structure and Function program, Hospital for Sick Children, Toronto, ON, Canada; Department of Molecular Genetics, University of Toronto, Toronto, ON, Canada; Department of Biochemistry, University of Toronto, Toronto, ON, Canada*
YUNHUANG YANG • *Department of Chemistry and Biochemistry and the Northeast Structural Genomics Consortium, Miami University, Oxford, OH, USA*
CHIN-PANG BENNY YIU • *33/F, Shui On Centre, Wan Chai, Hong Kong*
SHIGEYUKI YOKOYAMA • *RIKEN Systems and Structural Biology Center, Yokohama, Japan; RIKEN Structural Biology Laboratory, Yokohama, Japan*
HEPING ZHENG • *Department of Molecular Physiology and Biological Physics, University of Virginia, Charlottesville, VA, USA*
MATTHEW D. ZIMMERMAN • *Department of Molecular Physiology and Biological Physics, University of Virginia, Charlottesville, VA, USA*

Part I

Cloning, Expression and Protein Production

Chapter 1

DisMeta: A Meta Server for Construct Design and Optimization

Yuanpeng Janet Huang, Thomas B. Acton, and Gaetano T. Montelione

Abstract

Intrinsically disordered or unstructured regions in proteins are both common and biologically important, particularly in regulation, signaling, and modulating intermolecular recognition processes. From a practical point of view, however, such disordered regions often can pose significant challenges for crystallization. Disordered regions are also detrimental to NMR spectral quality, complicating the analysis of resonance assignments and three-dimensional protein structures by NMR methods. The DisMeta Server has been used by Northeastern Structural Genomics (NESG) consortium as a primary tool for construct design and optimization in preparing samples for both NMR and crystallization studies. It is a meta-server that generates a consensus analysis of eight different sequence-based disorder predictors to identify regions that are likely to be disordered. DisMeta also identifies predicted secretion signal peptides, transmembrane segments, and low-complexity regions. Identification of disordered regions, by either experimental or computational methods, is an important step in the NESG structure production pipeline, allowing the rational design of protein constructs that have improved expression and solubility, improved crystallization, and better quality NMR spectra.

Key words Intrinsically disorder protein prediction, Construct design, Construct optimization, Hydrogen–deuterium exchange with mass spectrometry (HDX-MS)

1 Introduction

Intrinsically disordered or unstructured regions in proteins are both common and biologically important, particularly in modulating intermolecular recognition processes in cellular regulation and signaling. Intrinsically disordered proteins also have broad associations with human diseases [1–4]. Intrinsic disorder, and or disorder-to-order structural transitions, is an important feature of many transcription factors [5, 6], signaling scaffold proteins [7, 8], stress-related proteins including protein chaperones [9], and hub proteins involved in protein–protein interactions [10, 11].

From a practical point of view, highly homogeneous protein samples with minimal amounts of intrinsic disorder are generally more amenable for successful protein crystallization and structure

Yu Wai Chen (ed.), *Structural Genomics: General Applications*, Methods in Molecular Biology, vol. 1091, DOI 10.1007/978-1-62703-691-7_1, © Springer Science+Business Media, LLC 2014

determination by X-ray crystallography [12–14]. While NMR can often be used successfully to study even fully disordered proteins, disordered segments of proteins can promote aggregation and deleteriously affect NMR spectral quality. In addition, a large percentage of targets for NMR studies, particularly human and other eukaryotic proteins, are multidomain proteins, which often misfold in prokaryotic systems [15]. Such multi-domain proteins also often exceed the size limitations for NMR structure determination techniques. Domain parsing can be used to circumvent these significant issues.

It can be extremely challenging to predict the protein subsequence that will produce a soluble well-behaved protein, particularly in studies of domains for which the three-dimensional structure is not yet known. In particular, it can be challenging to accurately predict domain boundaries and locations of disordered residues. This information is critical for identifying open reading frames that can be expressed in soluble form in bacterial or even in eukaryotic expression systems.

The DisMeta Server has been developed by the Northeast Structural Genomics (NESG) consortium as our primary tool for design and optimization of protein constructs expressed for both NMR and crystallization studies (http://www-nmr.cabm.rutgers.edu/bioinformatics/disorder/, Fig. 1). It is a meta-server that generates a consensus analysis of eight sequence-based disorder predictors to identify regions of the protein that are likely to be disordered (Fig. 2b). The DisMeta server uses a consensus approach for disorder prediction. The consensus analysis is conservative in identifying disordered regions, minimizing the possibility of cutting into an ordered region of the protein due to an inaccurate prediction that it is disorder, i.e., low false-positive rates. The server also identifies predicted secretion signal peptides using SignalP [16], transmembrane segments by TMHMM [17], low-complexity SEG regions [18], secondary structure by PROFsec [19] and PSIPred [20], and ANCHOR [21] (Fig. 2a).

The data from these disorder consensus of prediction servers, along with multiple sequence alignments of homologous proteins and hidden Markov models characteristic of the targeted protein domain families [22, 23], are used to predict possible structural domain boundaries. Based on this information, the user can generate nested sets of alternative constructs for full-length proteins, multi-domain constructs, and single-domain constructs. These alternative constructs often possess significantly better expression, solubility, and biophysical behavior than their full-length parent sequences, increasing the likelihood of success in crystallization and the efficiency of structure production. The success of this multiple construct strategy has been reported by the NESG and others [24–27]. NESG has developed an automated construct design software, which generates nested sets of constructs based on the DisMeta

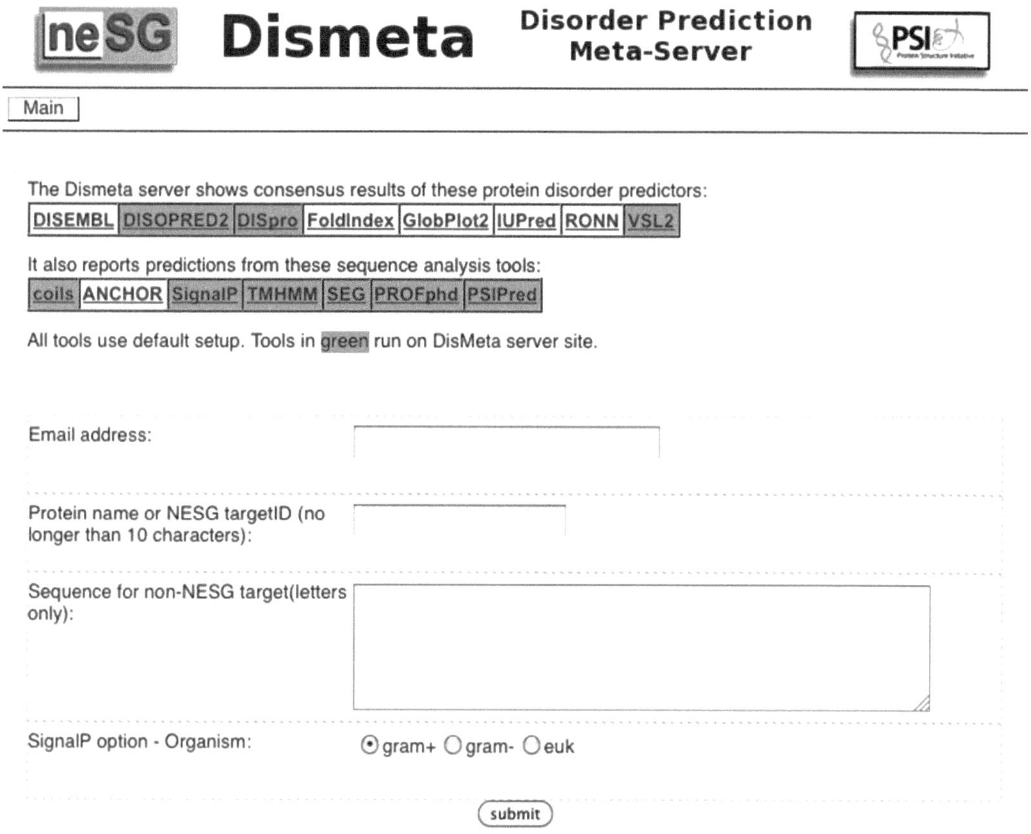

Dismeta Disorder Prediction Meta-Server

Fig. 1 The DisMeta Web server interface. The interface is designed to be simple and easy to use. All tools use default setup parameters. The programs DISOPRED2, DISpro, VSL2, coils, SignlP, TMHMM, SEG, PROFphd, and PSIPred are installed locally and run on DisMeta server site. Other results are retrieved over the Internet from the corresponding Web servers

report, to assist this construct design process for large-scale protein sample production (unpublished results) [27]. In addition, the server identifies domain-sized regions of the protein that are amenable to high-throughput NMR studies, allowing structural studies of proteins that would otherwise be too large to study by NMR.

The results from the DisMeta Server have been compared with biophysical data on protein disorder, including HSQC NMR, ^{15}N nuclear relaxation rate, and hydrogen–deuterium exchange mass spectroscopy (HDX-MS) data [14] for many NESG protein targets. We have found that consensus disordered region, regions predicted to be signal peptide, and/or transmembrane regions and low-complexity regions identified by DisMeta are often disordered in these experimental studies of protein samples. As shown in some examples in Subheading 3, truncated constructs lacking residues from these disordered regions have been successfully generated and used to provide diffraction quality crystals and/or good NMR spectra suitable for determining the 3D protein structure [14, 26–29].

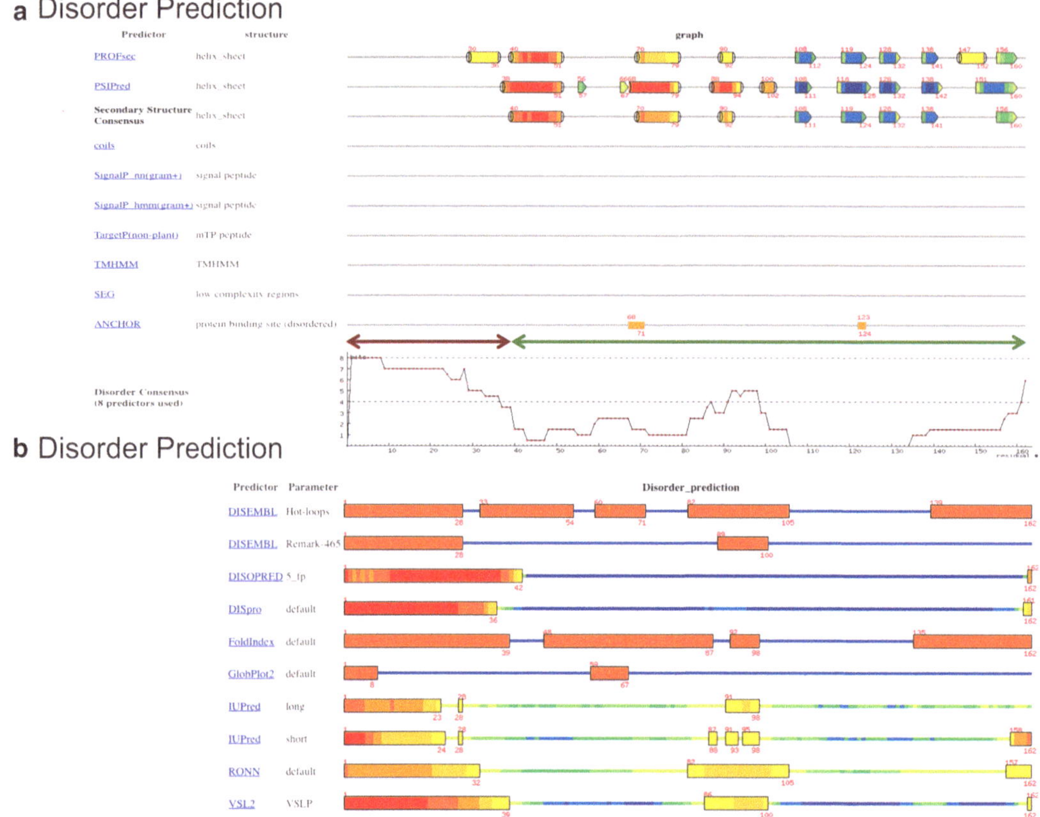

Fig. 2 The DisMeta report for the *Escherichia coli* Spr lipoprotein, NESG target ER541. The full-length protein provided NMR data of marginal quality and no crystals in HTP crystallization screens. The DisMeta report contains two parts: (**a**) sequence-based bioinformatic prediction for construct design and (**b**) disorder predictions from eight different servers. In this case, the disorder prediction programs provide a clear consensus result, with strong evidence for disorder in the N-terminal region of the protein (*red double-head arrow*). On the basis of the disorder consensus results and secondary structure prediction, several truncated constructs lacking residues from this region were generated, ultimately leading to the production of construct Spr (37–162) (*green double-headed arrow*) whose solution NMR structure was solved in NESG consortium (PDB ID, 2K1G) [28]. Sample preparation and NMR data collection for NESG target ER541 are described in ref. 28

2 Materials

2.1 Software Used by the DisMeta Server

The DisMeta Server runs several different disorder prediction software, including DISEMBL [30], DISOPRED2 [31], DISPro [32], FoldIndex [33], GlobPlot2 [34], IUPred [35], RONN [36], and VL2 [37]. For more detailed reviews on disorder prediction methods, *see* ref. 3.

The DisMeta Server also provides sequence-based structural prediction results from other bioinformatics software, including PROF [19], PSIPred [20], SignalP [16], TMHMM [17], Coils [38],

SEG [18], and ANCHOR [21]. This information is used together with the disorder predictions for construct design and optimization for both NMR and crystallization studies.

2.2 Amide Hydrogen–Deuterium Exchange with Mass Spectrometry Detection

HDX-MS measurements are based on the concept that backbone amide protons in disordered regions are solvent accessible and therefore exchange with solvent deuterium (^2H_2O) at a faster rate than backbone amide protons in less solvent-accessible ordered regions where they are generally involved in hydrogen bonds. The degree of exchange over various time intervals is assessed by quenching the exchange kinetics by lowering the pH and temperature, fragmenting the protein by pepsin proteolysis, and measuring the mass of the resulting fragments by mass spectrometry. Peptides with greater mass (higher deuterium exchange) compared to the fully protonated control are identified. The protocols used in this work for HDX-MS studies have been described in detail elsewhere [14]. The results are depicted graphically as a heat map (an example is shown in Fig. 4b); residues in peptides with the greatest amount of mass increase (disordered regions) are represented with red boxes. Residues in peptides showing little or no mass increase (ordered regions) are represented by blue boxes.

3 Methods

Identification of disordered regions of the protein by the DisMeta server, and elimination of these residues from the protein construct, is one of the keys to the NESG construct design and optimization process. Over the past decade, the use of DisMeta for construct design and optimization has greatly increased the efficiency of protein sample and structure production for this large-scale structural genomics project.

3.1 Excluding Consensus Disorder Regions from Construct Design

Figure 2 shows the DisMeta report for the *Escherichia coli* Spr lipoprotein (NESG target ER541), which originally provided NMR data of marginal quality and no crystals in HPT crystal screens. The DisMeta report contains two parts: (A) sequence-based bioinformatics prediction for construct design and (B) disorder predictions from eight different servers. In this case, the disorder prediction programs provide a clear consensus result, namely, strong evidence for disorder in the N-terminal region of the protein (red double-head arrow). On the basis of the disorder consensus results and secondary structure prediction, several truncated constructs lacking residues from this region were generated, ultimately leading to the production of Spr(37–162) (green double-headed arrow) whose solution NMR structure was subsequently determined by the NESG consortium (PDB ID, 2K1G) [28].

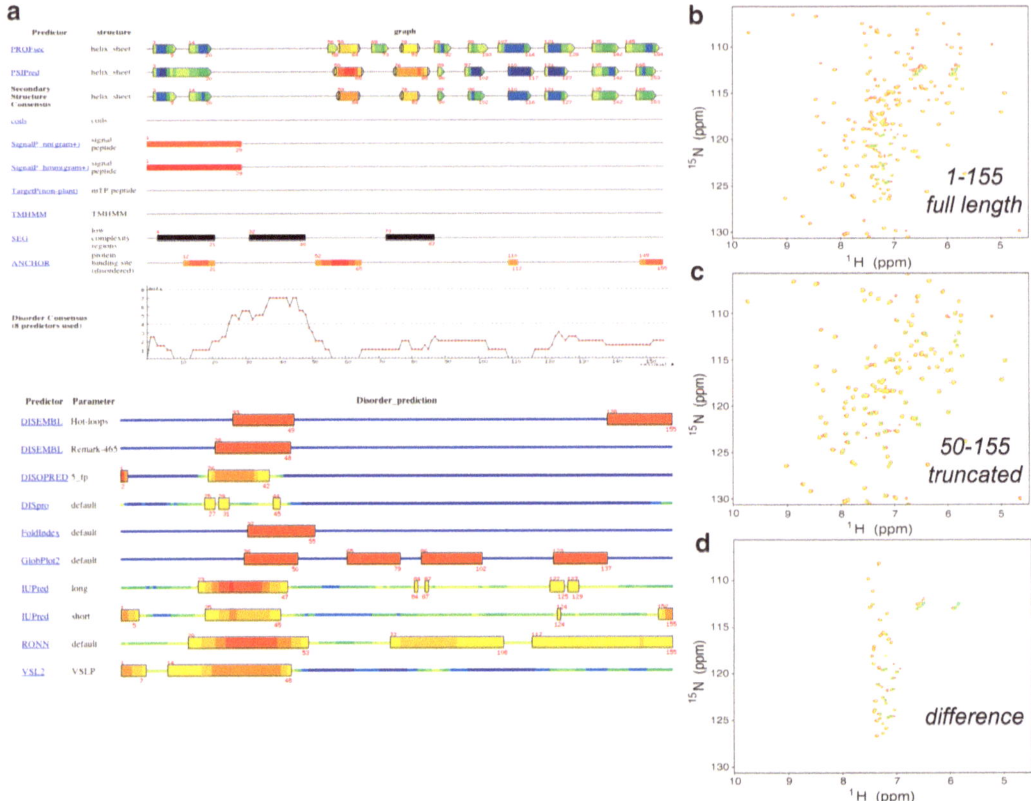

Fig. 3 (**a**) The DisMeta report for NESG target SyR11(full length 155), a bacterial secretory antigen. Both SignalP_nn and SignalP_hmm methods predict residues 1–29 to be a signal peptide. Disorder consensus predicts that the polypeptide segment region 1–29 is ordered and the region around 25–49 is disordered. (**b**) The NMR ^{1}H–^{15}N HSQC spectra are shown for the full-length 1–155 protein construct and (**c**) truncated (50–155) protein construct. (**d**) The differences between the two HSQC spectra (**b**) and (**c**). Sample preparation and NMR data collection for NESG target SyR11 are described in ref. 29

3.2 Excluding the TMHMM and SignalP Regions from Construct Design

The TMHMM and SignalP regions are often disordered for proteins expressed and purified in bacterial expression systems, although they are ordered based on disorder consensus prediction. For example, NESG target SyR11(full length 155, PDB ID: 2K3A) [29] is a bacteria putative secretory antigen. Both SignalP_nn and SignalP_hmm methods [16] predict polypeptide segment 1–29 to be a signal peptide. Disorder consensus predicts that the region of residues 1–29 is ordered and the region around 25–49 is disordered (Fig. 3a). The NMR ^{1}H–^{15}N HSQC spectra are shown for the full-length 1–155 protein construct (Fig. 3b) and truncated (residues 50–155) protein construct (Fig. 3c). The truncated (residues 50–155) construct was selected by removing both the signal peptide and the disorder consensus regions. Figure 3d shows the differences of two HSQC spectra (Fig. 3b, c). Many overlapping peaks in the full-length protein with chemical shift values typical of

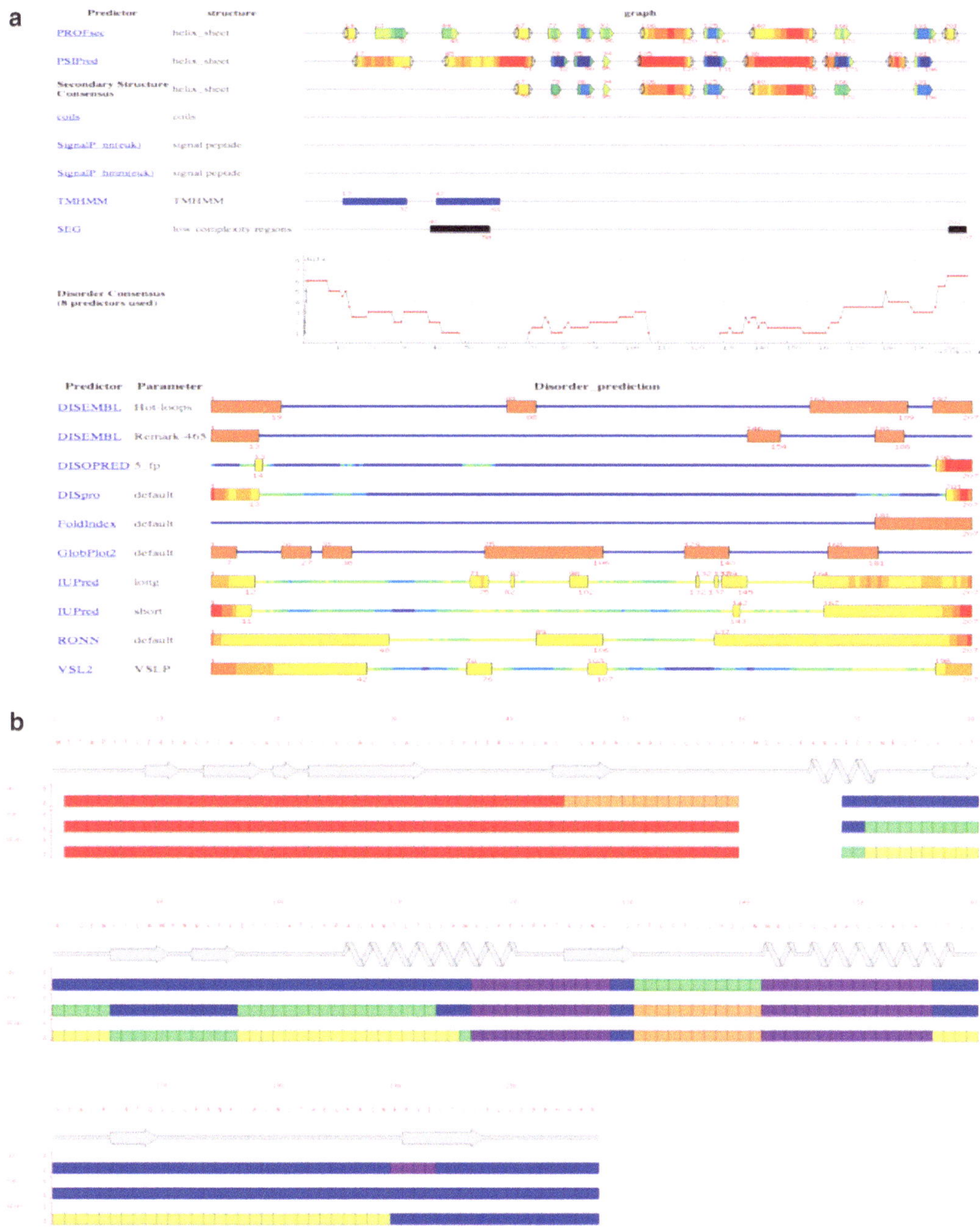

Fig. 4 (**a**) The DisMeta report for *E. coli* inner membrane lipoprotein YiaD (NESG target ER553). It has two TMHMM regions: 13–32 and 42–63. The disorder consensus predicts that the polypeptide segment region 1–13 is disordered and the TMHMM regions 13–32 and 42–63 are ordered. (**b**) HDX-MS analysis [14] shows that polypeptide segment region ~1–59 is in fact disordered. This is consistent with the DisMeta results combining predictions of the TMHMM regions and the disorder consensus region

disordered residues are absent in the truncated protein construct (Fig. 3d). The amide ^{15}N and ^{1}H resonance frequencies for the remaining ordered residues are identical in both spectra, confirming that the SignalP region 1–29 is disordered and deletion of the disordered region 1–49 does not disturb the structure of the remaining protein.

E. coli inner membrane lipoprotein YiaD (NESG target ER553) has two TMHMM regions: 13–32 and 42–63. The DisMeta disorder consensus predicts that the region around polypeptide segment 1–13 is disordered and the TMHMM regions 13–32 and 42–63 are ordered (Fig. 4a). HDX-MS analysis [14] shows that region ~1–59 is in fact disordered, which is consistent with the DisMeta prediction by combining both the TMHMM regions and the disorder consensus regions.

3.3 Low-Complexity SEG Regions, When Used Together with the Disorder Consensus Prediction, Are Good Disorder Indicators

Studies have shown that sequence regions with low complexity nearly always correspond to polypeptide segments that do not fold into ordered structures or to regions of proteins that form fibrous or extended structures [39], whereas intrinsically disordered regions do not always possess low sequence complexity [39, 40]. Both SEG analysis [18] for complexity and order–disorder prediction are useful and complementary in the analysis of protein sequences [41] and in construct design and optimization.

In our experience, many low-complexity SEG regions are next to or overlapped with disorder consensus regions, and those SEG regions next to the disorder consensus regions in the sequence are often indeed disordered. Examples are shown in Figs. 3, 4, and 5. In Fig. 3a, the low-complexity regions are in polypeptide segments 4–21, 32–48, and 73–87. Except for polypeptide segment 73–87, the two other low-complexity regions are overlapped with either the SignalP region or the disorder consensus regions. Figure 4a shows that the low-complexity region polypeptide segment 40–58 overlaps with the TMHMM region, which is disordered in protein samples produced in our *E. coli* expression system, as shown by the HDX-MS [14]. The low-complexity region polypeptide segment 202–207 also overlaps with disorder consensus region at the C-terminal.

C. elegans TPPP family protein CE32E8.3 (NESG target WR33) has been solved by the NESG consortium. It consists of five helices with an intrinsically disordered region in the C-terminal one-third of the protein sequence. HDX-MS data [14] (Fig. 5b) shows that regions ~1–12 (6× His tag) and ~121–190 are disordered. Figure 5a shows that polypeptide segments 3–25 and 125–136 are low-complexity regions. The region polypeptide segment 3–15 is overlapped with the disorder consensus. The region polypeptide segment 125–136 is very close to the C-terminal disorder consensus. Combing the SEG prediction with the disorder consensus agrees well with the HDX-MS data, which shows that the

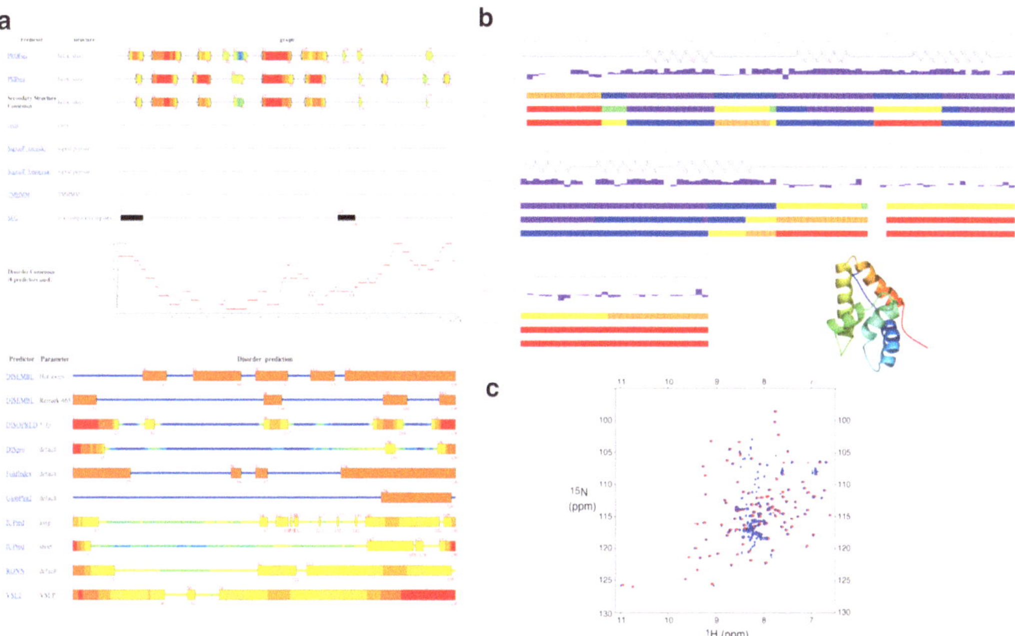

Fig. 5 (**a**) The DisMeta report for *C. elegans* TPPP family protein CE32E8.3 (NESG target WR33), whose structure has been solved by the NESG consortium (PDB ID 1PUL). It consists of five helices with an intrinsically disordered region in the C-terminal one-third of the protein sequence. The regions 3–25 and 125–136 are identified as the SEG regions. (**b**) The HDX-MS data [14] shows that regions ~1–12 (6× His tag) and ~121–190 are disordered. The region 3–15 is overlapped with the disorder consensus. The region 125–136 is very close to the C-terminal disorder consensus. The combined result of the low-complexity SEG regions and the disorder consensus agrees well with the HDX-MS data, which shows that the polypeptide region 121–190 is disordered [14]. (**c**) Comparison of the NMR ^{1}H–^{15}N HSQC spectra for the full-length CE32E8.3 protein, residues 1–190, and the truncated protein construct used for the solution structure determination, residues 1–125. This comparison shows the presence of many overlapping peaks in the full-length protein with chemical shift values typical of disordered residues. These peaks are absent in the truncated protein construct. The amide ^{15}N and ^{1}H resonance frequencies for the remaining ordered residues are identical in both spectra, confirming that deletion of the disordered sequence does not disturb the structure of the remaining protein. Sample preparation, NMR data collection, and HDX-MS data collection for NESG target WR33 are described in ref. 14

region of polypeptide segment 121–190 is disordered (Fig. 5b) [14]. Figure 5c compares the NMR ^{1}H–^{15}N HSQC spectra for the full-length CE32E8.3 protein 1–190 and the truncated protein construct residues 1–125 that was used for the solution structure determination. This comparison shows the presence of many overlapping peaks in the full-length protein with chemical shift values typical of disordered residues. These peaks are absent in the truncated protein construct. The amide ^{15}N and ^{1}H resonance frequencies for the remaining ordered residues are identical in both spectra, confirming that deletion of the disordered sequence does not disturb the structure of the remaining protein [14].

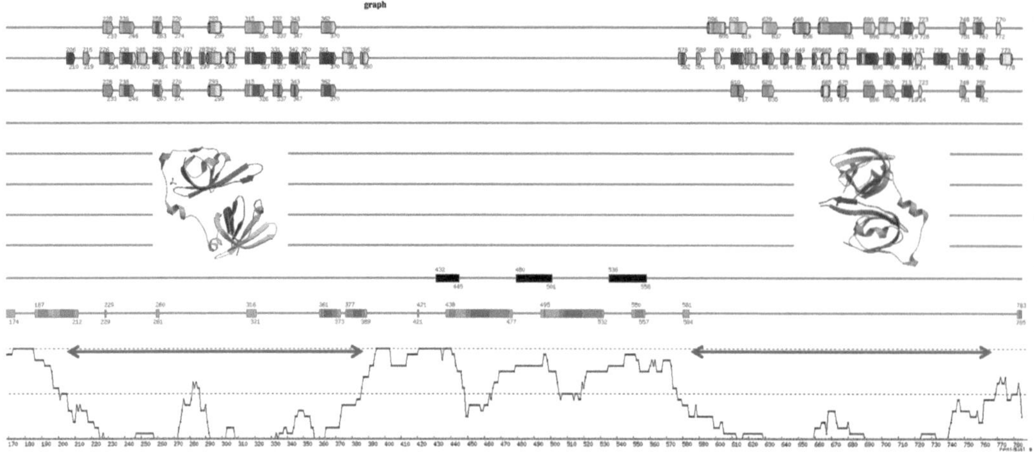

Fig. 6 Domain parsing based on disorder consensus prediction. The C-terminal part of the DisMeta report of gamma-interferon-inducible protein 16 (NESG target HR4626) from *Homo sapiens* is shown. There are two ordered regions, corresponding to residues ~206–385 and 577–768 (*double-headed arrows*), linked by a flexible disordered linker. The DisMeta consensus ordered regions are consistent with the domain constructs for which structures were determined and deposited in the PDB (i.e., HR4626A, PDB 2OQ0, 200–390 and HR4626B, PDB 3B6Y, 576–761). Indeed, the ordered regions identified by DisMeta have better agreement with the domain boundaries revealed by these experimental structures than the domains predicted by PFAM [42], polypeptide segments 201–370 and 575–740

3.4 Domain Parsing Based on the Disorder Consensus Prediction

Multi-domain proteins often misfold in prokaryotic system [15] and also often exceed the size limitations for high-throughput NMR structure determination techniques. Domain parsing can be used to circumvent these significant issues. The data from these disorder consensus of prediction servers, along with multiple sequence alignments of homologous proteins and hidden Markov models characteristic of the targeted protein domain families [22, 23], are used to predict possible structural domain boundaries.

Disorder consensus prediction can be used alone to predict the domain boundary in some cases. An example is shown in Fig. 6. The C-terminal part of the DisMeta report of gamma-interferon-inducible protein 16 (NESG target HR4626) from *Homo sapiens* is shown in Fig. 6. There are two ordered regions, corresponding approximately to polypeptide segments 206–385 and 577–768 (double-headed arrows), linked by a flexible disordered linker. The DisMeta consensus ordered regions are consistent with the domain constructs that provided these 3D structures (i.e., HR4626A, PDB 2OQ0, 200–390 and HR4626B, PDB 3B6Y, 576–761). Indeed, the DisMeta predictions are in better agreement with the experimentally determined domain parsing than the domain region polypeptide segments of residues 201–370 and 575–740 predicted by PFAM [42].

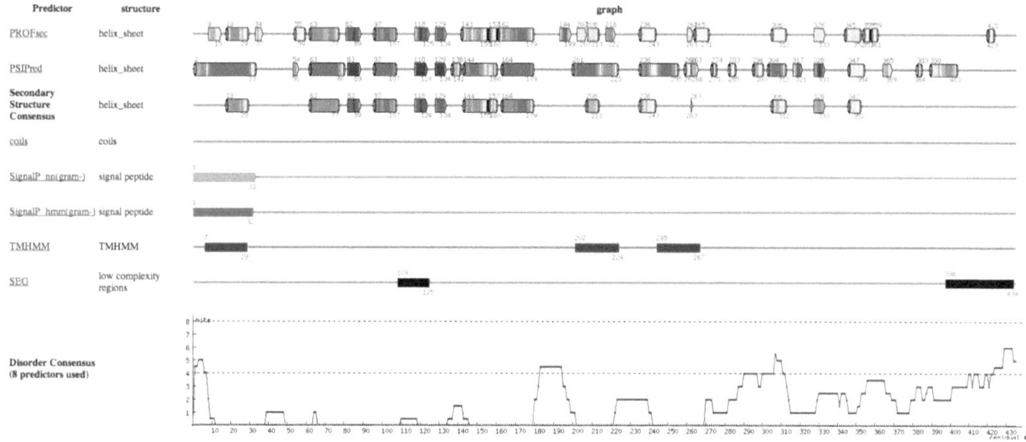

b Alternative Constructs

Fig. 7 (**a**) The DisMeta report for the *Porphyromonas gingivalis* protein Q7MX54 (NESG ID: PgR37). (**b**) Schematic representation of construct optimization for NESG protein target PgR37 from PFAM domain family DUF477, including full-length, residues 54–187, 59–182, and 35–182. Only the 35–182 construct produced a soluble-expressed protein and ultimately an NMR structure (Protein Data Bank ID: 2KW7)

3.5 Multiple Constructs as a Strategy for Obtaining Samples Suitable for Structural Analysis

It is extremely challenging to predict the protein subsequence that will produce a soluble well-behaved protein, particularly for protein containing domains for which the three-dimensional structure is not yet known. This arises from problems with accurately predicting the domain boundaries and locations of disordered residues, as this information is critical for designing an open reading frame that results in high-level expression and solubility in bacterial expression system. Currently, our approach is to take advantage of our high-throughput cloning and expression platform [26, 27] and produce several alternative constructs, varying the termini of a targeted domain based on the DisMeta predictions, followed by small-scale expression and solubility screening as well as NMR and crystallization screening to identify the protein subsequence with the best behavior [24–27].

An example of our domain parsing and multiple construct approach is shown in Fig. 7a. Consensus analysis of several disorder prediction algorithms (*see* Disorder Consensus panel) suggests that the C-terminal half of the 434-residue protein from *Porphyromonas gingivalis* Q7MX54 (NESG ID: PgR37) contains disordered regions. Cloning and expression analysis of the full-length protein in our bacterial expression system result in no detectable expression, supporting the disorder prediction based on the fact that proteins with significant disorder are often degraded in the *E. coli* cell. DisMeta analysis was then used for construct optimization, and alternative constructs were designed using these data together with information for the database of protein families that includes their annotations and multiple sequence alignments (DUF477 region 60–179). These alternative constructs are depicted schematically in Fig. 7b. The two expression constructs comprising residues 54–187 and 59–182 also did not express at detectable levels. However, a slightly longer construct (residues 35–182) was highly expressed and soluble and ultimately allowed the structure of this targeted domain to be solved by NMR (Protein Data Bank ID: 2KW7). Interestingly, the three-dimensional structure reveals the presence of two short helical regions between residues 38–48 and a β-strand for residues 54–56. These helices and the β-strand are tightly packed against each other and to other regions of the protein. The loss of these interactions in the shorter constructs likely destabilizes the protein, leading to degradation in the expression host [27].

4 Conclusion

The DisMeta server provides a consensus analysis of eight disorder predictors as well as predictions of secondary structure, signal peptides, transmembrane helical regions, and low-complexity regions of the protein sequence using publicly available servers. These data are combined into a single simple to read report. DisMeta analyses have been used for protein construct design and optimization in the large-scale sample and structure production pipeline of the NESG consortium of the Protein Structure Initiative. These disorder predictions have allowed production of many protein samples that have been used in many successful NMR and X-ray crystallography studies, some examples of which are illustrated in this chapter. The server is freely available online to the scientific community and should be useful both to small laboratories focused on specific biological problems and to large-scale protein sample production efforts, including antigen sample production projects.

Acknowledgements

We thank H. Zheng, S. Sharma, A. Ertekin, and R. Xiao for providing the HDX-MS data illustrated in this chapter and J. Aramini for providing the NMR spectrum shown in Fig. 5. The NMR data shown in Fig. 3 were recorded by P. Rossi. This work was supported by a grant from the National Institute of General Medical Sciences Protein Structure Initiative U54-GM074958 (to G.T.M.).

References

1. Dyson HJ, Wright PE (2002) Coupling of folding and binding for unstructured proteins. Curr Opin Struct Biol 12:54–60
2. Iakoucheva LM, Brown CJ, Lawson JD et al (2002) Intrinsic disorder in cell-signaling and cancer-associated proteins. J Mol Biol 323:573–584
3. Radivojac P, Iakoucheva LM, Oldfield CJ et al (2007) Intrinsic disorder and functional proteomics. Biophys J 92:1439–1456
4. Kovacs D, Szabo B, Pancsa R et al (2012) Intrinsically disordered proteins undergo and assist folding transitions in the proteome. Arch Biochem Biophys 531:80–89
5. Liu J, Perumal NB, Oldfield CJ et al (2006) Intrinsic disorder in transcription factors. Biochemistry 45:6873–6888
6. Minezaki Y, Homma K, Kinjo AR et al (2006) Human transcription factors contain a high fraction of intrinsically disordered regions essential for transcriptional regulation. J Mol Biol 359:1137–1149
7. Balazs A, Csizmok V, Buday L et al (2009) High levels of structural disorder in scaffold proteins as exemplified by a novel neuronal protein, CASK-interactive protein1. FEBS J 276:3744–3756
8. Buday L, Tompa P (2010) Functional classification of scaffold proteins and related molecules. FEBS J 277:4348–4355
9. Tompa P, Csermely P (2004) The role of structural disorder in the function of RNA and protein chaperones. FASEB J 18:1169–1175
10. Dosztanyi Z, Chen J, Dunker AK et al (2006) Disorder and sequence repeats in hub proteins and their implications for network evolution. J Proteome Res 5:2985–2995
11. Haynes C, Oldfield CJ, Ji F et al (2006) Intrinsic disorder is a common feature of hub proteins from four eukaryotic interactomes. PLoS Comput Biol 2:e100
12. Pantazatos D, Kim JS, Klock HE et al (2004) Rapid refinement of crystallographic protein construct definition employing enhanced hydrogen/deuterium exchange MS. Proc Natl Acad Sci USA 101:751–756
13. Spraggon G, Pantazatos D, Klock HE et al (2004) On the use of DXMS to produce more crystallizable proteins: structures of the T. maritima proteins TM0160 and TM1171. Protein Sci 13:3187–3199
14. Sharma S, Zheng H, Huang YJ et al (2009) Construct optimization for protein NMR structure analysis using amide hydrogen/deuterium exchange mass spectrometry. Proteins 76:882–894
15. Netzer WJ, Hartl FU (1997) Recombination of protein domains facilitated by co-translational folding in eukaryotes. Nature 388:343–349
16. Emanuelsson O, Brunak S, von Heijne G et al (2007) Locating proteins in the cell using TargetP, SignalP and related tools. Nat Protoc 2:953–971
17. Krogh A, Larsson B, von Heijne G et al (2001) Predicting transmembrane protein topology with a hidden Markov model: application to complete genomes. J Mol Biol 305:567–580
18. Wootton JC, Federhen S (1996) Analysis of compositionally biased regions in sequence databases. Methods Enzymol 266:554–571
19. Rost B, Yachdav G, Liu J (2004) The PredictProtein server. Nucleic Acids Res 32:W321–W326
20. McGuffin LJ, Bryson K, Jones DT (2000) The PSIPRED protein structure prediction server. Bioinformatics 16:404–405
21. Dosztanyi Z, Meszaros B, Simon I (2009) ANCHOR: web server for predicting protein binding regions in disordered proteins. Bioinformatics 25:2745–2746
22. Dessailly BH, Nair R, Jaroszewski L et al (2009) PSI-2: structural genomics to cover protein domain family space. Structure 17:869–881
23. Hunter S, Jones P, Mitchell A et al (2012) InterPro in 2011: new developments in the family and domain prediction database. Nucleic Acids Res 40:D306–D312
24. Graslund S, Sagemark J, Berglund H et al (2008) The use of systematic N- and C-terminal deletions to promote production and structural studies of recombinant proteins. Protein Expr Purif 58:210–221
25. Chikayama E, Kurotani A, Tanaka T et al (2010) Mathematical model for empirically optimizing large scale production of soluble protein domains. BMC Bioinformatics 11:113
26. Xiao R, Anderson S, Aramini J et al (2010) The high-throughput protein sample production platform of the Northeast Structural Genomics Consortium. J Struct Biol 172:21–33
27. Acton TB, Xiao R, Anderson S et al (2011) Preparation of protein samples for NMR structure, function, and small-molecule screening studies. Methods Enzymol 493:21–60
28. Aramini JM, Rossi P, Huang YJ et al (2008) Solution NMR structure of the NlpC/P60 domain of lipoprotein Spr from Escherichia coli: structural evidence for a novel cysteine peptidase catalytic triad. Biochemistry 47:9715–9717

29. Rossi P, Aramini JM, Xiao R et al (2009) Structural elucidation of the Cys-His-Glu-Asn proteolytic relay in the secreted CHAP domain enzyme from the human pathogen Staphylococcus saprophyticus. Proteins 74:515–519
30. Linding R, Jensen LJ, Diella F et al (2003) Protein disorder prediction: implications for structural proteomics. Structure 11:1453–1459
31. Ward JJ, Sodhi JS, McGuffin LJ et al (2004) Prediction and functional analysis of native disorder in proteins from the three kingdoms of life. J Mol Biol 337:635–645
32. Cheng JSM, Baldi P (2005) Accurate prediction of protein disordered regions by mining protein structure data. Data Min Knowl Discov 11:213–222
33. Prilusky J, Felder CE, Zeev-Ben-Mordehai T et al (2005) FoldIndex: a simple tool to predict whether a given protein sequence is intrinsically unfolded. Bioinformatics 21:3435–3438
34. Linding R, Russell RB, Neduva V et al (2003) GlobPlot: Exploring protein sequences for globularity and disorder. Nucleic Acids Res 31:3701–3708
35. Dosztanyi Z, Csizmok V, Tompa P et al (2005) The pairwise energy content estimated from amino acid composition discriminates between folded and intrinsically unstructured proteins. J Mol Biol 347:827–839
36. Yang ZR, Thomson R, McNeil P et al (2005) RONN: the bio-basis function neural network technique applied to the detection of natively disordered regions in proteins. Bioinformatics 21:3369–3376
37. Vucetic S, Brown CJ, Dunker AK et al (2003) Flavors of protein disorder. Proteins 52:573–584
38. Lupas A, Van Dyke M, Stock J (1991) Predicting coiled coils from protein sequences. Science 252:1162–1164
39. Romero P, Obradovic Z, Li X et al (2001) Sequence complexity of disordered protein. Proteins 42:38–48
40. Romero P, Obradovic Z, Dunker AK (1999) Folding minimal sequences: the lower bound for sequence complexity of globular proteins. FEBS Lett 462:363–367
41. Weathers EA, Paulaitis ME, Woolf TB et al (2007) Insights into protein structure and function from disorder-complexity space. Proteins 66:16–28
42. Finn RD, Mistry J, Tate J et al (2010) The Pfam protein families database. Nucleic Acids Res 38:D211–D222

Chapter 2

Stable Expression Clones and Auto-Induction for Protein Production in *E. coli*

F. William Studier

Abstract

Inducible production of proteins from cloned genes in *E. coli* is widely used, economical, and effective. However, common practices can result in unintended induction, inadvertently generating cultures that give poor or variable yields in protein production. Recipes are provided for (1) defined culture media in which expression strains grow to saturation without induction, thereby ensuring stable frozen stocks and seed cultures with high fractions of fully inducible cells, and (2) defined or complex media that maintain the same high fraction of inducible cells until auto-induction in late log phase to produce fully induced high-density cultures at saturation. Simply inoculating a suitable auto-inducing medium from such a seed culture and growing to saturation generally produces much higher levels of target protein per volume of culture than monitoring culture growth and adding IPTG or other inducer at the appropriate cell density. Many strains may be conveniently screened in parallel, and burdensome inoculation with fresh colonies, sometimes employed in hopes of assuring high yields, is entirely unnecessary. These media were developed for the T7 expression system using pET vectors in BL21(DE3) but are suitable or adaptable for other inducible expression systems in *E. coli* and for labeling proteins with selenomethionine for X-ray crystallography or with stable isotopes for NMR.

Key words Auto-induction, T7 expression system, Stable inducible cultures, Protein production, Protein labeling

1 Introduction

The T7 expression system is widely used for inducible production of target proteins from cloned genes in *E. coli*. The gene for T7 RNA polymerase in the chromosome of BL21(DE3) under the control of the *lacUV5* promoter is induced to express the target gene under the control of a T7 promoter and strong T7 translation start in a multi-copy plasmid. T7 polymerase is so selective, active, and processive that most resources of the cell can become directed to producing a great variety of target proteins [1]. From the first use, instability of inducible strains was encountered as a problem, because even slight basal expression

Yu Wai Chen (ed.), *Structural Genomics: General Applications*, Methods in Molecular Biology, vol. 1091, DOI 10.1007/978-1-62703-691-7_2, © Springer Science+Business Media, LLC 2014

of T7 RNA polymerase can generate enough basal expression of some target proteins to stress uninduced cells. Reduction of basal expression by supplying small amounts of T7 lysozyme, which inhibits transcription by binding T7 RNA polymerase [2] or, more effectively, reducing transcription of the target gene by placing a binding site for *lac* repressor immediately after the T7 promoter (referred to as a T7*lac* promoter) [3], allows maintenance and expression of clones for expressing a wide variety of target proteins [4]. A more recent strategy is to place the gene for T7 RNA polymerase in the chromosome of BL21 under the control of the pBAD promoter, which is inducible by arabinose and is thought to have very low basal expression (BL21-AI from Life Technologies). When the target gene is controlled by a T7*lac* promoter, both arabinose and a *lac* inducer are required for production of the target protein in BL21-AI.

It was also recognized early on that strains capable of expressing target proteins that stress the host cell should not be grown to saturation, because expression-competent cells could become overgrown by cells that had lost plasmid or mutants that were poorly inducible [1–4]. An explanation for why this precaution was advisable became apparent with the discovery that growth of expression strains in some complex media (but not others) produced high-level induction of target protein upon approach to saturation [5]. Investigation of how composition of the growth medium affects growth, saturation cell density, and expression of target protein produced a likely explanation for this unintended induction and a rationale for developing defined, non-inducing growth media and high-density auto-inducing media [6].

Amino acids and small peptides provide the primary carbon and energy sources in commonly used complex media such as LB, which contains enzymatic digests of the milk protein casein (e.g., tryptone or N-Z-amine) and yeast extract. Since milk is rich in lactose, an inducer of the T7 expression system, variable amounts of residual lactose may be present in different lots of these enzymatic digests. These small amounts of lactose do not promote appreciable induction during log-phase growth, but even minute amounts are sufficient to cause induction on approach to saturation, particularly at lower rates of aeration, which allow induction at lower lactose concentration and promote higher levels of induction [6]. The presence of glucose prevents such induction [5], but finding a concentration of glucose that reliably prevents induction in complex media without also causing cultures to become undesirably acidic at saturation proved to be difficult if not impossible [6]. Recent work found that small amounts of galactose present in complex media derived from plant sources also cause unintended induction in BL21(DE3) [7]. This can happen because BL21 strains lack galactokinase [8], thereby preventing galactose from being metabolized and allowing the intracellular galactose concentration to reach levels high enough for robust induction of the *lac* operon by this weak inducer. However, the galactose transporters necessary for such induction are also strongly inhibited by glucose [9].

Formulation of non-inducing and auto-inducing media is based on experimentation reported in [6]. Non-inducing media are made entirely from purified ingredients to minimize potential contamination by inducing agents. Glucose is the primary carbon and energy source because it is highly effective at preventing induction of operons responsible for metabolizing sugars such as lactose, galactose, and arabinose by a combination of catabolite repression and inducer exclusion. However, to grow cultures to high cell densities (OD_{600nm} ~10 and cell concentrations greater than 10^{10}/ml), glucose concentration must be adjusted so that the pH of the culture does not fall much below ~6 before metabolism of another component of the defined medium (typically aspartate, succinate, and/or a mixture of amino acids) increases the final pH at saturation to ~7. This metabolic balancing of pH requires that the culture be well aerated. Even expression strains that produce target proteins highly toxic to the host cell retain plasmid and remain viable when grown to saturation in these non-inducing media, indicating that little expression of target protein occurs at any stage of growth.

Auto-inducing media can be made either with purified or complex ingredients, because the inducing sugar is intentionally present in the medium throughout growth. The principle of auto-induction is that glucose in the growth medium completely prevents uptake and metabolism of inducing sugar also present in the growth medium. However, if the glucose concentration is such that all of the glucose is metabolized before saturation of the culture, other sugars present in the medium can be transported into the cell and induce the operons for metabolizing them. Lactose, arabinose, and galactose are all subject to this glucose effect, and auto-inducing media have been formulated for protein expression systems induced by them. Auto-induction is potentially applicable for any expression system having an inducer that is subject to this type of regulation.

Expression strains suitable for auto-induction must have functional transporters for the appropriate sugar. Induction by lactose requires active β-galactosidase to convert lactose to allolactose, the actual inducer, and a functional LacY transporter. Induction by galactose would not require active β-galactosidase, but the host strain must lack galactokinase activity and transport galactose well enough to achieve an intracellular concentration sufficient for inducing promoters blocked by *lac* repressor. IPTG is not suitable for use in auto-induction because it can enter the cell and induce expression without a specific transporter, and cultures cannot grow uninduced in the presence of IPTG.

A reliable carbon and energy source in addition to amino acids is needed to maintain metabolic activities in support of high-level expression of target protein when glucose becomes depleted during growth in auto-inducing media. Metabolism of the inducing sugar may not be sufficient because the operon for metabolism of that sugar may not be well induced in competition with the highly active T7 RNA polymerase-directed expression of target mRNA and protein. Furthermore, galactose cannot be metabolized at all in BL21 strains. Therefore, glycerol is provided in auto-inducing media as a good carbon and energy source that does not prevent glucose depletion during growth, glucose exclusion of inducing sugars, or the uptake of inducing sugars upon glucose depletion.

A glucose concentration is chosen so that its depletion causes auto-induction to begin in the mid-to-late log phase of growth, as aeration is diminishing and induction of target protein is robust. Glucose is highly effective at preventing induction by other sugars present in the medium, and even strains that express highly toxic target proteins grow well and maintain a high fraction of inducible cells before induction begins. Our comparisons using stained SDS-PAGE gels to detect target protein found auto-induction to comparable levels by lactose or galactose at the same molar concentration in MDA-505: no target protein was detected at 10 μM but detectable accumulation began around 20–50 μM and increased with inducer concentration to a maximum accumulation between about 1 and 10 mM. The standard 0.2 % lactose selected initially for auto-inducing media [6] corresponds to 5.6 mM, which equates to 0.1 % galactose. Induction of the pBAD promoter in these media is effective at 0.05 % arabinose.

Non-inducing and auto-inducing media make production of proteins from cloned genes in *E. coli* reliable and convenient and are adaptable for applications from small-scale laboratory testing to large-scale screening and protein production. These growth media and protocols were developed for use with the T7 expression system, but the same methods are applicable to existing expression systems inducible by IPTG or arabinose and potentially to any expression system with an inducer whose action is blocked by glucose.

2 Materials

2.1 Stock Solutions

Stock solutions are autoclaved for 15 min and stored at room temperature unless specified otherwise. Deionized distilled water is used for all solutions. If the final solution is not to be autoclaved, sterile water (autoclaved for 15 min) is used in making the solution and the final solution is filter sterilized. Dissolve components sequentially in the order given, usually in water stirred in a beaker.

Brief heating in a microwave oven can be effective in speeding up the process (plastic-covered magnetic stirring bars need not be removed in the microwave). High concentrations of sugars usually have to be heated to dissolve in a reasonable time.

1. 50× M: 80 ml water, 17.75 g Na_2HPO_4, 17.0 g KH_2PO_4, 13.4 g NH_4Cl, 3.55 g Na_2SO_4. 1× concentration: 25 mM Na_2HPO_4, 25 mM KH_2PO_4, 50 mM NH_4Cl, 5 mM Na_2SO_4, pH ~6.7 (*see* **Note 1**).
2. 40 % glucose: 74 ml water, 40 g glucose.
3. 80 % glycerol (v/v) = 100 % (w/v): 100 g glycerol (weigh in beaker), 20 ml water.
4. 50× 5052: 25 g glycerol (weigh in beaker), 73 ml water, 2.5 g glucose, 10 g α-lactose monohydrate. 1× concentration: 0.5 % glycerol, 0.05 % glucose, 0.2 % α-lactose.
5. 50× 5051: 25 g glycerol (weigh in beaker), 73 ml water, 2.5 g glucose, 5 g galactose. 1× concentration: 0.5 % glycerol, 0.05 % glucose, 0.1 % galactose.
6. 100× 505: 50 g glycerol (weigh in beaker), 57 ml water, 5 g glucose. 1× concentration: 0.5 % glycerol, 0.05 % glucose.
7. 25 % aspartate: 84 ml water, 25 g aspartic acid, 8 g NaOH (pH should be near neutral).
8. 17aa (no C,Y,M) (10 mg/ml each): 90 ml water stirred in beaker, add 1 g of each pure amino acid in the order NaGlu, Asp, Lys-HCl, Arg-HCl, His-HCl-H_2O, Ala, Pro, Gly, Thr, Ser, Gln, Asn-H_2O, Val, Leu, Ile, Phe, Trp. Stir until everything dissolves, and heat in microwave if necessary. Filter sterilize rather than autoclave. Keep working stock in refrigerator, and store aliquots in −20 °C freezer (*see* **Note 2**).
9. 18aa (no C,Y) (7.14 mg/ml each): 10 ml 17aa (10 mg/ml each), 4 ml methionine (25 mg/ml, autoclaved). Do not autoclave the final mixture. Keep working stock in refrigerator, and store aliquots in −20 °C freezer. Incorporating 280 μl of 18aa in a total of 10 ml of medium gives 200 μg/ml of each amino acid for a total mixed concentration of 0.36 % (*see* **Notes 2** and **3**).
10. 1 M $MgSO_4$: 87 ml water, 24.65 g $MgSO_4$–$7H_2O$.
11. 0.1 M $FeCl_3$ in ~0.12 M HCl: 99 ml water, 1 ml concentrated HCl (~12 M), 2.7 g $FeCl_3$–$6H_2O$, do not autoclave (*see* **Note 4**).
12. 1,000× metals: 50 mM $FeCl_3$, 20 mM $CaCl_2$, 10 mM $MnCl_2$, 10 mM $ZnSO_4$, 2 mM $CoCl_2$, 2 mM $CuCl_2$, 2 mM $NiCl_2$, 2 mM Na_2MoO_4, 2 mM Na_2SeO_3, 2 mM H_3BO_3, do not autoclave (*see* **Note 4**).
13. ZY: 1 l water, 10 g N-Z-amine AS, 5 g yeast extract (*see* **Note 5**).

2.2 Growth Media

Recipes are given for a total volume of 500 ml for ~25 agar plates, 800 ml for labeling with selenomethionine (SeMet), and 10 ml for other growth media to give convenient multiples for scaling up or down. Growth media contain 50 mM phosphate, which provides significant buffering and supports growth to high densities. As little as 25 mM phosphate is sufficient if a lower concentration is desirable. The effectiveness of kanamycin as a selective agent against the growth of BL21(DE3) decreases with increasing concentration of phosphate in rich media [6]: 100 μg/ml of kanamycin is needed to assure killing in the media given here. BL21(DE3) and BL21-AI grow well in these media. The recipes generally do not contain selective antibiotics or nutrients that may be essential for growth of other hosts, which must be added as appropriate.

2.2.1 Non-inducing Media

1. MDAG-11 non-inducing agar plates for isolating transformants (*see* **Note 6**): 5 g agar, 475 ml H_2O, autoclave for 15 min, mix well, let cool for ~10 min on bench or equilibrate in a 50 °C water bath. Add 1 ml 1 M $MgSO_4$, 100 μl 1,000× metals, 1.25 ml 40 % glucose, 2 ml 25 % aspartate, 10 ml 50× M, 14 ml 18aa, and any nutrients required by the host cell (e.g., 50 μl 10 mM thiamine for XL1Blue-MR) or selective antibiotics (e.g., 2 ml of kanamycin stock solution, 25 mg/ml). Mix well, and pour ~20 ml per standard plastic Petri plate (pouring slowly until liquid just covers the bottom usually gives about the right amount per plate). This recipe makes ~25 plates with final composition of 1 % agar, 25 mM Na_2HPO_4, 25 mM KH_2PO_4, 50 mM NH_4Cl, 5 mM Na_2SO_4, 2 mM $MgSO_4$, 0.2× metals, 0.1 % glucose, 0.1 % aspartate, 200 μg/ml of each of 18 amino acids (no C,Y) and optionally 1 μM thiamine, and 100 μg/ml kanamycin (*see* **Note 7**).
2. MDAG-11, non-inducing growth medium for suspending colonies, making dilutions or growing standing cultures (*see* **Note 8**): 9.43 ml water, 20 μl 1 M $MgSO_4$, 2 μl 1,000× metals, 25 μl 40 % glucose, 40 μl 25 % aspartate, 200 μl 50× M, 280 μl 18aa. Final composition: 25 mM Na_2HPO_4, 25 mM KH_2PO_4, 50 mM NH_4Cl, 5 mM Na_2SO_4, 2 mM $MgSO_4$, 0.2× metals, 0.1 % glucose, 0.1 % aspartate, 200 μg/ml of each of 18 amino acids (no C,Y).
3. MDAG-135, non-inducing medium for growing high-density freezer stocks, working or seed cultures, or cultures for isolating plasmids (*see* **Note 9**): 9.37 ml water, 20 μl 1 M $MgSO_4$, 2 μl 1,000× metals, 87.5 μl 40 % glucose, 40 μl 25 % aspartate, 200 μl 50× M, 280 μl 18aa. Final composition: 25 mM Na_2HPO_4, 25 mM KH_2PO_4, 50 mM NH_4Cl, 5 mM Na_2SO_4, 2 mM $MgSO_4$, 0.2× metals, 0.35 % glucose, 0.1 % aspartate, 200 μg/ml of each of 18 amino acids (no C,Y).

4. MDA-505, non-inducing medium for testing auto-induction by different concentrations of inducers (*see* **Note 10**): 9.36 ml water, 20 μl 1 M $MgSO_4$, 2 μl 1,000× metals, 100 μl 100× 505, 40 μl 25 % aspartate, 200 μl 50× M, 280 μl 18aa. Final composition: 25 mM Na_2HPO_4, 25 mM KH_2PO_4, 50 mM NH_4Cl, 5 mM Na_2SO_4, 2 mM $MgSO_4$, 0.2× metals, 0.5 % glycerol, 0.05 % glucose, 0.1 % aspartate, 200 μg/ml of each of 18 amino acids (no C,Y).
5. MDG, non-inducing minimal medium (*see* **Note 11**): 9.55 ml water, 20 μl 1 M $MgSO_4$, 2 μl 1,000× metals, 125 μl 40 % glucose, 100 μl 25 % aspartate, 200 μl 50× M. Final composition: 25 mM Na_2HPO_4, 25 mM KH_2PO_4, 50 mM NH_4Cl, 5 mM Na_2SO_4, 2 mM $MgSO_4$, 0.2× metals, 0.5 % glucose, 0.25 % aspartate.

2.2.2 Auto-Inducing Media

All of these recipes except MD-5051 are formulated for auto-induction with 0.2 % lactose, which can be replaced for auto-induction with 0.1 % galactose, 0.05 % arabinose, or other sugars subject to glucose inhibition, as appropriate. Lactose and galactose can be exchanged by exchanging 50× 5052 (which provides 0.2 % lactose) and 50× 5051 (which provides 0.1 % galactose) in the recipes for auto-inducing media. Auto-induction of target genes under the control of the T7*lac* promoter in BL21-AI requires both arabinose, to induce production of T7 RNA polymerase, and either lactose or galactose, to unblock the T7*lac* promoter in the expression plasmid.

1. ZYM-5052 complex auto-inducing medium: 9.57 ml ZY, 20 μl 1 M $MgSO_4$ (2 μl 1,000× metals, optional), 200 μl 50× 5052, 200 μl 50× M. Final composition: 1 % N-Z-amine AS, 0.5 % yeast extract, 25 mM Na_2HPO_4, 25 mM KH_2PO_4, 50 mM NH_4Cl, 5 mM Na_2SO_4, 2 mM $MgSO_4$ (0.2× metals, optional), 0.5 % glycerol, 0.05 % glucose, 0.2 % α-lactose.
2. MDA-5052 defined auto-inducing medium: 9.26 ml water, 20 μl 1 M $MgSO_4$, 2 μl 1,000× metals, 200 μl 50× 5052, 40 μl 25 % aspartate, 200 μl 50× M, 280 μl 18aa. Final composition: 25 mM Na_2HPO_4, 25 mM KH_2PO_4, 50 mM NH_4Cl, 5 mM Na_2SO_4, 2 mM $MgSO_4$, 0.2× metals, 0.5 % glycerol, 0.05 % glucose, 0.2 % α-lactose, 0.1 % aspartate, 200 μg/ml of each of 18 amino acids (no C,Y).
3. MD-5051 minimal auto-inducing medium for flexible labeling of target proteins (*see* **Note 12**): 9.48 ml water, 20 μl 1 M $MgSO_4$, 2 μl 1,000× metals, 200 μl 5051, 100 μl 25 % aspartate, 200 μl 50× M. Final composition: 25 mM Na_2HPO_4, 25 mM KH_2PO_4, 50 mM NH_4Cl, 5 mM Na_2SO_4, 2 mM $MgSO_4$, 0.2× metals, 0.5 % glycerol, 0.05 % glucose, 0.1 % galactose, 0.25 % aspartate.

4. MDASM-5052 for labeling proteins with SeMet, 800 ml (*see* **Note 13**): 746 ml sterile water, 1.6 ml 1 M $MgSO_4$, 160 µl 1,000× metals, 16 ml 50× 5052, 3.2 ml 25 % aspartate, 16 ml 50× M, 16 ml 17aa (no C,Y,M), 320 µl Met (25 µg/ml), entire 100 mg bottle of SeMet, 800 µl 1 mM vitamin B_{12}. Do not autoclave but use immediately. Final composition: 25 mM Na_2HPO_4, 25 mM KH_2PO_4, 50 mM NH_4Cl, 5 mM Na_2SO_4, 2 mM $MgSO_4$, 0.2× metals, 0.5 % glycerol, 0.05 % glucose, 0.2 % α-lactose, 0.1 % aspartate, 200 µg/ml of each of 17 amino acids (no C,Y,M), 10 µg/ml Met, 125 µg/ml SeMet, 1 µM vitamin B_{12}.

2.2.3 General-Purpose Complex Medium for Rapid Growth to High Density

1. ZYM-505: 9.68 ml ZY, 20 µl 1 M $MgSO_4$ (2 µl 1,000× metals, optional), 100 µl 100× 505, 200 µl 50× M. Final composition: 1 % N-Z-amine AS, 0.5 % yeast extract, 25 mM Na_2HPO_4, 25 mM KH_2PO_4, 50 mM NH_4Cl, 5 mM Na_2SO_4, 2 mM $MgSO_4$ (0.2× metals, optional), 0.5 % glycerol, 0.05 % glucose.

3 Methods

3.1 Assembly and Storage of Growth Media

Stock solutions were designed for convenience and flexibility in assembling different growth media and to avoid combining components that are incompatible upon autoclaving. Media are usually assembled from stock solutions immediately before use. However, the growth media can be stable for extended periods in the refrigerator if not contaminated.

3.2 Media in Routine Use

1. MDAG-11 non-inducing plates are used for transformation to isolate clones. Transformants of some clones that express target genes highly toxic to the host have been obtained on MDAG-11 plates but only poorly or not at all on plates containing complex media (which may also differ from lot to lot). Liquid culture is used to make dilutions for colony PCR or to titer cultures.
2. MDAG-135 non-inducing medium is used for growing cultures for temporary or long-term storage at −70 °C, for growing working or seed cultures, or for isolating plasmids (*see* **Notes 9** and **14**).
3. ZYM-5052 is the auto-inducing medium used routinely for screening pET clones for expression and solubility of target proteins in BL21(DE3) and for producing target protein for purification.
4. ZYM-505 is used for rapid growth of high-density cultures and for isolating plasmids from strains that do not supply T7 RNA polymerase. Because of the potential for unintended induction in this medium, plasmids from BL21(DE3) and BL21-AI are usually isolated from cultures grown on the non-inducing MDAG-135.

3.3 Special-Purpose Media

1. MDA-505 is used to test auto-induction as a function of inducer concentration (*see* **Note 10**).
2. MDA-5052 auto-inducing medium is used instead of ZYM-5052 when a defined medium is desired. The two media typically produce comparable levels of target protein.
3. MDASM-5052 is designed for labeling proteins with SeMet for crystallography (*see* **Note 13**).
4. The MD-5051 auto-inducing minimal medium is designed to be useful for specific labeling of target proteins for NMR analysis (*see* **Note 12**).
5. The MDG non-inducing minimal medium may be useful in labeling experiments and for defining nutritional requirements (*see* **Note 11**).

3.4 Growth of Cultures, Aeration, and Scale-Up

Reasonably good aeration is important for maintaining metabolically balanced pH near neutral and obtaining growth to high cell densities. We typically grow cultures in a temperature-controlled rotary shaker at ~350 rpm, using vessels and volumes of culture that give approximately equivalent levels of aeration. For auto-induction of many cultures in parallel to test expression and solubility, we grow 0.5 ml cultures in 13 × 100 mm glass culture tubes with plastic caps in the rotary shaker-incubator upright in plastic racks that hold up to 72 tubes. At the scale we work, this is much more convenient and controllable than using multi-well microtiter plates. However, microtiter plates can be used in high-throughput automation if aeration is properly addressed. Usually only a few microliters from such cultures are needed for reading densities (*see* **Note 15**), analysis by gel electrophoresis (*see* **Note 16**), and any other likely testing. Up to 2.5 ml of culture in non-inducing media is grown in 18 × 150 mm glass culture tubes in the same way to make freezer stocks, plasmid preps, or seed stocks for moderate-scale auto-induction (*see* **Note 14**). Seed stocks for larger scale auto-induction are grown in Erlenmeyer flasks, the culture occupying approximately 5–10 % of the flask volume. Moderate-scale auto-induction can use 400–500 ml of culture in 1.8-l baffled Fernbach flasks (Bellco).

In general, auto-inducing cultures are inoculated with one-thousandth volume from a culture grown to saturation in non-inducing MDAG-135 from a freezer stock. The high dilution allows the entire culture to be growing uniformly by the time auto-induction begins. Cultures grown at 37 °C are typically grown overnight for 12–16 h, well past saturation. Saturation in ZYM-5052 at 37 °C typically reaches an OD_{600nm} around 7–10 but can reach 20–30 in some highly expressing strains. Incubation for several hours at saturation after auto-induction usually has little effect on accumulation or solubility of target proteins.

Auto-induction works over the entire temperature range from 18 to 37 °C, an advantage because some target proteins are substantially more soluble when expressed at lower temperatures. However, care must be taken to ensure that auto-induction is complete before harvesting cultures grown at the lower temperatures. Saturation densities are usually significantly higher than at 37 °C, presumably because of the higher solubility of oxygen at the lower temperatures. To shorten the total incubation time, we typically grow cultures for a few hours at 37 °C until they become lightly turbid (less than OD_{600nm} ~1) and then transfer them to the lower temperature. It is a good idea to continue incubation of cultures grown overnight at low temperature and to read the culture density a few hours apart to be sure that they are saturated. I learned this lesson when a colleague harvested a low-temperature culture because it had reached a high density overnight only to find that it actually saturates at a considerably higher density and had not yet induced. Sometimes incubation over two nights may be required.

In general, increasing the rate of aeration increases the density at which auto-induction begins and the density at which the culture saturates. Higher aeration also increases the minimum concentration of lactose needed for good induction. The standard 0.2 % lactose was chosen to be well in excess of that needed for good auto-induction over the range of conditions tested. Auto-induced culture densities greater than OD_{600nm} ~50 have been attained by using higher levels of glycerol, higher levels of aeration, and appropriate metabolic balancing of pH with aspartate or succinate. Properly constituted auto-inducing media should be capable of producing even higher densities of fully induced cells in batch culture in fermentors, where high levels of oxygenation and near-neutral pH can be maintained to even higher densities.

3.5 Expression Clones, Selective Antibiotics, and Toxic Target Proteins

We use expression clones made by inserting the coding sequence for the target protein in a pET vector plasmid under the control of a T7*lac* promoter and the strong upstream translation signals for the T7 major capsid protein (Novagen), but clones in any of a wide range of vectors inducible by IPTG or arabinose are suitable. Initial clones are isolated by transformation into a host that does not supply T7 RNA polymerase (XL1Blue-MR in our work), and clones are usually verified by DNA sequencing. Expression plasmids are then transformed into BL21(DE3) or BL21-AI (usually selected on MDAG-11 plates), freezer stocks and working cultures are grown in MDAG-135, and target proteins are produced by auto-induction, usually in ZYM-5052 or MDA-5052 (*see* **Notes 6**, **7**, and **14**). Compatible plasmids that supply rare tRNAs are also included in the expression strain if there are likely to be issues with codon usage.

Plasmids used to clone and express target genes usually carry a gene that confers resistance to an antibiotic to allow selection of the desired clones in transformation and to help maintain cultures in which the vast majority of cells are capable of inducing production of the target protein. However, proper practice is important if cultures are not to become overgrown by cells that have lost plasmid. If basal expression is sufficient to stress the cell or if unintended induction occurs, the unwary can end up trying to produce target protein from cultures in which only a small fraction of cells remain competent to express it. Clones in some early vectors for expression by T7 RNA polymerase had significant basal expression and the problem was discussed in some detail [1–4]. The antibiotic ampicillin is degraded by a secreted enzyme, β-lactamase, and is usually destroyed by the time turbidity becomes apparent in a culture, at which point cells that have lost plasmid can begin to overgrow the culture. Furthermore, enough β-lactamase can be produced and secreted that even a 200-fold dilution to grow a subculture can bring along enough enzyme to destroy the ampicillin present in the fresh medium and allow continued overgrowth of the culture. Kanamycin also had unanticipated problems. As pointed out in the first paragraph of Subheading 2.2, kanamycin loses the ability to restrict the growth of BL21(DE3) (and presumably other *E. coli* strains as well) in rich media with commonly used phosphate concentrations. Once recognized, such problems can be avoided.

Problems due to basal expression of target protein are much reduced for the great majority of proteins when expressed from a T7*lac* promoter. In the few cases we have examined, equivalently high accumulation of target protein was obtained in MDA-5052 auto-inducing medium whether selective antibiotic was present in the medium or not, indicating that basal expression of target protein in the early stage of growth in auto-inducing media is low enough not to stress the cell significantly.

However, occasional target proteins are highly toxic to the cell at extremely low concentrations. In the limit case where a single transcript of the target gene can generate enough protein to kill the cell, an expression plasmid could not be maintained in a culture unless the stochastic bursts of target protein from all the plasmids in the cell occurred at a frequency significantly lower than an average of once per cell division. Basal expression of target protein will be reduced if basal expression of T7 RNA polymerase is reduced. This seems to be the case for BL21-AI, where T7 RNA polymerase appears to be produced from the uninduced pBAD promoter at a lower rate than in BL21(DE3) from the uninduced *lacUV5* promoter. Indeed, some toxic target genes we have worked with were easier to establish and maintain in BL21-AI than in BL21(DE3), although both hosts showed signs of stress.

Some tools are available to help in trying to obtain, maintain, and express clones that express highly toxic target proteins. As pointed out in **Note 6**, some clones that cannot be obtained on commonly used plates containing complex media such as LB can be obtained by selection on non-inducing MDAG-11 plates. Colony PCR with appropriate primers is a rapid way to screen many colonies from a transformation plate for the presence of inserts. Touching a colony with a sterile pipettor tip and dispersing the cells in 0.5 ml MDAG-11 non-inducing medium by vortexing produces a suspension that can be used directly for PCR (1 μl in a 25 μl PCR reaction) or after further dilution. The composition of a culture at any point in growth can also be determined by titering on four different plates [1, 4]: (1) all viable cells will form colonies on an appropriate nutrient plate; (2) only cells that retain plasmid will form colonies on a plate containing the selective antibiotic; (3) cells that have lost plasmid or that cannot induce the target protein will give colonies on a plate with strong inducing capacity, such as an inducing concentration of IPTG where repression is maintained by *lac* repressor; and (4) mutant cells that retain plasmid but cannot induce target protein will give colonies on plates containing both the selective antibiotic and the inducing capacity. Using such a plating assay can help to determine where problems lie.

4 Notes

1. Occasionally 50× M has showered crystals upon standing at room temperature. They can be redissolved in the microwave.
2. The mixture of 17 amino acids is quite acidic and may have to be neutralized when using final concentrations greater than 200 μg/ml of each. The effect of neutralizing the stock solution has not been explored. Trp and His slowly oxidize, producing a slightly yellow color with time. Tyr and Cys are not included because Tyr has low solubility and Cys oxidizes with time to precipitate as insoluble cystine. Met is not included so that SeMet can be used for labeling proteins for crystallography.
3. All 18 amino acids could be dissolved together at 10 mg/ml of each, but the solution did not remain soluble upon storage in the refrigerator.
4. The trace metal mix was assembled from autoclaved stock solutions of the individual components except for $FeCl_3$, which was added from the 0.1 M solution in ~0.12 M HCl. Defined media made with purified components usually will not have a sufficient supply of trace metals for growth to high density and auto-induction. The trace metal mix was designed to supply all of the trace metals known to be needed: 0.2× metals is sufficient for growth to high density and auto-induction in the media we use, and 1× metals attempts to supply sufficient amounts to saturate most metal-binding target proteins whose metal requirements may

not be known. As much as 5× metals in the growth medium can be tolerated with little effect on saturation density. The most critical need is for iron: less than 5 μM limited growth in minimal media and less than 10 μM limited growth in defined media containing amino acids. If a trace metal mixture is not available, 100 μM $FeCl_3$ supported growth in a defined medium almost as well as the total metal mix. The highest iron concentration tested, 800 μM, remained soluble in 1 mM citrate and was well tolerated. Citrate at a concentration of 1 mM in growth media may prevent a light turbidity due to added iron or metal mix but is not necessary for their beneficial effects. Trace metals are generally not needed in complex media, but 0.2× metals could be added to ensure that metal requirements are met.

5. N-Z-amine AS, a soluble enzymatic digest of casein in 100-lb barrels, and yeast extract (HY-YEST 444 in a 55-lb barrel) were obtained from Quest International, 5515 Sedge Blvd., Hoffman Estates, IL 60192, telephone 800-833-8308. These or equivalent materials (e.g., tryptone) are also available in various quantities from Difco, Sigma, Fisher, or other biochemical and chemical suppliers.
6. MDAG-11 plates are used for transformation of clones into BL21(DE3) (competent cells from Novagen), BL21-AI (competent cells from Life Technologies), or XL1Blue-MR, a host that does not supply T7 RNA polymerase and which requires thiamine for growth in minimal media (competent cells from Stratagene). Colonies appear after overnight incubation at 37 °C almost as rapidly on plates containing MDAG-11 as on plates containing complex media. Transformants of some clones that express target genes highly toxic to the host have been obtained on MDAG-11 plates but only poorly or not at all on plates containing complex media (which may also differ from lot to lot).
7. Plates should dry at room temperature for a day or two before using them or placing in a sealed plastic bag and storing in the refrigerator. Remove condensed moisture inside the lids with a Kimwipe. For use on the same day as pouring, allow the agar to set and then place in a 37 °C incubator with lids removed for 30–60 min or until the agar surface begins to show the fine lines or creases that indicate drying. To prevent small bubbles from appearing, plates that have been stored in the refrigerator should be separated on a bench top and allowed to warm gradually to room temperature for several hours before incubating at 37 °C.
8. The lower concentration of glucose in MDAG-11 is to prevent standing cultures or colonies on agar plates from becoming too acidic from metabolism of excess glucose at low dissolved oxygen concentration.

9. The combination of 0.35 % glucose, 0.1 % aspartate, and 200 μg/ml of each of 18 amino acids (0.36 %) in MDAG-135 was arrived at experimentally to provide metabolic balancing of pH at relatively high aeration so that cultures grow to high cell density and arrive at saturation near-neutral pH. Poor aeration should be avoided, as such cultures may become quite acidic.
10. The fully defined non-inducing MDA-505 contains the same mixture of carbon and energy sources as the fully defined auto-inducing MDA-5052 except for lack of an inducing sugar. An expression strain capable of producing large amounts of target protein grows to saturation in MDA-505 with no detectable induction, making this medium suitable for testing the effectiveness of different concentrations of inducing sugars.
11. The only carbon and energy sources in MDG are 0.5 % glucose and 0.25 % aspartate for metabolic balancing of pH. Succinate can replace aspartate to make NH_4 the only source of nitrogen in this medium, and glycerol could replace glucose if desirable for labeling. BL21(DE3) grows well in this minimal medium.
12. The only carbon and energy sources available to BL21(DE3) in the minimal auto-inducing MD-5051 are 0.5 % glycerol, 0.05 % glucose, and 0.25 % aspartate for metabolic balancing of pH, since galactose cannot be metabolized by BL21 strains. Succinate can replace aspartate to make NH_4 the only source of nitrogen in this medium, and glucose should have been exhausted when production of target protein gets under way. BL21(DE3) grows well in this medium, which could be adapted as needed for labeling target proteins with various stable isotopes for NMR analysis.
13. MDASM-5052 is a reformulation of PASM-5052, previously used for SeMet labeling of target proteins in BL21(DE3) with greater than 90 % replacement of Met by SeMet [6], to reduce the phosphate concentration from 100 to 50 mM. Methionine at 10 μg/ml facilitates growth and auto-induction in 125 μg/ml SeMet, which would otherwise be too toxic. Vitamin B_{12} stimulates the *metH* enzyme, which should regenerate SeMet from selenohomocysteine generated in methylation reactions and thereby help to make efficient use of SeMet. Vitamin B_{12} is present at the relatively high concentration of 1 μM so as to be taken up in spite of the BL21 deficiency in the *btuB* transporter [8]. Cultures grown well aerated from a 1,000-fold dilution of uninduced expression strain should reach saturation fully induced in 14–16 h at 37 °C. We use 400 ml of medium in a 1.8-l baffled flask (Bellco) in a rotary shaker at ~350 rpm. Yields of SeMet-labeled target protein have been comparable to yields of unlabeled target protein obtained in the absence of SeMet.

14. Freezer stocks for storage at −70 °C are typically made by placing in a 2-ml plastic freezer tube 1 ml of fresh overnight culture grown at 37 °C in non-inducing MDAG-135, adding 0.1 ml of 80 % glycerol, mixing well by vortexing, and placing the tube directly in a storage box in a −70 °C freezer. Working cultures or seed cultures are inoculated from the frozen stock as needed by scraping material from the frozen surface with a sterile pipettor tip without thawing the remainder. An advantage of these defined, pH-balanced, non-inducing media is that working cultures and colonies on plates retain plasmid and high viability when stored for weeks in the refrigerator, much longer than is typical in complex media.
15. Cultures grow to such high densities in these media that saturated cultures are routinely diluted 100-fold to read an accurate OD_{600nm} in a spectrophotometer with a 1 cm path length. Because the reading is due to light scattering, an accurate reading requires dilution to an OD lower than ~0.200.
16. Protein patterns of whole-cell lysates, soluble portion, and pelleted fraction are analyzed by SDS-polyacrylamide gel electrophoresis. Cells pelleted from a sample of culture in a 1.5-ml microcentrifuge tube are resuspended in 40 μl of lysis solution at a concentration equivalent to OD_{600nm} ~5 (the volume of culture centrifuged, in microliters, equals $200/OD_{600nm}$). The lysis solution is Bugbuster Protein Extraction Reagent containing 25 units/ml Benzonase Nuclease and 3 KU/ml recombinant lysozyme, all from Novagen. After allowing lysis for at least 30 min at room temperature, 20 μl is removed to a second tube and centrifuged for 1 min to separate soluble and pellet fractions, and the supernatant is carefully removed from the pellet with a pipettor and deposited in a new tube. Samples for electrophoresis are made by adding 10 μl of 3× SDS loading buffer to the 20 μl whole-cell lysate and supernatant samples, and the lysate pellet is suspended in 30 μl of 1× SDS sample buffer, all three being well mixed by vortexing. The three samples are heated for 2 min in a boiling water bath, and 10 μl of each is resolved by electrophoresis on a 4–20 % gradient gel, which is then stained with Coomassie brilliant blue.

Acknowledgements

Work was supported by the Office of Biological and Environmental Research of the US Department of Energy, the Protein Structure Initiative of the National Institute of General Medical Sciences of the National Institutes of Health, as part of the New York Structural Genomics Research Consortium, and by internal research funding from Brookhaven National Laboratory.

References

1. Studier FW, Moffatt BA (1986) Use of bacteriophage T7 RNA polymerase to direct selective high-level expression of cloned genes. J Mol Biol 189:113–130
2. Studier FW (1991) Use of bacteriophage T7 lysozyme to improve an inducible T7 expression system. J Mol Biol 219:37–44
3. Dubendorff JW, Studier FW (1991) Controlling basal expression in an inducible T7 expression system by blocking the target T7 promoter with lac repressor. J Mol Biol 219:45–59
4. Studier FW, Rosenberg AH, Dunn JJ, Dubendorff JW (1990) Use of T7 RNA polymerase to direct expression of cloned genes. Methods Enzymol 185:60–89
5. Grossman TH, Kawasaki ES, Punreddy SR, Osburne MS (1998) Spontaneous cAMP-dependent derepression of gene expression in stationary phase plays a role in recombinant expression instability. Gene 209:95–103
6. Studier FW (2005) Protein production by auto-induction in high-density shaking cultures. Protein Expr Purif 41:207–234
7. Xu J, Banerjee A, Pan S-H, Li ZJ (2012) Galactose can be an inducer for production of therapeutic proteins by auto-induction using *E. coli* BL21 strains. Protein Expr Purif 83:30–36
8. Studier FW, Daegelen P, Lenski RE, Maslov S, Kim JF (2009) Understanding the differences between genome sequences of *Escherichia coli* B strains REL606 and BL21(DE3) and comparison of the *E. coli* B and K-12 genomes. J Mol Biol 394:653–680
9. Adhya S, Echols H (1966) Glucose effect and the galactose enzymes of *E. coli*: Correlation between glucose inhibition of induction and inducer transport. J Bacteriol 92:601–608

Chapter 3

High-Throughput Expression Screening and Purification of Recombinant Proteins in *E. coli*

Natalie J. Saez and Renaud Vincentelli

Abstract

The protocols outlined in this chapter allow for the small-scale test expression of a single or multiple proteins concurrently using several expression conditions to identify optimal strategies for producing soluble, stable proteins. The protocols can be performed manually without the need for specialized equipment, or can be translated to robotic platforms. The high-throughput protocols begin with transformation in a 96-well format, followed by small-scale test expression using auto-induction medium in a 24-well format, finishing with purification in a 96-well format. Even from such a small scale, there is the potential to use the purified proteins for characterization in pilot studies, for sensitive micro-assays, or for the quick detection of and differentiation of the expected size and oxidation state of the protein by mass spectrometry.

Key words *E. coli*, Bacteria, Expression, Recombinant, High-throughput, Purification, Auto-induction, Immobilized metal affinity chromatography (IMAC), TEV cleavage

1 Introduction

Traditionally, protein production approaches have centered on the case-by-case exploration of proteins of particular interest. With advances in genomics and thousands of novel and interesting proteins being discovered at such an accelerated rate, these production strategies have become outdated, causing a bottleneck in structural and functional studies. Parallelization of these traditional approaches into high-throughput pipelines at a small scale allows the screening for optimal expression conditions, enabling the testing of various parameters on soluble expression levels. This may include, but is not limited to, using varying expression strains [1, 2], temperature [3, 4], media [2, 3], target variants [5], fusion partners [6–12], co-expression with chaperones [13, 14], cytoplasmic or periplasmic expression [15], and purification buffer components [3]. Testing all of these variables using traditional methods would be highly inefficient. However, by implementing high-throughput approaches,

Yu Wai Chen (ed.), *Structural Genomics: General Applications*, Methods in Molecular Biology, vol. 1091, DOI 10.1007/978-1-62703-691-7_3, © Springer Science+Business Media, LLC 2014

up to 96 expressions and purifications can be tested in parallel, and multiple can be performed in any 1 week. This means that many variables can be tested on a few targets or a few variables can be tested on many targets with a high level of efficiency. The strategy also provides good reproducibility upon scale-up as the same culture and purification conditions are utilized at both stages.

Our high-throughput strategy utilizes *E. coli*, taking advantage of its ease of use, fast growth rates, relatively low cost of production, and adaptability to scaling up cultures for large-scale expression once optimal conditions are identified. With the range of *E. coli* strains available, the system is also applicable to a wide range of targets, even those that are not codon optimized for expression in *E. coli* (using strains that compensate for rare codons) or for targets with complex folds, containing multiple disulfide bonds (using strains that modify the reducing environment of the cytoplasm). Given that a large proportion of constructs are generally cloned directly without codon optimization, for the protocol described herein we have chosen to utilize the Rosetta 2 (DE3) pLysS strain, which carries tRNAs for rare codons that are not highly expressed in *E. coli*. However, it is possible for the high-throughput protocol to be trialled using different strains to check for variances in the soluble expression levels of target proteins and to continue optimization on the most desirable strain. We also use auto-induction [16], which simplifies expression, eliminating the need for manual induction. One downfall of expression in *E. coli* is that it does not allow for posttranslational modifications, other than disulfide bonding, and for these types of targets alternative production methods need to be sought (either by expression in eukaryotes or by chemical synthesis for small proteins (<70 residues)).

Apart from the expression strain, several other factors are important for the optimization of expression. Temperature is also a factor that influences the rate of expression, and may affect the solubility levels of the target protein. For poorly soluble proteins it may be beneficial to reduce the rate of expression by decreasing the temperature in order to prevent aggregation issues arising from too high a rate of expression [3, 4]. While there is evidence that auto-induction medium can give higher or equivalent levels of expression than standard media (e.g., LB, TB, 2YT) with IPTG induction [6], for cases where soluble expression is low or not detected, alternative media can also be trialled during screening as it may have an influence on soluble expression for particular proteins. The choice of fusion partner is another important factor in protein expression. Various fusion tags can be used to aid in the solubilization or proper folding of the targets (e.g., thioredoxin (TRX) [9], small ubiquitin-like modifier (SUMO) [6, 10], maltose binding protein (MBP) [8, 11], N-utilizing substance A (NusA) [7], and glutathione *S*-transferase (GST) [12]). Several tags should be tested for each target as their effect on expression levels may be target-dependent. For example, tags with some redox activity may be more beneficial for target proteins containing multiple disulfide bonds [17]. For proteins containing complex folds, other strategies to be considered include the effect of periplasmic expression [18] compared to cytoplasmic expression and the effect of co-expression of chaperones to assist folding [13, 14]. For targets composed of individual domains from larger proteins, different constructs with variable N- and C-termini could be tested to determine the optimal domain boundaries (the multi-construct approach) [5]. For purification, there are several important factors to be taken into consideration including ionic strength (by varying salt concentration) and incorporation of protein structure stabilizing agents such as glycerol (for proteins with poor stability). A detailed review of all potential variables is out of the scope of this communication, but has been analyzed in further detail previously [2, 3].

The high-throughput strategy is also desirable for reasons other than efficiency. Cultures are grown in deep well 24 (DW24) format, which means that they can be grown using regular shaking incubators, in contrast to cultures grown in deep well 96 (DW96) format which require the use of specific, shaking incubators with a high speed of shaking to allow for reasonable culture aeration. The handling of cultures in this side-by-side manner also minimizes the variables involved in expression and purification, allowing for a more effective and simplified comparison of results. Specifically, because only one variable is changed well-to-well (such as the type of fusion partner), all the other parameters remain strictly the same (temperature, shaking, aeration, medium) preventing artificial batch-to-batch variations. Even for laboratories with defined expression and purification conditions in place, it is possible to move directly into this system to aid side-by-side comparison of individual variables or in order to simply increase throughput, without needing to completely change their methods.

The strategy utilized at AFMB [19] is simple and generally applicable to a wide range of targets. We use these expression screening pipelines and high-throughput protocols in a semi-automated way (using a Tecan liquid handling robot and Caliper GXII LabChip system for analysis of results) for up to 1,152 (12×96) cultures in parallel over 1 week [19, 20]. However, it is important to note that the same methods are suitable for a high-throughput manual approach. This means that these protocols can be used in laboratories with a basic setup and without the need of expensive equipment. In fact, depending on the detection system used for analyzing solubility, the procedure can be done manually with a throughput of 96 (using SDS-PAGE detection) or 384 (4×96; using dot blot and SDS-PAGE [20]) cultures per week. Using the recommended protocol (*see* Fig. 1), there is a reasonable

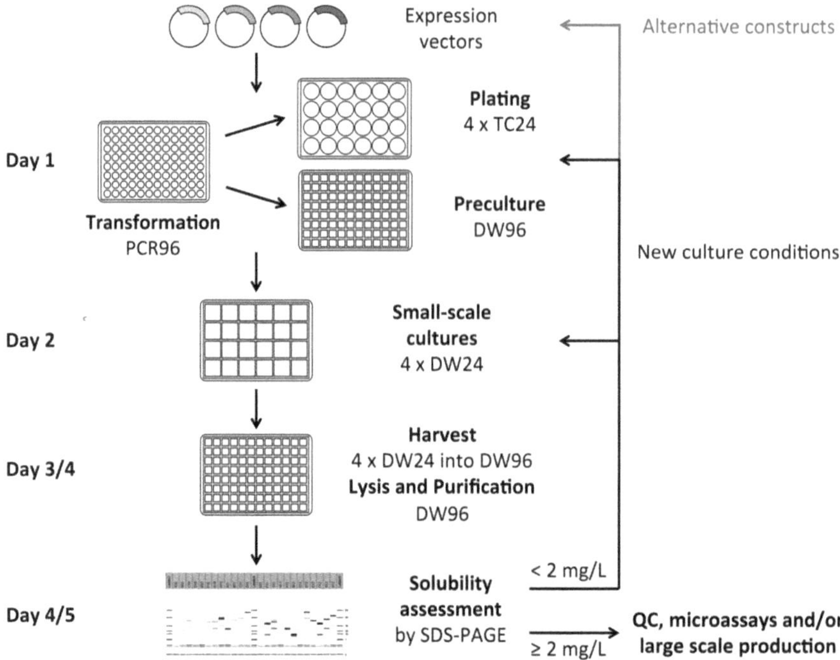

Fig. 1 Schematic representation of the high-throughput expression screening protocol. The variables to be tested in the expression screening protocols (culture conditions: strain, temperature, medium; and constructs: fusion partner, target construct, tag position, co-expression, periplasmic expression) are discussed further in the Introduction text and in Pipeline 1 and 2 (Fig. 2a, b, respectively). Using this protocol, 96–384 conditions can be tested in 1 week using manual methods [19], or up to 1,152 conditions can be tested in an automated manner [19]

chance that sufficient quantities of soluble proteins will be obtained for the majority of targets in the time frame of 1 week. For further details regarding the choice of conditions in this strategy *see* ref. 19.

Using this strategy there are two possible pipelines to be considered, the choice of which depends on the number of targets to be tested. For optimal plate layout, the number of targets should be restricted to 8, 16, 24, 32, 48 or 96 (one column, one sixth, a quarter, a third, half or a full plate, respectively). For 48 targets and above Pipeline 1 is recommended, while for 1–24 targets Pipeline 2 is recommended (*see* Fig. 2a, b, respectively, explained in further detail below). In both cases, for a new project, expression screening is initially carried out in only one culture condition. The cultures are grown in ZYP5052 auto-induction medium [16] at 37 °C for 4 h, at which point the temperature is then dropped to 17 °C and left overnight for another 18–20 h. The change in temperature corresponds to the glucose depletion time and the induction of expression by lactose. Because we usually produce several proteins concurrently, we use the Rosetta 2 (DE3) pLysS strain as the default strain without considering the protein origin or codon optimization, however, if codon optimization has been performed then BL21 (DE3) pLysS is also suitable. For cloning, all

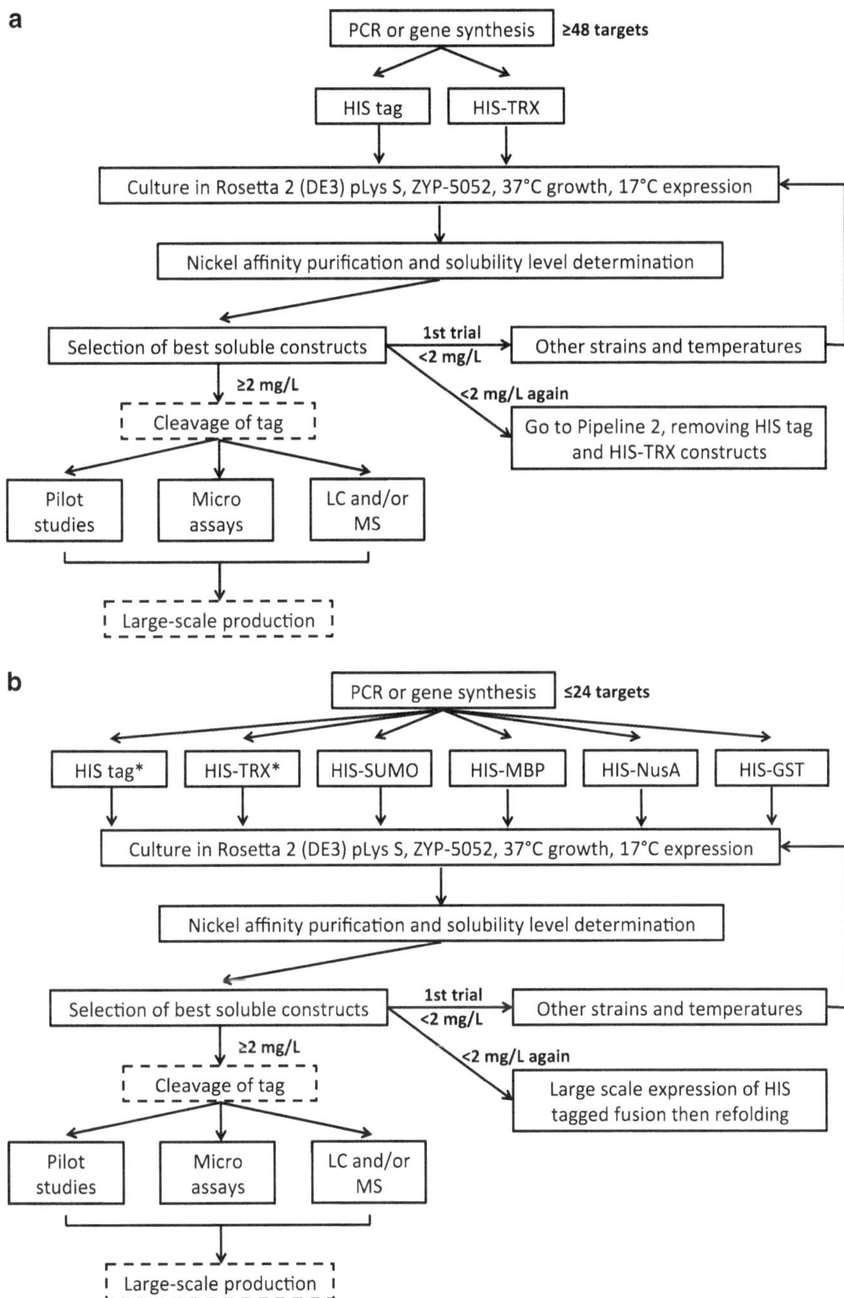

Fig. 2 Constructs are initially expressed in Rosetta 2 (DE3) pLysS at 37/17 °C. Purification is performed on nickel resin followed by detection of soluble constructs via SDS-PAGE (or dot-blot then SDS-PAGE or on a Caliper Lab Chip). If the first round of expression screening is unsuccessful, alternative culture conditions are trialled. If constructs produce soluble proteins in high enough yields, microassays and quality control can be performed and, if required, large-scale production can be pursued. Cleavage of the tag and large-scale production are optional steps. (**a**) Pipeline 1 is recommended for 48 or more constructs, where initially only HIS and HIS-TRX constructs are trialled. If Pipeline 1 is not successful, expression screening is resumed at Pipeline 2. (**b**) Pipeline 2 is recommended for 1–24 targets, where up to 6 expression constructs are trialled in parallel. Note, if Pipeline 1 has already been trialled, the HIS tag and HIS-TRX constructs (marked with an *asterisk* (*)) can be removed in Pipeline 2

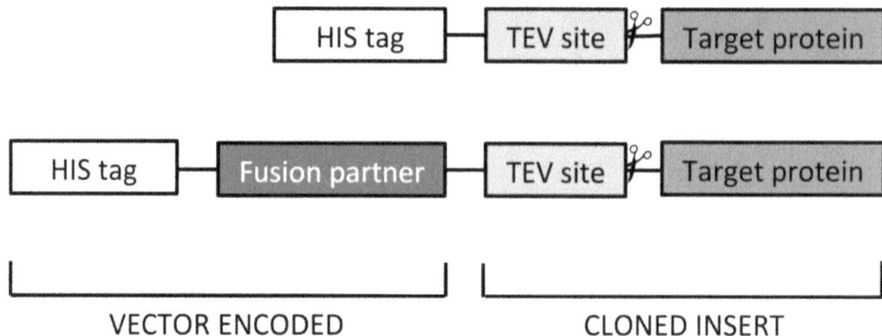

Fig. 3 Constructs used in Pipeline 1 and Pipeline 2 (Fig. 2a, b, respectively). The HIS tag is used for nickel affinity purification. The fusion partner is used to increase solubility and/or aid folding of the target protein. The fusion partners suggested in this protocol are thioredoxin (TRX) [9], small ubiquitin-like modifier (SUMO) [6, 10], maltose binding protein (MBP) [8, 11], N-utilizing substance A (NusA) [7], and glutathione *S*-transferase (GST) [12], however alternative fusion partners, or a different number of fusion partners, can be selected at the user's discretion. The TEV site enables cleavage of the HIS tag and fusion partner if required, leaving only a single glycine or serine residue at the N-terminus of the native target protein. In these constructs the target protein does not need to be codon optimized as expression is generally performed in Rosetta 2 (DE3) pLysS, however, if the target protein is codon-optimized expression can be performed in BL21 (DE3) pLysS instead. Stop codons should be added after the target protein sequence prior to cloning into the expression vectors

proteins are directly preceded by a TEV protease cleavage site (ENLYFQ/[G or S]) to produce native protein (with a single vestigial glycine or serine) after cleavage (Fig. 3). While not essential for this protocol, we utilize the Gateway system for cloning [21]. This system is very efficient and gives us access to numerous clone collections from private or public libraries. Furthermore, with its versatility and the hundreds of vectors available, it allows us to work on the same clones as collaborators that may be working with divergent techniques, some examples being expression in eukaryotic cells, yeast two hybrid or in vivo localization.

For Pipeline 1 (suitable for 48 or more constructs; Fig. 2a) all proteins are cloned directly into two plasmids [19] containing an N-terminal HIS-tag (pDEST17OI) [22] or a HIS-TRX fusion-tag [9]. If the HIS-tag construct is soluble (above 2 mg/L), then production is carried out with this construct. Alternatively, if the HIS-tag construct is not soluble or produces less than 2 mg/L of protein, and if the TRX construct is soluble above 2 mg/L, production is carried out using this construct. In all other cases, constructs are screened using 3 alternative strains (BL21 (DE3) pLysS, Origami (DE3) pLysS, and C41 (DE3) pRos) at three induction temperatures (37/25/17 °C) in ZYP5052. The same thresholds are applied and for recalcitrant targets alternative fusion partners are pursued as in Pipeline 2.

Due to the fewer number of targets in Pipeline 2 (1–24 targets; Fig. 2b), more fusion partners can be tested initially to identify those that produce higher soluble yields. In this case the targets are cloned in up to six vectors: one containing an N-terminal HIS-tag [22], and the others containing a HIS-tag with additional fusion partner. The recommended fusion partners in this case are: thioredoxin (TRX) [9], small ubiquitin-like modifier (SUMO) [6, 10], maltose binding protein (MBP) [8, 11], N-utilizing substance A (NusA) [7], and glutathione *S*-transferase (GST) [12], however alternative fusion partners, or a different number of fusion partners, can be selected at the user's discretion. As in Pipeline 1, if the HIS-tag construct is soluble (above 2 mg/L), then production is carried out with this construct. Alternatively, if the HIS-tag construct is not soluble or produces less than 2 mg/L of protein, and if at least one of the fusion constructs is soluble above 2 mg/L, production is carried out using the construct producing the highest yield. In all other cases, constructs are screened using 3 alternative strains (BL21 (DE3) pLysS, Origami (DE3) pLysS, and C41 (DE3) pRos) at three induction temperatures (37/25/17 °C) in ZYP5052. The same thresholds are applied at this stage. For recalcitrant targets, the next step would be to purify the insoluble HIS-tagged target from inclusion bodies, solubilize, and refold (this is out of the scope of this protocol and will not be discussed here, *see* ref. 23).

The amount of resin used in the purification steps can be adjusted (either 50 or 200 μL, as discussed further in Subheadings 3 and 4) allowing simple high-throughput screening useful for the quick comparison of expression conditions [17], or (with the larger amount of resin) there is even the potential to purify tagged proteins directly from small-scale expressions for characterization by sensitive functional assays, binding assays (e.g., Systematic Evolution of Ligands by Exponential Enrichment (SELEX) for DNA-binding proteins [19, 24]) or in pilot studies where tens of micrograms of sample is sufficient. Yields are also generally sufficient for the quick detection of and differentiation of the expected size and oxidation state of the protein by mass spectrometry or to confirm homogeneity by chromatographic methods [25]. In many cases it is unnecessary or even undesirable to cleave the fusion tag. Cleavage can be a limiting step in terms of yield, due to suboptimal cleavage efficiency, and poor recovery after further purification, therefore if the tag does not interfere with the structure or function of the target it is advisable to leave the fusion protein intact. For proteins that are poorly soluble and prone to aggregation, often the fusion protein is used to maintain solubility and should not be removed, for example during crystallization in structural studies [26, 27]. If cleavage of the fusion tag is desired, it is also possible to perform TEV cleavage from the small-scale purification to analyze the efficiency of cleavage, optimize cleavage conditions if necessary and obtain a reliable estimate of yields for future scale-up experiments, at which time the native protein can be purified.

Once the optimal conditions for soluble expression have been defined (for example strain, temperature, media and fusion partner), production can be scaled up for the production of milligram quantities of purified proteins for further structural and functional studies. By extrapolating the yield from small-scale expression, the culture volume required at large scale can be inferred (typically 1–5 L). When growing scale-up culture, in auto-induction medium, to get optimal aeration we limit the volume of medium to 800 mL in a 2 L flask. For 800 mL of culture, a 20 mL preculture is generally inoculated from the glycerol stock or LB agar plate produced at expression screening (of the best fusion in the best cell line). The culture is grown at the same temperature and in the same medium as the optimal condition from expression screening. Lysis is achieved in the same way as at the small scale, but in the presence of a suitable protease inhibitor (e.g., PMSF). The whole cell lysate is first clarified by centrifugation to obtain the soluble cell lysate.

The general large-scale strategy employed at AFMB [28] is to purify the protein in a semi-automated manner on an AKTA Express. In the first step nickel affinity purification is performed on the soluble cell lysate using the same buffers as in the test purification with a HisTrap FF Crude column (GE Healthcare Life Sciences, Product code: 17–5286). After this step we have our target protein purified in a high concentration of imidazole. The next step is dependent on whether tag cleavage is required. If the HIS-tagged version is soluble or the fusion is more stable, there is no need to remove the tag and the second step is gel filtration (on a Superdex 200 preparative grade column, GE Healthcare Life Sciences, Product code: 28-9893-35) for the purposes of oligomeric state characterization as well as buffer exchange into an appropriate buffer for subsequent applications. If the tag is to be removed, desalting must be performed first. Here, the protein is desalted (on a HiPrep 26/10 desalting column, GE Healthcare Life Sciences, Product code: 17-5087-01) into a buffer with no imidazole. Tag cleavage is then performed using the same conditions identified at small scale. The tag and protease (which also carries a HIS-tag) are removed by reapplying the cleavage mixture over the regenerated and re-equilibrated nickel column. This time the column flowthrough, containing the purified target protein, is collected. An additional gel filtration step, as described above for tagged protein, can be performed to assess oligomerization and for buffer exchange into an appropriate buffer for subsequent applications. Quality control by LC-MS can be performed at various stages during the purification in order to confirm the integrity of the protein produced. Alternatively, purification can be performed manually with gravity flow nickel columns followed by buffer exchange using desalting columns (such as PD-10 desalting columns, GE Healthcare Life Sciences, reference 17-0851-01) if an AKTA express is not available.

A case study utilizing the high-throughput screening approach for the expression of DNA-binding domains from *Ciona intestinalis* transcription factors and the subsequent Systematic Evolution of Ligands by Exponential Enrichment (SELEX) DNA-binding assay is given in [19, 24] and these protocols have continued to be used successfully for various other projects since their publication.

2 Materials

Material quantities are given for one set of 96 transformations, cultures and purifications. If more than 96 are being performed at one time, please adjust the values accordingly. Reference numbers for the author's preferred choice of materials are provided where relevant, however equivalent products may also be suitable. A regular shaking incubator can be used for all culture steps (at a speed of 200 rpm) for DW96 (LB preculture only) and DW24 and will provide sufficient aeration, however if a short orbital specialized incubator for higher speeds is available (such as an Infors Incubation Multitron Shaker with 3 mm throw, model number AJ103) this can be used at a speed of 800 rpm for DW96 and 400 rpm for DW24. The procedures are most efficient when using multichannel pipettes with variable span such as Matrix Equalizer Pipettes, Thermo Scientific. If there are none already available and the equipment budget is limited, it is possible to limit the purchase of these pipettes to the 15–1,250 μL pipette only (reference 2034), which is the most versatile pipette for these techniques.

2.1 Transformation Materials

1. 4×24-well sterile tissue culture (TC24) plates (Greiner Bio-One, reference 662160).
2. Antibiotics: Ampicillin (100 mg/mL in water), Chloramphenicol (34 mg/mL in ethanol), store stocks at −20 °C and use a 1 in 1,000 dilution.
3. (a) LB agar: Dissolve 10 g tryptone, 5 g yeast extract, and 10 g NaCl in ~950 mL water. Adjust the pH of the medium to 7.0 using 1 M NaOH. Add 15 g of agar and make up to 1 L. Dispense into 500 mL Schott bottles containing no more than 450 mL of LB agar each. Autoclave.

   (b) Preparation of agar plates: Melt a bottle of LB agar in a microwave set to low (ensuring that it does not boil over). Once thoroughly melted, allow it to cool to approximately 45 °C and add the required antibiotics. Pour onto plates immediately. The 24-well tissue culture plates should contain 2 mL of LB agar per well (*see* **Note 1**).
4. Chemically competent Rosetta 2 (DE3) pLysS *E. coli* strain (Novagen, Millipore).

After the initial purchase of competent cells, the cells are cultivated and made competent in-house, in order to be more cost-efficient. New batches of chemically competent cells are shock frozen (using liquid nitrogen) in 1 mL aliquots and stored at −80 °C.

5. 96-well PCR (PCR96) plate (Greiner Bio-One, reference 652270).
6. Multichannel pipettes with variable span (suitable for dispensing 1, 25, 60 and 100 μL volumes) to dispense reagents into a 96- and 24-well format (Matrix Equalizer Pipettes, Thermo Scientific: 1–30 μL, reference 2130-11 and 5–250 μL, reference 2032-11) and 100 mL disposable reagent reservoirs, sterile (Thermo Scientific, reference 8085).
7. Expression plasmids.
8. Adhesive tape pads (Qiagen, reference 19570).
9. PCR machine suitable for 96-well PCR plates.
10. Deep-well 96 (DW96) plate, autoclaved for sterility, with 2 mL volume capacity (Greiner Bio-One, reference 780270).
11. LB medium: Prepare in advance and store at room temperature. Dissolve 10 g tryptone, 5 g yeast extract, and 10 g NaCl in ~950 mL water. Adjust the pH of the medium to 7.0 using 1 M NaOH and make up to 1 L. Autoclave in volumes of less than 500 mL.
12. Repeat pipettor (Eppendorf, reference 22 26 020-1) and 50 mL Combitips (Eppendorf, reference 22 26 660-8).
13. Shaking incubator set to 37 °C.
14. Plate incubator set to 37 °C.
15. Breathseal breathable adhesive film (Greiner Bio-One, reference 676050).

2.2 Test Expression Materials

1. ZY medium: Dissolve 10 g of N-Z-amine AS (or any tryptic digest of casein, e.g., tryptone) and 5 g of yeast extract in 925 mL of water and autoclave. A final volume of 1 L ZYP-5052 medium will be achieved with the addition of the remaining components ($MgSO_4$, 5052, and NPS, provided in *no. 2*, *3*, and *4*, respectively).
2. 2 M $MgSO_4$ stock: Dissolve 49.3 g of $MgSO_4 \cdot 7H_2O$ in water to a final volume of 100 mL. Autoclave.
3. 5052 50× stock: In a beaker weigh out 250 g of glycerol. To this, add 730 mL water, and while stirring, add 25 g of glucose and 100 g of α-lactose. Lactose dissolves slowly; stirring over low heat will hasten the process. Autoclave once dissolved.
4. NPS 20× stock: To 900 mL of water in a beaker, add (in the following order) 66 g of $(NH_4)_2SO_4$, 136 g of KH_2PO_4, and 142 g of Na_2HPO_4. Stir until dissolved, then autoclave. A 20-fold dilution in water should have a pH of around 6.75.

5. Antibiotics: Ampicillin (100 mg/mL in water), Chloramphenicol (34 mg/mL in ethanol), store stocks at −20 °C and use a 1 in 1,000 dilution.
6. 4× Deep well 24 (DW24) plates autoclaved for sterility, with 10 mL volume capacity (Whatman UNIPLATE, reference 7701–5102).
7. Repeat pipettor (Eppendorf, reference 22 26 020–1) and 50 mL Combitips (Eppendorf, reference 22 26 660–8).
8. Multichannel pipettes with variable span (suitable for dispensing 50, 100, 150 µL and 1 mL volumes) (Matrix Equalizer Pipettes, Thermo Scientific: 5–250 µL reference 2032–11, 15–1,250 µL reference 2034) and 1,250 µL pipette tips (Matrix, Thermo Scientific, reference 8041–11).
9. Breathseal breathable adhesive film (Greiner Bio-One, reference 676050).
10. Shaking incubator set to 37 °C that can be adjusted to 17 °C.
11. Flat-bottomed, clear microtiter plate (Greiner Bio-One, reference 655101).
12. 96-well plate reading spectrophotometer for measuring OD_{600nm} (optical density) of bacterial cultures.
13. Centrifuge with rotor for deep well plates (3,800 × *g*).
14. Bactinyl (Orapi Group) or equivalent microbial disinfectant.
15. Lysozyme stock (50 mg/mL): Dissolve 0.5 g lysozyme in water to a final volume of 10 mL. Store in 0.5 mL aliquots at −20 °C.
16. Imidazole ACS grade (Merck, reference 104716). A high quality grade of imidazole must be used so that it will not interfere with A_{280nm} readings for calculating protein yield.
17. Lysis/binding buffer 10× stock (*see* **Note 2**): Prepare 1 L of buffer containing 500 mM Tris pH 8, 3 M NaCl and 100 mM Imidazole ACS grade (Merck, reference 104716) in advance, filter through a 0.22 µm filter and store at 4 °C.

    Preparation of lysis buffer: On the day of use, dilute 10 mL of 10× stock into a final volume of 100 mL. Add lysozyme stock to a final concentration of 0.25 mg/mL.
18. Deep-well 96 (DW96) plate, with 2 mL volume capacity (Greiner Bio-One, reference 780270).

2.3 Test Purification Components

1. Water bath.
2. Shaking incubator.
3. DNase stock (2 mg/mL): Dissolve 100 mg of DNase in 50 mL of water. Filter-sterilize and divide into 1 mL aliquots. Store aliquots at −20 °C.

4. 2 M $MgSO_4$ stock: Dissolve 49.3 g of $MgSO_4 \cdot 7H_2O$ in water to a final volume of 100 mL. Autoclave.
5. Multichannel pipettes with variable span (suitable for dispensing 5, 10, 15, 25 and 200 or 600 μL and 1.2 mL volumes) (Matrix Equalizer Pipettes, Thermo Scientific: 5–250 μL reference 2032–11, 15–1,250 μL reference 2034), 1,250 μL pipette tips (Matrix, Thermo Scientific, reference 8041–11), and 100 mL disposable reagent reservoirs (Thermo Scientific, reference 8085).
6. Adhesive tape pads (Qiagen, reference 19570).
7. 4× SDS-PAGE sample buffer (*see* **Note 3**): Prepare 10 mL of 250 mM Tris–HCl pH 6.8, 8 % SDS, 300 mM DTT, 30 % glycerol, 0.02 % bromophenol blue. Divide into 1 mL aliquots and store at −20 °C.
8. 4× 96-well PCR (PCR96) plates (Greiner Bio-One, reference 652270).
9. Ni Sepharose 6 Fast Flow resin (GE Healthcare, reference 17-5318-02): The resin is supplied in 20 % ethanol. Put aliquots of the resin in 15 mL falcon tubes. To equilibrate the resin, wash twice in water and then twice in binding buffer (*no. 10*). This is done by first centrifuging at 500 ×*g* for 1 min, discarding the supernatant by inverting the tubes and resuspending in water or buffer. Repeat at each step of equilibration. After the final wash, resuspend in binding buffer as a 25 % (v/v) (25:75 mL) or 33 % (v/v) (35:70 mL) (resin–buffer) slurry (*see* **Note 4**). Store the equilibrated resin at 4 °C when not in use.
10. Lysis/binding buffer 10× stock (*see* **Note 2**): Prepare 1 L of buffer containing 500 mM Tris–HCl pH 8, 3 M NaCl and 100 mM Imidazole ACS grade (Merck, reference 104716) in advance, filter through a 0.22 μm filter and store at 4 °C.

    Preparation of binding buffer: On the day of use, dilute 20 mL of 10× stock into a final volume of 200 mL.
11. Wash buffer 10× stock (*see* **Note 2**): Prepare 1 L of buffer containing 500 mM Tris–HCl pH 8, 3 M NaCl and 500 mM Imidazole ACS grade (Merck, reference 104716) in advance, filter through a 0.22 μm filter and store at 4 °C.

    Preparation of wash buffer: On the day of use, dilute 25 mL of 10× stock into a final volume of 250 mL.
12. Elution buffer 5× stock (*see* **Note 2**): Prepare 1 L of buffer containing 250 mM Tris–HCl pH 8, 1.5 M NaCl and 1.25 M Imidazole ACS grade (Merck, reference 104716) in advance, filter through a 0.22 μm filter and store at 4 °C.

    Preparation of wash buffer: On the day of use, dilute 20 mL of 5× stock into a final volume of 100 mL.

13. CHROMABOND® MULTI 96 vacuum manifold (Macherey-Nagel, reference 738630.M) and vacuum pump.
14. Macherey-Nagel 96-well Receiver/Filter Plate 20 μm, 1.5 mL capacity (Macherey-Nagel, reference 740686.4).
15. 4× Deep-well 96 (DW96) plates, with 2 mL volume capacity (Greiner Bio-One, reference 780270).
16. PCR machine suitable for 96-well PCR plates for boiling SDS-PAGE or Caliper samples.
17. SDS-PAGE Equipment (*see* **Note 3**): Electrophoresis apparatus and choice of gel type is at the user's discretion.
18. Spectrophotometer and cuvettes for measuring absorbance at 280 nm (A_{280nm}) to calculate yield of soluble proteins.
19. Optional: Deep-well 96 (DW96) plates, with 2 mL volume capacity (Greiner Bio-One, reference 780270) for elution 2 and collecting the soluble fraction after cleavage. 96-well PCR (PCR96) plates (Greiner Bio-One, reference 652270) for SDS-PAGE samples of elution 2, cleavage mixture and soluble fraction after cleavage. Tobacco Etch Virus (TEV) protease, 2 mg/mL. A 96-well 0.22 μm filter plate (Millipore, reference MSGV N22 10) to filter the soluble fraction after cleavage.

3 Methods

NB: All the steps below have been performed manually as described here or with slight variations in a fully automated manner using a liquid handling robot (TECAN Freedom EVO series) [19, 28]. A regular shaking incubator can be used for all culture steps (at a speed of 200 rpm) for DW96 (LB preculture only) and DW24 and will provide sufficient aeration, however if a short orbital specialized incubator for higher speeds is available (such as an Infors Incubation Multitron Shaker with 3 mm throw, model number AJ103) this can be used at a speed of 800 rpm for DW96 and 400 rpm for DW24. For simplicity, only the speed for a regular shaker is given in the protocols. If a high-speed shaker is used, please adjust the shaking speed accordingly.

3.1 Transformation into E. coli Rosetta 2 (DE3) pLysS

1. For each set of 96 constructs to be transformed, prepare four TC24 plates containing 2 mL of LB agar supplemented with 100 μg/mL ampicillin (or an alternative antibiotic that the expression vector is resistant to) and 34 μg/mL chloramphenicol (for pLysS strains). Allow to set and dry (*see* **Note 1**).
2. Thaw 3 × 1 mL of competent Rosetta 2 (DE3) pLysS strain on ice then aliquot 25 μL of competent cells into each well of a PCR96 plate using a multichannel pipette. Keep the plate on ice until the thermal shock in **step 4**.

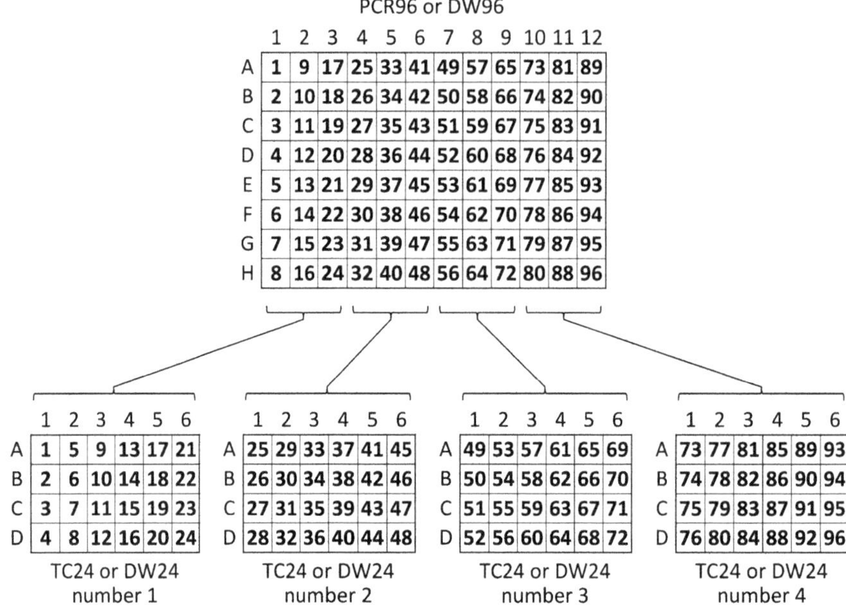

Fig. 4 Schematic for transferring from a single 96-well plate into four 24-well plates

3. Add 1 μL of the expression plasmids (at a concentration of ~10 ng/μL for pure plasmids) with a multichannel pipette (*see* **Note 5**). Ensure that the plasmid is dispensed into the cells but do not mix by pipetting. Cover the plate with plastic film to avoid contamination.
4. Incubate on ice for 30 min, then place the plate at 42 °C for 45 s (thermal shock), then transfer back to ice for 3 min (*see* **Note 6**). Add 100 μL of LB medium using a multichannel pipette and a reagent reservoir and incubate for 60 min at 37 °C.
5. In the meantime, prepare a sterile DW96 containing 1 mL LB (with the appropriate antibiotic(s)) in each well using a repeat pipettor and seal with plastic adhesive to prevent contamination.
6. At the end of the transformation, dispense 60 μL of transformed cells onto the pre-prepared 24-well LB agar plates (*see* **Note 7** and Fig. 4). Place in a shaker for 10 min to spread and leave plates open to dry for 10 min under a hood (or in the incubator). Close the plates and leave them inverted at 37 °C, overnight. *See* **Note 8**.
7. Dilute 60 μL of transformed cells into the DW96 containing the medium (*see* **Note 7**). Seal the deep well plate with a breathable film to allow culture aeration. Place in a 37 °C shaking incubator at maximum speed overnight (200 rpm). This is the preculture for the expression screening.

8. The next day, the preculture is used to inoculate the test expression in auto-induction medium (*see* **Note 9**). The remaining preculture is used to prepare glycerol stocks if desired (*see* **Note 10**).

3.2 Test Expression

1. Make up 500 mL ZYP-5052 auto-induction medium (464 mL ZY medium, 250 μL 2 M $MgSO_4$, 10 mL 50× 5052, 25 mL 20× NPS, in that order) supplemented with the appropriate antibiotics. Dispense 4 mL into each well of 4× DW24 plates with a repeat pipettor.
2. Use 100 μL of preculture (1 in 40 dilution) to inoculate the test expression cultures (using a multichannel pipette with variable span to transfer the preculture from the DW96 into DW24) using the scheme provided in Fig. 4. Incubate at 37 °C with shaking (200 rpm) for 4 h (the growth phase, during which time glucose from the medium will preferentially be depleted) before reducing the temperature to 17 °C (after the 4 h the glucose will have been depleted and lactose will start to be metabolized, leading to induction of expression). Leave the cells to express overnight. *See* **Note 9**.
3. To determine the OD_{600nm} take 50 μL of each culture and dispense into a flat-bottomed, clear microtiter plate containing 150 μL of medium. Measure the OD_{600nm}, taking into account the 4-fold dilution.
4. Centrifuge the 4× DW24 plates at 3,800 × *g* for 10 min then discard the supernatant into a waste container with diluted Bactinyl for decontamination before disposal. Tap the plates, upside-down, onto absorbent paper to remove any excess medium.
5. In the meantime, prepare 100 mL of lysis buffer containing lysozyme (*see* **Note 11**).
6. Add 1 mL of lysis buffer to each well and resuspend the pellets by shaking them at 17 °C and 200 rpm for 15 min. *See* **Note 12**. For purification on the same day or short-term freezing, store at −80 °C for a minimum of 1 h, otherwise store at −20 °C.

3.3 Test Purification

1. Thaw the frozen cell suspensions in a water bath (at room temperature) for approximately 15 min and if any pelleting has occurred resuspend in the shaking incubator for an additional 10 min (*see* **Note 12**). The cultures should become viscous (*see* **Note 11**).
2. Take 500 μL of DNase stock and mix it with 1 mL of $MgSO_4$ stock. Dispense 15 μL into each well of the DW96, to give a final concentration of 10 μg/mL of DNase and 20 mM $MgSO_4$. Re-seal the plate with plastic tape and shake for a further 15 min, after which stage the cultures should be non-viscous (*see* **Note 13**). Check carefully (by visual examination)

that all the cultures are no longer viscous. This is the most critical point of the whole procedure, if some cultures are still viscous (for example, if the DNase was accidentally forgotten in some wells), the filter will clog, generating an uneven pressure on the samples and contamination or total clogging of the filter plate could happen during the purification.

3. Aspirate 10 μL of the whole cell lysate and dispense into a 96-well PCR plate containing 10 μL of 4× SDS-PAGE sample buffer (*see* **Note 3**) and 20 μL of water. Denature for 3 min at 95 °C and freeze until analysis (Total fraction).
4. Add 600 μL of 33 % (v/v) Ni sepharose resin/binding buffer suspension (or 200 μL of 25 % (v/v) Ni sepharose resin/binding buffer suspension; *see* **Notes 4** and **14**) to each well of lysate, mixing before aspiration to ensure that the resin is suspended evenly and an equal amount of resin is dispensed into each well. Seal the plate with plastic tape.
5. Incubate with shaking for 10 min to allow for optimal binding.
6. In the meantime, assemble the vacuum manifold according to the Manufacturer's instructions.
7. Place the filter plate on top of the vacuum manifold, with a DW96 below to collect the flowthrough. Transfer the lysate/bead mixture to the Macherey-Nagel filter/receiver plate (20 μm) using a multichannel pipette, mixing before aspiration otherwise the resin will be retained at the bottom of the DW96 (*see* **Note 14**).
8. Turn the vacuum on for approximately 60 s to filter the lysate through the plate into the DW96 to collect the flowthrough, taking care not to dry out the resin (*see* **Note 15**). Turn the vacuum off.
9. Remove the DW96 containing the flowthrough and replace it with the waste reservoir. Keep the flowthrough aside until the end of the purification.
10. Wash the resin with 1 mL of binding buffer, turn the vacuum on until the buffer has passed through. Switch the vacuum off. Repeat once more.
11. Remove the waste reservoir, discarding the waste and replace with a fresh DW96 to collect the 50 mM imidazole wash.
12. Add 150 μL of wash buffer, turn the vacuum on until the buffer has passed through. Switch the vacuum off. Remove the DW96 containing the wash sample and replace it with the waste reservoir. Keep the wash sample aside until the end of the purification.

13. Wash the resin with 1 mL of wash buffer, turn the vacuum on until the buffer has passed through. Switch the vacuum off. Repeat once more.
14. Remove the waste reservoir, discarding the waste and replace with a fresh DW96 to collect the elution.
15. Add 500 μL of elution buffer (for 200 μL resin; or 150 μL buffer for 50 μL resin; *see* **Note 4**) and incubate in situ for 3 min. Turn on the vacuum until all buffer has passed through.
16. Optional: A second elution can be performed into a fresh DW96 as in **step 15**.
17. Take samples of the flowthrough, wash and elutions for SDS-PAGE analysis (*see* **Note 3**). For the flowthrough dispense 10 μL into a 96-well PCR plate containing 10 μL of 4× SDS-PAGE sample buffer and 20 μL of water. For the wash and elutions dispense 30 μL into PCR plates containing 10 μL of 4× SDS-PAGE sample buffer. Denature for 3 min at 95 °C and freeze until analysis.
18. Identify the constructs expressing soluble protein by analyzing the elution samples on SDS-PAGE (*see* **Note 16**).
19. Quantification can be performed for positive samples by measuring the absorbance at 280 nm (A_{280nm}), taking the extinction co-efficient of the protein into account and using the elution buffer as a blank, to provide an estimate of protein yield (*see* **Note 17**) in order to identify the highest expressing soluble constructs.
20. Optional: If tag cleavage is desired, TEV protease should be added to the eluted protein in a ratio of 1:10 (w/w, after measurement of the A_{280nm}, or alternatively v/v) (*see* **Note 2**). The TEV cleavage is, to an extent, construct-dependent, however initial attempts should be performed for a duration ranging from overnight to two nights and at room temperature or at 4 °C, depending on the stability of the protein (the colder the temperature, the longer the incubation should be). At the end of cleavage dispense 30 μL into a PCR plate containing 10 μL of 4× SDS-PAGE sample buffer. The remaining cleavage mixture can be filtered through a 96-well 0.22 μm filter plate on the vacuum manifold to collect the soluble protein after cleavage in DW96 and remove any precipitated protein. After filtration, dispense 30 μL into a PCR plate containing 10 μL of 4× SDS-PAGE sample buffer. This allows the comparison of the protein before cleavage, the mixture after cleavage and the soluble protein remaining after cleavage and gives good indications of the expected results in subsequent scale-up experiments.

3.4 Quality Control

We recommend performing an additional quality control step after soluble constructs have been identified. This allows confirmation of the expected size and oxidation state of the target protein. Samples can be analyzed directly from the elution or cleavage sample by liquid chromatography–mass spectrometry (LC/MS), or they can be desalted first, to remove imidazole, using ZipTip pipette tips (Millipore) followed by analysis using matrix-assisted laser desorption/ionization time-of-flight (MALDI-TOF) or electrospray ionization (ESI) mass spectrometry.

4 Notes

1. Plates can be made ahead of time and stored for up to 2 weeks at 4 °C. They should be pre-warmed and dried to room temperature or 37 °C prior to use. This can be done during the 1-h incubation of the transformations. To dry the plates, leave them inside a hood (or in a plate incubator) with their lids off until all moisture has evaporated.
2. These buffers can be replaced by your usual buffers for expression screening and scale-up. Hydrophobic or membrane proteins may also require the addition of detergents or other solubilizing agents. Unstable proteins may benefit from the addition of 10 % glycerol to buffers. Reducing agents can be added for proteins that need to stay in their reduced form. Ensure that all buffer additives are compatible with the purification resin (check the Manufacturer's instructions for buffer additive compatibility). For TEV cleavage, the addition of reducing agents is often recommended. We have found these to be unnecessary and they are often incompatible with subsequent purification steps. If they are used, a buffer exchange step may be required before further purification.
3. If a Caliper LabChip system is available, the samples can be analyzed on the Caliper instead. In this case, follow the Manufacturer's recommended protocol for sample preparation. Alternatively, to increase throughput over SDS-PAGE alone, samples can first be run on a dot blot, and then only the positive samples can be selected to be run on SDS-PAGE.
4. If the aim of the experiment is only detection of soluble protein then a 25 % resin slurry is suitable, so that the final volume of resin in the purification is 50 μL (this protocol has been successfully used previously for the high-throughput screening of recombinant expression conditions for small disulfide-rich peptides in *E. coli* [17]). However, if you want to be able to capture as much protein as possible (to purify for pilot assays or MS, or to extrapolate for scale-up yields) then a 33 % resin slurry should be used so that the final volume of resin in the purification is 200 μL. The downside

to using the larger volume of resin is that vacuum steps during the purification may take slightly longer than when using the smaller volume of resin.

5. A positive and negative control should be included in order to assess the success of the transformation. These can be performed in separate individual PCR tubes if there is no additional room on the 96-well plate. A positive control would be a plasmid that is known to have worked successfully in the past, while a negative control would be a lack of plasmid. If the positive control transformation does not work, or if the negative control does work, the transformation should be repeated.
6. For 96 samples (or less), the whole transformation protocol (incubation on ice, heat shock and 37 °C growth) is always done on a PCR machine (with 96 block) with one manual step (addition of LB medium). This gives more consistent results than the water bath. For more than 96 samples (up to 1,152 transformations; 12 plates) we incubate on ice, then use the PCR machine only for the heat-shock step and a 37 °C incubator for the 1-h incubation. All 1,152 transformations can be done in one afternoon.
7. This is most easily done using a multichannel pipette with variable span, using only four consecutive pipette tips at a time. Four transformations can be aspirated at once from the 96-well PCR plate, then the pipette span can be extended to fit over the 24-well plate and the culture dispensed. If the pipette has step-based programming (as for the Matrix Equalizer Pipettes, Thermo Scientific), then **steps 6** and 7 can be performed in one step by aspirating 120 μL and dispensing 60 μL onto the LB agar plate and 60 μL into the medium, changing the tip span between each dispensing step.
8. The agar plates are only back-ups. They can be stored at 4 °C for the scale-up production or for any cases where the liquid preculture does not grow but there are colonies on the plates. In that case the expression screening is postponed 24 h and the precultures are redone by a dilution in fresh medium of the original preculture and the picking of colonies for the few missing precultures to complete the plate.
9. If less than 80 % of the precultures and LB agar plates grow the transformation should be started again. If less than 80 % of the test expression cultures grow they should be done again starting from the preculture step, either directly from the LB agar plates or from the glycerol stocks.
10. Glycerol stocks can be stored at −80 °C and used to inoculate precultures for subsequent rounds of expression. Glycerol stocks should be made in replicates.

    Preparation of glycerol stocks: Dispense 30 μL of 100 % glycerol using a multidispensing pipette set to slow speed into each well of a 96-well microtiter plate. Transfer 120 μL of each culture into the corresponding well of the microtiter plate and mix by pipetting slowly and gently. Seal with plastic adhesive tape and store at −80 °C.

11. While it is possible to include DNase and $MgSO_4$ in the lysis buffer, we recommend not to. That way when the cells are thawed the lysis will be visible because the cell suspension will be viscous. If the lysozyme was accidentally omitted and DNase and $MgSO_4$ are also present in the lysis buffer then it will be impossible to discriminate whether the lysis was successful.
12. When only working on 96 proteins at a time, the samples can be frozen in DW24 format as they will be quicker to thaw on the day of lysis. In that case, once thawed on the day of lysis, the lysate should be transferred back to DW96 using a multichannel pipette with variable span. When working on multiple lots of 96, to save freezer space, transfer back into DW96 after resuspension on the day of harvesting, before freezing.
13. With this protocol, depending of the time of incubation, approximately 80–100 % of the cells are lysed. To speed up the process and achieve 100 % lysis an additional sonication step can be performed if desired using a plate sonicator (Ultrasonic processor XL, Misonix Inc., USA).
14. A slow aspiration speed should be used for pipetting all resin suspensions, as the suspensions are quite thick. In the protocol with 600 μL of 33 % (v/v) beads, the volume of lysate/resin has to be transferred in two steps, so as not to cause an overflow on the filter plate (using vacuum to remove the first lot of flowthrough between the two transfers).
15. Take care not to over-dry the resin, which will result in a reduction in binding capacity and viability. However, some wells may empty faster than others.
16. Elution samples can be run on SDS-PAGE first to identify constructs producing soluble protein. Only for constructs where no soluble protein can be detected is it necessary to then run the whole cell lysate to see if the protein was expressed. If the protein was not expressed but the cells did grow (reached an $OD_{600nm} \geq 6.0$ at Subheading 3.2, **step 3**) a new expression strategy must be pursued, or if the OD_{600nm} was not high enough the culture can be regrown and reanalyzed. If the protein was expressed it is advisable to run the flowthrough and wash samples on SDS-PAGE to see if it was insoluble or did not bind to the resin.
17. For the most reliable comparison of soluble yields, it is recommended to normalize the yields by the density of the culture (using the OD_{600nm} measurement), which was taken at Subheading 3.2, **step 3**.

References

1. Berrow NS, Bussow K, Coutard B et al (2006) Recombinant protein expression and solubility screening in Escherichia coli: a comparative study. Acta Crystallogr D 62:1218–1226
2. Correa A, Oppezzo P (2011) Tuning different expression parameters to achieve soluble recombinant proteins in E. coli: advantages of high-throughput screening. Biotechnol J 6:715–730
3. Graslund S, Nordlund P, Weigelt J et al (2008) Protein production and purification. Nat Methods 5:135–146
4. Vera A, Gonzalez-Montalban N, Aris A et al (2007) The conformational quality of insoluble recombinant proteins is enhanced at low growth temperatures. Biotechnol Bioeng 96:1101–1106
5. Graslund S, Sagemark J, Berglund H et al (2008) The use of systematic N- and C-terminal deletions to promote production and structural studies of recombinant proteins. Protein Expr Purif 58:210–221
6. Bird LE (2011) High throughput construction and small scale expression screening of multi-tag vectors in Escherichia coli. Methods 55:29–37
7. Davis GD, Elisee C, Newham DM et al (1999) New fusion protein systems designed to give soluble expression in Escherichia coli. Biotechnol Bioeng 65:382–388
8. Kapust RB, Waugh DS (1999) Escherichia coli maltose-binding protein is uncommonly effective at promoting the solubility of polypeptides to which it is fused. Protein Sci 8:1668–1674
9. LaVallie ER, Lu Z, Diblasio-Smith EA et al (2000) Thioredoxin as a fusion partner for production of soluble recombinant proteins in Escherichia coli. Methods Enzymol 326:322–340
10. Marblestone JG, Edavettal SC, Lim Y et al (2006) Comparison of SUMO fusion technology with traditional gene fusion systems: enhanced expression and solubility with SUMO. Protein Sci 15:182–189
11. Sachdev D, Chirgwin JM (2000) Fusions to maltose-binding protein: control of folding and solubility in protein purification. Methods Enzymol 326:312–321
12. Smith DB (2000) Generating fusions to glutathione S-transferase for protein studies. Methods Enzymol 326:254–270
13. de Marco A, Deuerling E, Mogk A et al (2007) Chaperone-based procedure to increase yields of soluble recombinant proteins produced in E. coli. BMC Biotechnol 7:32
14. Hatahet F, Nguyen VD, Salo KE et al (2010) Disruption of reducing pathways is not essential for efficient disulfide bond formation in the cytoplasm of E. coli. Microb Cell Fact 9:67
15. Katzen F, Beckwith J (2002) Disulfide bond formation in periplasm of Escherichia coli. Methods Enzymol 348:54–66
16. Studier FW (2005) Protein production by auto-induction in high density shaking cultures. Protein Expr Purif 41:207–234
17. Nozach H, Fruchart-Gaillard C, Fenaille F et al (2013) High throughput screening identifies disulfide isomerase DsbC as a very efficient partner for recombinant expression of small disulfide-rich proteins in E. coli. Microb Cell Fact 12:37
18. Klint JK, Senff S, Saez NJ et al (2013) Recombinant production of disulfide-rich peptides in the periplasm of E. coli for structural and functional analysis. PLoS One 8(5):e63865
19. Vincentelli R, Cimino A, Geerlof A et al (2011) High-throughput protein expression screening and purification in Escherichia coli. Methods 55:65–72
20. Vincentelli R, Canaan S, Offant J et al (2005) Automated expression and solubility screening of His-tagged proteins in 96-well format. Anal Biochem 346:77–84
21. Katzen F (2007) Gateway (R) recombinational cloning: a biological operating system. Expert Opin Drug Discov 2:571–589
22. Bornhorst JA, Falke JJ (2000) Purification of proteins using polyhistidine affinity tags. Methods Enzymol 326:245–254
23. Vincentelli R, Canaan S, Campanacci V et al (2004) High-throughput automated refolding screening of inclusion bodies. Protein Sci 13:2782–2792
24. Jolma A, Yan J, Whitington T et al (2013) DNA-binding specificities of human transcription factors. Cell 152:327–339
25. Sala E, de Marco A (2010) Screening optimized protein purification protocols by coupling small-scale expression and mini-size exclusion chromatography. Protein Expr Purif 74:231–235
26. Moon AF, Mueller GA, Zhong X et al (2010) A synergistic approach to protein crystallization: combination of a fixed-arm carrier with surface entropy reduction. Protein Sci 19:901–913
27. Zanier K, Charbonnier S, Sidi AO et al (2013) Structural basis for hijacking of cellular LxxLL motifs by papillomavirus E6 oncoproteins. Science 339:694–698
28. Vincentelli R, Bignon C, Gruez A et al (2003) Medium-scale structural genomics: strategies for protein expression and crystallization. Acc Chem Res 36:165–172

Chapter 4

Medium-Throughput Production of Recombinant Human Proteins: Ligation-Independent Cloning

Claire Strain-Damerell, Pravin Mahajan,
Opher Gileadi, and Nicola A. Burgess-Brown

Abstract

Structural genomics groups have identified the need to generate multiple truncated versions of each target to improve their success in producing a well-expressed, soluble, and stable protein and one that crystallizes and diffracts to a sufficient resolution for structural determination. At the SGC, we opted for the Ligation-Independent Cloning (LIC) method which provides the medium throughput we desire to produce and screen many proteins in a parallel process. Here, we describe our LIC protocol for generating constructs in a 96-well format and provide a choice of vectors suitable for expressing proteins in both *E. coli* and the baculovirus expression vector system (BEVS).

Key words PCR, Gene, Ligation-independent cloning (LIC), Construct, Protein, Crystallography

1 Introduction

The knowledge base resulting from sequencing of the human genome has provided a strong foundation for identifying and understanding the role of genes encoding various proteins involved in health and disease as well as in physiological processes. Determining three-dimensional (3D) structures of the proteins is important to understand the biochemical reactions they catalyze at the molecular level. According to the latest estimate by the International Human Genome Sequencing Consortium, the human genome seems to encode 20,000–25,000 proteins [1]. However, there is a major gap between the number of protein sequences and experimentally determined 3D protein structures. The Structural Genomics Consortium (SGC) is a not-for-profit organization that is addressing this gap by solving the structures of medically relevant proteins and placing them into the public domain without restriction (http://www.thesgc.org/).

Yu Wai Chen (ed.), *Structural Genomics: General Applications*, Methods in Molecular Biology, vol. 1091, DOI 10.1007/978-1-62703-691-7_4,

Determining protein structures by X-ray crystallography on the genome scale creates a number of bottlenecks, the first being expression and purification of the large number of soluble, homogeneous, and stable proteins in heterologous systems. We have developed robust protocols for medium-throughput cloning, expression testing, and protein production in *E. coli* and in insect cells which have resulted in a portfolio of hundreds of protein domains. We have used *E. coli* as the primary expression system for producing our soluble target proteins; however, for expression of more challenging proteins such as kinases and integral membrane proteins (IMPs), the baculovirus expression system is our first choice. The recombinant proteins expressed globally at SGC have yielded more than 1,400 protein structures, but in addition, these proteins have provided a rich resource for functional genomics, small molecule inhibitor screens, and generation of antibodies.

Ligation-independent cloning (LIC) [2] was our method of choice as it provided a simple and cost effective tool for producing many constructs of a single target or multiple targets in parallel without the need to select specific restriction enzymes for each gene. Briefly, the process involves T4 DNA polymerase treatment of linearized vectors in the presence of a single deoxynucleotide (dNTP). PCR fragments of the gene of interest (GOI) with complementary overhangs are generated by adding appropriate 5′ extensions into the primers (LIC sequences) and treating the fragments with T4 DNA polymerase in the presence of the paired dNTP (*see* Fig. 1). At the SGC we have engineered many of our vectors to share the same LIC site which allows one LIC-prepared PCR fragment to be cloned into a range of vectors within the same and across different expression systems. Alternative efficient cloning methods are available including Gateway® [3–5], MAGIC [6], and In-Fusion® [7], the latter being the method preferred by our SGC node in Toronto. More recently, the LIC method has evolved to SLIC [8, 9] which removes sequence constraints. In this chapter, we begin the process of medium-throughput screening by describing in detail our methods for (1) identifying domain boundaries to increase the likelihood of producing a stable and correctly folded protein, (2) primer design, PCR, and vector preparation, (3) annealing and transformation into *E. coli*, and (4) confirmation of cloning success by colony PCR screening. In Chapters 5 and 6, we provide detailed protocols for expression testing using *E. coli* and baculovirus/insect cells and producing milligram quantities of protein of sufficient quality and purity for crystallization and functional screening. Although our cloning and expression testing protocols are described for a 96-well format, the whole process can easily be applied to generate and screen a smaller number of proteins. Handling 24 or more samples should be performed in block format rather than in individual tubes as described in the methods.

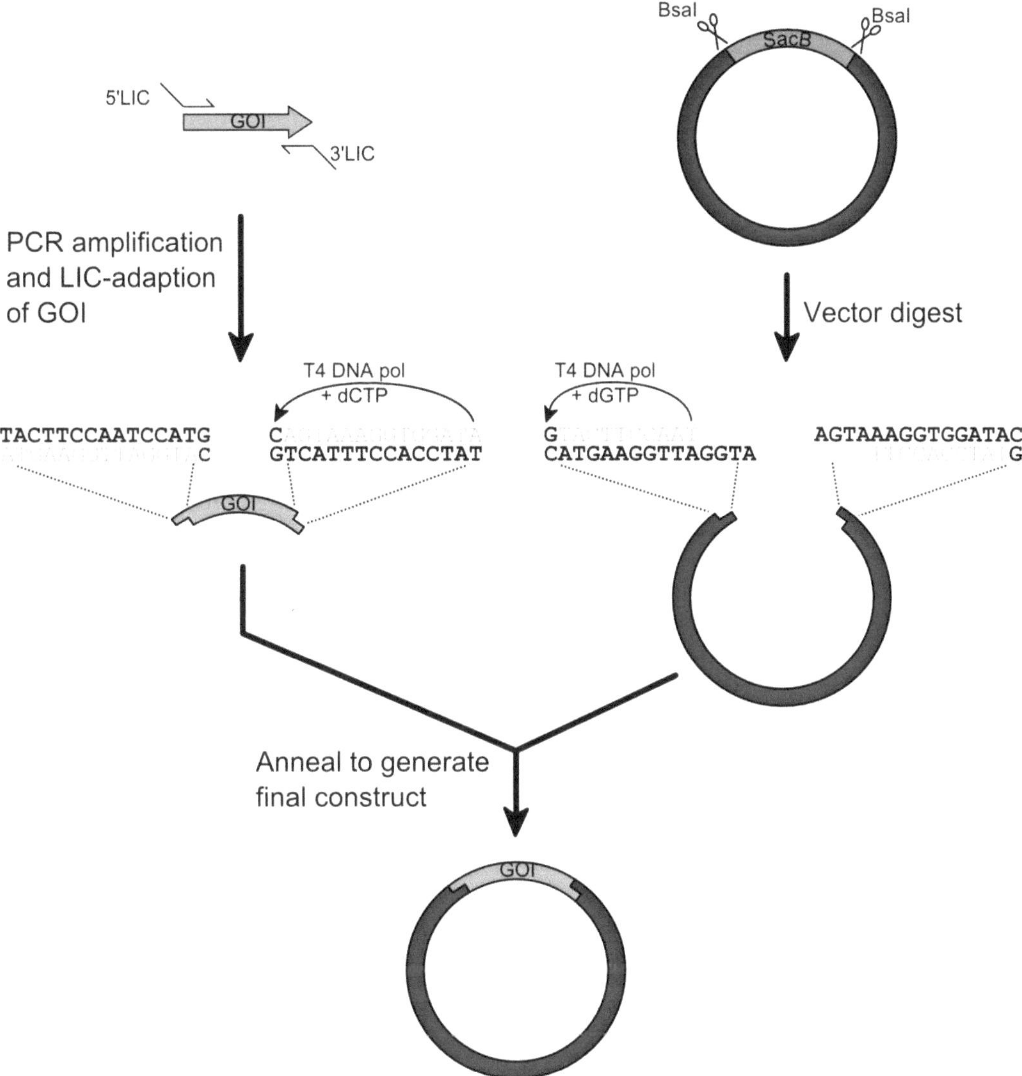

Fig. 1 Overview of the Ligation Independent Cloning (LIC) Process. The gene of interest (GOI) is amplified with primers that include the LIC sequence specific to the target vector. The vector is linearized by restriction digest, removing the *sacB* gene. Both insert and vector are then T4 DNA polymerase treated to resect 3′ ends, creating large overhangs, promoting efficient circularization without the need for T4 DNA ligase

2 Materials

Unless otherwise stated, molecular biology grade water (Thermo Scientific HyClone) is used for all dilutions and reactions set out below. Where ultrapure water is instead specified, it is prepared by purifying deionized water to reach a resistivity of 18 MΩ cm at 25 °C. All reagents should be of analytical grade or higher and all plasticware should be DNase-free.

2.1 PCR

1. Primers: Primers are supplied by either MWG-Biotech or Sigma-Aldrich and are HPSF purified at 0.01 μmol scale or DST purified at 0.025 μmol scale, respectively. Primer stocks are either supplied at or diluted (in 10 mM Tris–HCl buffer, pH 8.0) to 100 μM and stored at −20 °C.
2. Template library: Human cDNA clones were obtained from the IMAGE cDNA collection (currently distributed by Source BioScience, UK), from other commercial providers (OriGene, Invitrogen, FivePrime), or isolated in-house by PCR from human cDNA. Synthetic DNA clones, including either the natural cDNA or codon-optimized sequences, were synthesized to order by GenScript or Codon Devices.
3. Enzymes: Platinum Pfx DNA polymerase (2.5 units/μl, Invitrogen), alternatively Herculase II Fusion DNA polymerase (Agilent Technologies) for difficult to amplify targets, and BIOTAQ™ Red DNA polymerase (1 unit/μl, Bioline) for colony PCR screening. DpnI (20 units/μl, New England BioLabs (NEB)).
4. 10 mM dNTP solution: 10 mM dATP, 10 mM dTTP, 10 mM dGTP, and 10 mM dCTP (prepared from 100 mM dNTP set, Invitrogen) and stored at −20 °C.
5. TE Buffer: 10 mM Tris–HCl and 1 mM EDTA, pH 8.0. Filtered through a 0.20 μm syringe filter (Sartorius) and stored at room temperature (RT).
6. 50× TAE buffer (1 L): 242 g Tris base, 57.1 ml glacial acetic acid, and 100 ml of 0.5 M EDTA, pH 8.0, pH adjusted to 8.5. Filtered through a 0.2 μm membrane filter and used as a 1× solution.
7. 96-well 1.5 % TAE-agarose gels: 3 g agarose powder (Invitrogen), 200 ml of 1× TAE buffer, and 8 μl of SYBR-safe DNA gel stain (Invitrogen), cast in a Sub-cell Model 96 (Bio-Rad or similar) gel cast.
8. DNA ladders: For the E-Gel® system, the Low Range Quantitative DNA Ladder (Invitrogen), and for the colony PCR screen, the 1 kb Plus DNA ladder (Invitrogen) prepared in 1× BlueJuice™ (Invitrogen) are used.
9. QIAquick PCR purification kit (Qiagen).
10. MultiScreen PCR_{96} filter plate (Millipore).
11. 96-Well PCR plates (4titude Ltd. or similar).
12. Adhesive PCR seals (ABgene).
13. Adhesive tape pads (Qiagen).
14. V-bottomed microtiter plates.
15. Minisart syringe filters, 0.20 μm (Sartorius).
16. Express™ PLUS filter unit, 0.22 μm (Millipore).

17. Membrane filters, 0.2 μm and unit.
18. Reagent reservoirs for multichannel pipetting (Fisher).
19. Multichannel pipettes and repeat pipettors are used to dispense reagents into a 96-well format.
20. 96-Well PCR thermocycler with heated lid.
21. E-Gel® 96 Mother base and E-Gel® 96 1 % Agarose Gels (Invitrogen).
22. 96-Well gel cast and tank (Subcell Model 96 Bio-Rad or similar).
23. Centrifuge suitable for 96-well PCR plates (150 × *g*).
24. Microcentrifuge.
25. MultiScreen$_{HTS}$ Vacuum Manifold (Millipore).
26. Gel Logic 200 Imaging System (Kodak).
27. Water bath set at 37 °C.

2.2 Cloning

The following reagents, consumables, and equipment are required in addition to those listed above:

1. Competent cells: All cloning is performed in Mach1™ cells (originally purchased from Invitrogen), with chemically competent cells produced in-house using the RbCl method [10]. Other cell lines are suitable for cloning but we recommend using a *recA*$^-$ phage resistant strain, to promote plasmid stability and to reduce the risk of bacteriophage infection during *E. coli* expression, respectively.
2. All enzymes and their associated buffers are supplied by NEB; including T4 DNA Polymerase (3 units/μl), BsaI (10 units/μl), BfuAI (5 units/μl), and BseRI (4 units/μl).
3. 25 mM dGTP and 25 mM dCTP (prepared from 100 mM dNTP set, Invitrogen) and stored at −20 °C.
4. 100 mM DTT (Dithiothreitol): filtered through a 0.20 μm syringe filter (Sartorius) and stored as 1 ml aliquots at −20 °C.
5. Bovine serum albumin (BSA) (100× supplied with most NEB enzymes).
6. 25 % (w/v) sucrose: 250 g sucrose dissolved in 1 l of ultrapure water and filtered through a 0.22 μm filter unit (Millipore).
7. 60 % (v/v) glycerol autoclaved to sterilize.
8. Antibiotic stocks: Ampicillin (50 mg/ml); Kanamycin (50 mg/ml), filtered through a 0.20 μm syringe filter (Sartorius) and stored at −20 °C.
9. LB-agar: 22.5 g premixed LB-broth and 13.5 g agar dissolved in 800 ml of ultrapure water. Volume adjusted to 900 ml and autoclaved on the same day.

10. LB-agar plates: LB-agar melted slowly in a microwave and sucrose added to a final concentration of 5 % (w/v) (*see* **Note 1**). Once cooled to hand-hot, the appropriate antibiotic (*see* Table 1) is added and swirled vigorously to mix. 10 ml of the molten agar is poured into each 50 mm petri dish and once set, upturned and left open to dry. These can be prepared ahead of time and stored at 4 °C sealed in a plastic bag to prevent over-drying.
11. 1× LB: 22.5 g premixed LB-broth dissolved in 800 ml of ultrapure water. Volume adjusted to 900 ml and autoclaved on the same day.
12. SOC medium: 18 g tryptone, 4.5 g yeast extract, 0.45 g NaCl, and 2.25 ml of 1 M KCl dissolved in 800 ml of ultrapure water. Volume adjusted to 900 ml and autoclaved on the same day. Once cooled, 9 ml of 2 M $MgCl_2$ hexahydrate, and 18 ml of 1 M (18 %) glucose are added; both solutions are filtered through a 0.20 μm syringe filter (Sartorius) prior to use (*see* **Note 2**).
13. Virkon (Appleton Woods).
14. Montage Plasmid Miniprep$_{HTS}$ 96 Kit (Millipore).
15. 50 mm petri dishes.
16. Disposable sterile spreaders or 2 mm autoclaved glass beads (e.g., Sigma-Aldrich) for spreading as these are reusable and allow faster plating for the medium-throughput scale.
17. Disposable sterile inoculation loops (1 μl).
18. 96-Deep-well blocks (Thomson).
19. AirOtop porous seals (Thomson).
20. Centrifuge suitable for 96-deep-well blocks (3,000 × *g*).
21. Micro-Express Glas-Col shaker (Glas-Col, Indiana, USA) or similar set to 37 °C.
22. Water bath set at 42 °C.
23. Incubator set at 37 °C.
24. Heated block set at 50 °C.

3 Methods

3.1 Construct Design

In order to give the best possible chance of producing soluble protein with a high propensity for crystallization we opt for a multi-construct design approach [11–13]. Whilst we do include the full length protein in the initial target screen, only 8.6 % of our solved structures have arisen from such constructs. By repositioning the start and stop boundaries of our constructs by only five amino acids either side, our success increases to 13.3 % (unpublished data). By expanding the design out to include only certain domains of the protein, our success rate improves further meaning that structures that would have otherwise been missed make it

Table 1
LIC-adapted vectors for bacterial and baculovirus expression which are freely available from the SGC on request

Vector name	Antibiotic resistance marker	Tags for purification	Restriction site for LIC	dNTP for vector	dNTP for insert	5′ LIC primer extension	3′ LIC primer extension	Screening primers
Bacterial expression vectors								
pNIC28-Bsa4	Kanamycin	N-terminal His_6	BsaI	dGTP	dCTP	TACTTCCAATCCATG	TATCCACCTTTACTG*TCA*	pLIC-F+R
pGTVL2	Kanamycin	N-terminal His_6+GST	BsaI	dGTP	dCTP	TACTTCCAATCCATG	TATCCACCTTTACTG*TCA*	pLIC-F+R
pNH-TrxT	Kanamycin	N-terminal His_6+Trx	BsaI	dGTP	dCTP	TACTTCCAATCCATG	TATCCACCTTTACTG*TCA*	pLIC-F+R
pNIC-CTH0	Kanamycin	C-terminal His_6	BfuAI	dCTP	dGTP	TTAAGAAGGAGATAT ACTATG	GATTGGAAGTAGAGGTTC TCTGC	pLIC-F+R
pNIC-CTHF	Kanamycin	C-terminal His_6+Flag	BfuAI	dCTP	dGTP	TTAAGAAGGAGATAT ACTATG	GATTGGAAGTAGAGGTTC TCTGC	pLIC-F+R
pNIC-CT10HF	Kanamycin	C-terminal His_{10}+Flag	BfuAI	dCTP	dGTP	TTAAGAAGGAGATAT ACTATG	GATTGGAAGTAGAGGTTC TCTGC	pLIC-F+R
Baculovirus transfer vectors								
pFB-LIC-Bse	Ampicillin	N-terminal His_6	BseRI	dGTP	dCTP	TACTTCCAATCCATG	TATCCACCTTTACTG*TCA*	FBac-1+2
pFB-HGT-LIC	Ampicillin	N-terminal His_6+GST	BseRI	dGTP	dCTP	TACTTCCAATCCATG	TATCCACCTTTACTG*TCA*	FBac-1+2
pFB-CT6H-LIC	Ampicillin	C-terminal His_6	BfuAI	dCTP	dGTP	TTAAGAAGGAGATATA CTATG	GATTGGAAGTAGAGGTTC TCTGC	FBac-3+1
pFB-CT6HF-LIC	Ampicillin	C-terminal His_6+Flag	BfuAI	dCTP	dGTP	TTAAGAAGGAGATATA CTATG	GATTGGAAGTAGAGGTTC TCTGC	FBac-3+1
pFB-CT10HF-LIC	Ampicillin	C-terminal His_{10}+Flag	BfuAI	dCTP	dGTP	TTAAGAAGGAGATATA CTATG	GATTGGAAGTAGAGGTTC TCTGC	FBac-3+1

The antibiotic resistance cassette and purification tags are indicated, along with the details required for LIC: forward and reverse primer extension required to LIC-adapt the GOI, restriction enzyme to cut the vector with, and the dNTP required for the T4-treatment step of either vector or PCR product. Note that the start codon included by the 5′ LIC sequence is underlined and the stop codon included by the 3′ LIC sequence is italicized through to Protein Data Bank (PDB) submission using the multi-construct approach. Constructs are therefore designed based on available protein domain information, secondary structure predictions, and sequence alignments, as well as taking account of disordered regions to try to produce more stable proteins at the expression stage. Due to uncertainty in predictive methods and in our understanding of factors affecting protein behavior, we test a number of construct endpoints (2–5 on either end) closely spaced around the predicted domain boundaries.

3.2 Primer and Plate Design

Having identified appropriate construct boundaries in the previous step, we design primers for PCR amplification of the desired DNA segments. The primer sequences themselves typically include the appropriate LIC sequence (*see* Table 1) followed by ~20 bp from the construct sequence. In each case, the ATG underlined in Table 1 should be in-frame with the target sequence. Where the construct includes an N-terminal purification tag, the stop codon is incorporated by the 3′ LIC sequence marked in italics (*see* Table 1). For C-terminally tagged constructs, the reverse primer must not include a stop codon but must be in-frame with the 3′ LIC sequence, i.e., do not include additional nucleotides between the 3′ of the reverse LIC site and the codon encoding the C-terminal amino acid. As the primer sequences are dictated by the desired boundaries in the protein sequence, the corresponding DNA sequences may have properties (e.g., repetitions or biased nucleotide composition) that make it difficult to design optimal primers. Primers are thus designed with care to avoid mispriming or primer-dimers and to ensure compatible T_m values, determining the lengths and base composition accordingly. The arrangement of constructs in a 96-well format is done with the following constraints for ease of cloning: (1) constructs from the same entry clone are kept together, (2) constructs are arranged in order of size, (3) where possible, only one vector and or T4-treatment condition is used per plate, and (4) if the plate is mixed then like-vectors and T4-treatment conditions are kept together on the plate. Arrangement in this manner enables easy identification of correctly sized products and limits mistakes caused by erroneous pipetting. Once you have designed the plate format keep a record of what primers, template, and vector will be associated with each well and use this for all subsequent steps.

3.3 PCR

1. Using a multichannel pipette and reagent reservoir, add 180 μl of water to each well of a 96-well PCR plate. To this, add 10 μl each of the 100 μM forward and reverse primers (*see* Subheading 3.2) and mix well.
2. For each template (*see* Subheading 2.1, **step 2**) prepare a 2.5 ng/μl dilution in a 1.5 ml eppendorf tube, mix well, and

aliquot 20 µl of this into the appropriate wells of a second 96-well PCR plate.

3. Prepare a PCR master mix as follows: 250 µl of 10× Pfx reaction buffer (Invitrogen), 50 µl of 50 mM $MgSO_4$, 75 µl of 10 mM dNTP mixture, 20 µl of Platinum Pfx (Invitrogen), and 1.705 ml of water. Mix the solution well. Using a multichannel pipette or repeat pipettor, aliquot 21 µl into each well of a third 96-well PCR plate (*see* **Note 3**).
4. Using a multichannel pipette, transfer 1.5 µl of the diluted primers (**step 1**), followed by 2.5 µl of diluted template DNA (**step 2**), into the corresponding wells of the reaction plate (**step 3**). Mix well then seal the plate using an adhesive PCR seal, making sure to press down well in order to limit evaporation (*see* **Note 4**).
5. Place the reaction plate into the thermocycler and cycle with the following conditions—touchdown PCR (*see* **Note 5**):

   95 °C, 10 min.

   (95 °C, 30 s; 68 °C, 30 s; 68 °C, 1–3 min*) × 5 cycles.

   (95 °C, 30 s; 60 °C, 30 s; 68 °C, 1–3 min*) × 5 cycles.

   (95 °C, 30 s; 55 °C, 30 s; 68 °C, 1–3 min*) × 5 cycles.

   (95 °C, 30 s; 50 °C, 30 s; 68 °C, 1–3 min*) × 20 cycles.

   68 °C, 10 min.

   15 °C hold.

   *Extension time dependent on length of PCR product—e.g., 1 min per 1 kb.
6. Remove 3 µl of each reaction and dilute with 12 µl of water. Run on an E-Gel® (Invitrogen) against 20 µl of 2× diluted Low Range Quantitative DNA Ladder (Invitrogen) (Fig. 2).
7. Transfer the successful reactions into the corresponding wells of a fresh PCR plate and repeat any failed reactions using different cycling conditions or with additives such as the PCR enhancer supplied with the Platinum Pfx (Invitrogen) kit (*see* **Note 6**).
8. Any products amplified from templates containing the same antibiotic resistance cassette as the target vector, require DpnI-treatment to limit template carryover (*see* **Note 7**). Prepare a 1 in 20 dilution of DpnI (20 units/µl, NEB) in NEB buffer 2 and aliquot 1 µl into the appropriate wells of the PCR reaction plate. Incubate the plate in a 37 °C water bath for 1 h.
9. Purify the products (*see* **Note 8**) using a MultiScreen PCR_{96} purification plate (Millipore) following the manufacturer's instructions. Recover the DNA from the plate in 50 µl of TE buffer, transferring into a V-bottomed microtiter plate and store at −20 °C.

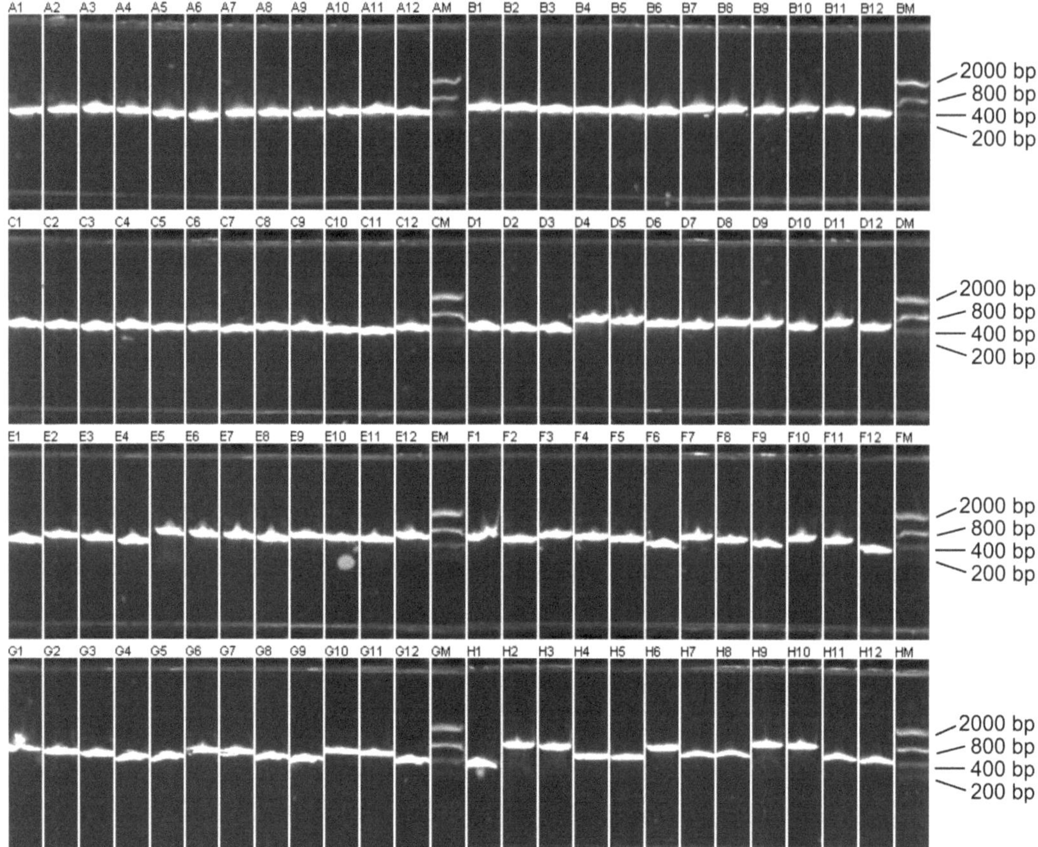

Fig. 2 Image of an initial PCR performed in 96-well format, analyzed using the E-Gel® system and Low Range Quantitative DNA Ladder. The sizes of the ladder are indicated. Due to the low resolution of these gels the products are judged based on the sizing of neighboring bands, e.g., the products of E5 to E8 should be in decreasing size order, which can be observed on the gel

3.4 Vector Preparation

1. Digest the target vector using the restriction enzyme indicated in Table 1 (*see* **Note 9** for alternative restriction enzymes), for example for BsaI vectors (*see* Fig. 3 for example vector) prepare the digest as follows: 5 μg vector, 10 μl of 10× NEB buffer 3, 1 μl of 100× BSA, 3 μl of BsaI (10 units/μl, NEB), make up to 100 μl with water and incubate at 50 °C for 2 h.
2. Mix 3 μl of the digested vector with 3 μl of 2× BlueJuice™ (Invitrogen) and analyze on a 1.5 % TAE-agarose gel to confirm complete digestion (*see* **Note 10**).
3. Purify the digested vector using a QIAquick PCR purification spin column (Qiagen), following the manufacturer's instructions, and elute in 50 μl.

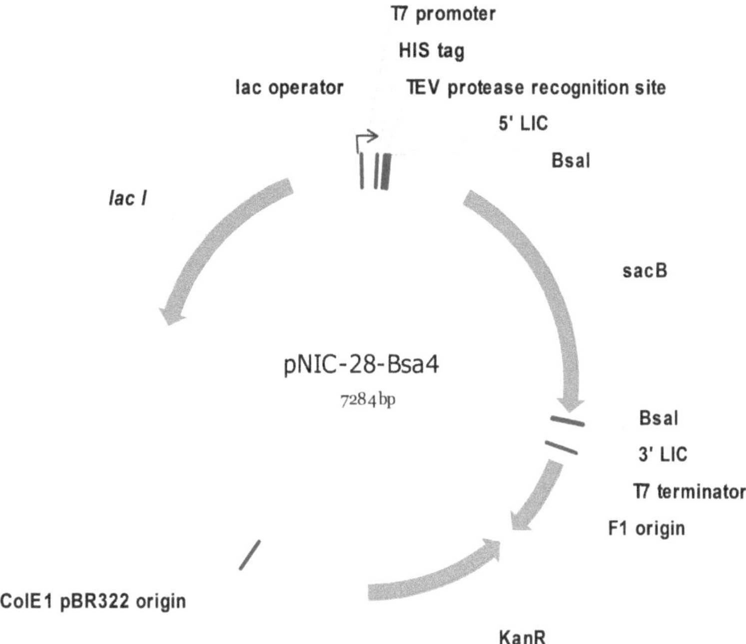

Fig. 3 Vector map of standard bacterial expression vector pNIC28-Bsa4. Digestion with BsaI excises the *sacB* gene and T4-treatment resects the 3′ ends of the LIC sites to provide complementary cohesive ends to the PCR products. The vector incorporates a His_6 tag at the N-terminus followed by TEV cleavage site in frame with the PCR product. This vector also includes the T7 promoter and terminator sequences for expression in the BL21 (DE3) strain and is under the control of the lac repressor for induction with IPTG during the expression stage (*see* Chapter 5)

3.5 T4 DNA Polymerase Treatment

1. To the purified vector (50 μl) add 21.5 μl of water, 10 μl of 10× NEB buffer 2, 10 μl of 25 mM dCTP or dGTP (*see* Table 1), 5 μl of 100 mM DTT, 1 μl of 100× BSA (NEB), and 2.5 μl of T4 DNA polymerase (3 units/μl, NEB). Place in a thermocycler with the following conditions: 22 °C for 30 min, 75 °C for 20 min, 15 °C hold (*see* **Note 11**).
2. For T4-treament of the PCR products prepare a master mix as follows: 215 μl of water, 100 μl of 10× NEB buffer 2, 100 μl of 25 mM dCTP or dGTP (*see* Table 1), 50 μl of 100 mM DTT, 10 μl of 100× BSA, and 25 μl of T4 DNA polymerase (3 units/μl, NEB). Using a repeat pipettor aliquot 5 μl into each well of a PCR plate. Using a multichannel pipette, transfer 5 μl of the purified PCR product (Subheading 3.3, **step 9**) into the corresponding wells of the T4 reaction mix, mixing as you dispense. Place in a thermocycler with the following conditions: 22 °C for 30 min, 75 °C for 20 min, 15 °C hold.

3.6 Annealing and Transformation

1. Using a repeat pipettor, aliquot 1 μl of the T4-treated vector into each well of a 96-well PCR plate and centrifuge briefly at 150×*g*. Confirm that there is liquid in each well before progressing to **step 2**.
2. Using a multichannel pipette, transfer 2 μl of T4-treated insert into the corresponding wells of the plate from **step 1**. Spin briefly and incubate the reaction at RT for at least 20 min before placing on ice (*see* **Note 12**).
3. Take two 1.5 ml eppendorf tubes, label one with "vector-only control" and the other with "insert-only control" (*see* **Note 13**). To the first add 1 μl of the T4-treated vector and to the other 2 μl of T4-treated insert from a well that has undergone DpnI-treatment (*see* Subheading 3.3, **step 8**). Place both tubes on ice.
4. Using a repeat pipettor, aliquot 50 μl of chemically competent sub-cloning efficiency cells (*see* **Notes 14** and **15**) into each well of the plate from **steps 1** and **2** and into the two tubes from **step 3**. Incubate on ice for 30 min.
5. Heat-shock the cells at 42 °C for 45 s, then return to ice briefly.
6. Using a multichannel, pipette 100 μl of SOC medium (*see* **Note 16**) into each well, seal with a porous seal and incubate at 37 °C for 1.5 h in a stationary incubator.
7. Plate 100 μl of the transformation mixture onto LB-agar plates containing 5 % sucrose (*see* **Note 1**) supplemented with either 50 μg/ml kanamycin or 200 μg/ml ampicillin (*see* Table 1). Spread the sample across the plate using either sterile spreaders or glass beads (*see* **Note 17**).
8. Incubate the plates at 37 °C for ~16 h, then store at 4 °C until the colony PCR screening step is complete.

3.7 Colony PCR Screening

1. Prepare a 96-deep well block containing 1 ml of LB and the appropriate antibiotic selection (*see* Table 1).
2. Set up a PCR master mix as follows: 200 μl of 10× NH_4 Reaction Buffer (Bioline), 60 μl of 50 mM $MgCl_2$, 60 μl of 100 % DMSO, 1.24 ml of water, 200 μl of 2 mM dNTPs, 200 μl of 10 μM screening primers (*see* Tables 1 and 2), and 40 μl of BIOTAQ™ Red DNA polymerase (1 unit/μl, Bioline). Using a repeat pipettor or a multichannel pipette, aliquot 20 μl into each well of a 96-well PCR plate.
3. Using a 1 μl sterile loop, pick one colony from each transformation plate and inoculate into the corresponding well of the PCR reaction plate (**step 2**) followed by the corresponding well of the deep-well block (**step 1**) (*see* **Note 18**).
4. Once all of the wells have been inoculated, seal the deep-well block with a porous seal and incubate at 37 °C overnight in a Glas-Col with shaking at 700 rpm, then store at 4 °C.

Table 2
Colony PCR screening primers for SGC vectors

Primer name	Primer sequence
pLIC-F	TGTGAGCGGATAACAATTCC
pLIC-R	AGCAGCCAACTCAGCTTCC
FBac-1	TATTCATACCGTCCCACCA
FBac-2	GGGAGGTTTTTTAAAGCAAGTAAA
FBac-3	TTAAAATGATAACCATCTCG

The screening primers are situated upstream of the LIC sites, allowing full sequencing of the purification tags incorporated by the vector sequence. The pLIC primers are for the bacterial expression vectors, whereas the FBac primers are for the baculovirus expression vectors (BEVs). Note that FBac-1 and -2 may be used to screen all BEVs but that FBac-1 is located too close to the start codon of the C-terminally tagged vectors to allow complete coverage during sequencing; FBac-3 is recommended for this purpose

5. Seal the PCR reaction plate with a adhesive PCR seal and set a thermocycler with the following conditions, making sure that the block is up to temperature before placing your sample plate in the instrument (*see* **Note 19**):

   95 °C, 10 min.

   (95 °C, 30 s; 50 °C, 30 s; 72 °C, 1–3 min*) × 25 cycles.

   72 °C, 5 min.

   15 °C hold.

   *Extension time dependent on length of PCR product—e.g., 1 min per 1 kb. Please note that ~200 bp will be added to your products due to the positioning of the screening primers.

6. Whilst the cycle is running, prepare a 96-well 1.5 % TAE agarose gel.
7. Using a multichannel pipette, load 10 μl of the PCR reaction mixtures directly onto the gel. Note that the spacing of the wells means that samples will be interleaved (Fig. 4). Load 6 μl of 1 kb DNA ladder and run the gel at 150 V for 1 h.
8. Confirm the sizing of the products and repeat the screen for additional clones if necessary (*see* **Note 20**).

3.8 Preparation of Glycerol Stocks and 96-Well Miniprep

1. Combine the correct clones into a single block by inoculating 20 μl of each culture (*see* Subheading 3.7, **step 4**) into 1 ml of LB (*see* **Note 21**) in a new 96-deep-well block, containing the same antibiotic selection as above. Grow overnight at 37 °C in a Glas-Col with shaking at 700 rpm.
2. To each well of a V-bottomed microtiter plate, add 30 μl of 60 % (v/v) glycerol, followed by 120 μl of culture. Mix well as you add the culture (*see* **Note 22**). Seal this with an adhesive tape pad and store at −80 °C.

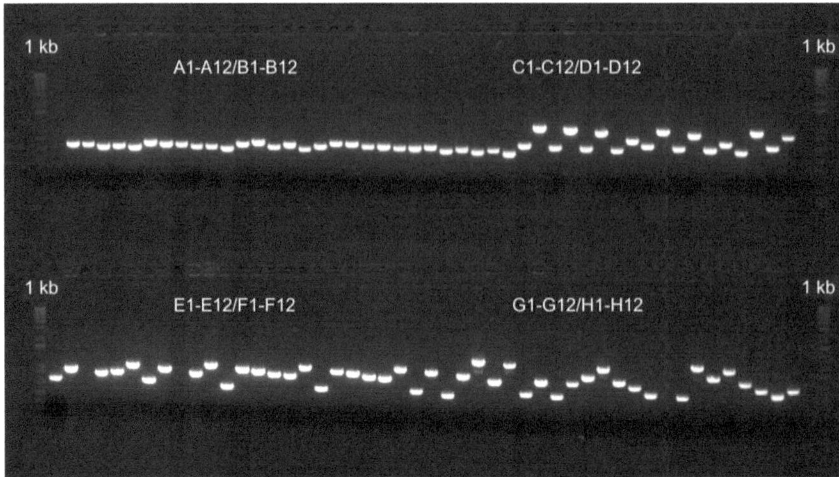

Fig. 4 Image of a colony PCR screen performed in a 96-well format, analyzed on a 1.5 % TAE agarose gel. The samples are interleaved, e.g., A1, B1, A2, B2. Note that the products are larger (~200 bp) at the colony screening stage due to the positioning of the screening primers

3. Centrifuge the remaining culture at 3,000 × *g* for 20 min.
4. Discard the supernatant into a waste pot containing 1 % Virkon and blot the excess liquid onto a clean paper towel.
5. Use a 96-well plasmid purification kit (Millipore) to purify the plasmids from these cell pellets following the manufacturer's instructions, with a few modifications (*see* **Note 23**).
6. Recover the DNA in 50 μl of Solution 5 (Millipore) and transfer into a V-bottomed microtiter plate. Seal with an adhesive tape pad and store at −20 °C.

4 Notes

1. The *sacB* gene product, expressed from our LIC-adapted vectors (Fig. 3), is capable of converting sucrose to a toxic by-product. By adding sucrose to the LB-agar plates we select for recombinant plasmids only, as these will lack the *sacB* gene, having been replaced by our GOI (Fig. 1).
2. It is advisable to prepare small volumes of SOC medium at a time as it is prone to contamination.
3. Note that if using the repeat pipettor to aliquot the PCR master mix, the volume added will actually be 20 μl (not 21 μl) but this will not affect the reaction.
4. Spend plenty of time sealing your PCR plate, applying a lot of pressure around the wells to ensure efficient adherence to prevent evaporation. It is important that your thermal cycler has a heated lid as this will again limit the amount of evaporation.

5. When dealing with a mixture of targets and primers on one 96-well plate, it is not always possible to optimize each reaction, therefore the best approach is to perform touchdown PCR as a first pass and then use a more tailored cycle for any missing products. As touchdown cycles through a range of annealing temperatures, it will cover the differences in melting temperatures of your primers across the plate.
6. If you get multiple bands from your PCR, try using a fixed annealing temperature instead which should be ~5 °C lower than the melting temperature of your primers. If you get no bands, try using additives such as the enhancer supplied with the Platinum Pfx kit (Invitrogen) or DMSO at a final concentration of 3 %, or test higher concentrations of $MgSO_4$ (1.5–3 mM). If all of the reactions for a certain target have failed, you can try an alternative polymerase such as Herculase II (Agilent Technologies); however, you may also want to sequence your template to check that it is what you think it is.
7. DpnI is a restriction endonuclease that can only cleave at its recognition sites when they have been methylated. Standard strains of *E. coli* (including Mach1™, Invitrogen) methylate their DNA, thus any entry clones propagated in them will be methylated. By DpnI-treating a PCR product, we specifically cleave the template DNA leaving only the product intact. This limits the chance of template carryover when the entry clone carries the same antibiotic resistance marker as the cloning vector.
8. It is important to purify the PCR products away from any unincorporated dNTPs in the reaction mixture as these will inhibit resection of the 3′ ends during the T4 DNA polymerase step.
9. Alternative restriction digest conditions are as follows: *BfuAI vectors*: 5 μg vector, 5 μl 10× NEB buffer 3, 1 μl BfuAI (5 units/μl, NEB), make up to 50 μl with water and incubate at 50 °C for 2 h. *BseRI vectors*: 5 μg vector, 10 μl 10× NEB buffer 2, 1 μl 100× BSA (NEB), 6.25 μl BseRI (4 units/μl, NEB), make up to 100 μl with water and incubate at 37 °C for 2 h. Add a further 20 μl of water, 2.5 μl 10× NEB buffer 2, 2.5 μl BseRI (4 units/μl, NEB) and incubate for an additional 1 h at 37 °C.
10. Check by agarose gel analysis that your vector has two clearly distinct bands; the top one is the vector backbone that you will ligate your fragment into and the lower band is the *sacB* fragment (~2 kb). You do not need to purify the lower fragment away from the top fragment as self-ligation is selected against by using sucrose in the medium (*see* **Note 1**).
11. It is important that you only add one dNTP to your reaction as this will determine the stop position of the 3′ resection (Fig. 1). For this reason it is also important that your dNTP stock is stored at −20 °C when not in use to ensure that it remains fresh. The same rule applies to the DTT.

12. The longer you give the annealing step the more successful your transformation will be. Give your samples no less than 20 min but give them longer whenever possible.
13. It is important to include a vector-only control during the transformation to check that the rate of insert-independent colonies is low. The sucrose will select against reinsertion of the *sacB* fragment and uncut vector but the vector backbone can occasionally close on itself. If there are many colonies on this plate then there may be an issue with your sucrose selection or with your T4-treatment step as self-ligation should be rare. Note that these will be distinguishable at the PCR screen step as they will produce a ~200 bp product. You should also include an insert-only control at the transformation step when your PCR products have required DpnI-treatment as this will indicate any template carryover from insufficient Dpn-I treatment.
14. When using a repeat pipettor to aliquot your cells, care should be taken to prevent cross-contamination between wells caused by splash-back.
15. It is important to use high cloning efficiency cells for the transformation; if you fail to get colonies this is normally the reason why. If you prepare your cells in-house, then check the efficiency is on the order of 1×10^6 CFU per μg by transforming 0.5 ng of vector. This test should be done every time new competent cells are prepared. You should also test for contamination by plating 50 μl of untransformed cells on plates containing either ampicillin or kanamycin. This test should be performed using aseptic techniques to ensure that the cells are the only potential source of contamination.
16. Other media can be used during this step (e.g., 1× or 2× LB); however, SOC gives a higher transformation efficiency when dealing with the low DNA concentrations that are used in this protocol.
17. To plate using sterile glass beads: Stack the plates, agar-side down, in order of row (e.g., A1 to A12) and add ~5 beads per plate. Working from one side of the transformation plate to the other transfer 100 μl of the culture to the relevant agar plate. When each row is completed, split the stack into two blocks of six and shake the plates from side to side to spread the culture. Once all wells have been plated, shake the plates once more and upturn to move the beads onto the lid. The beads can then be disposed of into a beaker containing 1 % (w/v) Virkon to be cleaned, autoclaved, and reused.
18. Give the inoculation loop a twirl in both the PCR mixture and the LB to transfer more material for the PCR and growth, respectively.
19. We have found BIOTAQ™ Red DNA polymerase (Bioline) reactions to be more successful when the samples are placed in

a thermocycler preheated to 95 °C, rather than allowing the enzyme to heat up to 95 °C. If using an alternative screening polymerase, check the conditions specified by the manufacturer; however, note that Bioline do not specify preheating with their product.

20. If your colony screen is not working there may be several reasons: If you get a smear on your agarose gels then it can often be remedied by cleaning your pipettes and gel tank before starting the screen. If you get no product, then check that your reagents and cycling conditions are working by including a small sample of your uncut vector (use 2 μl of a 2.5 ng/μl dilution for a 20 μl reaction) to act as a positive control. If this works but your screen does not, then there may be an issue with your cells (*see* **Note 15**). If the positive control fails, then you may want to try alternative reagents and/or cycling conditions, and if the initial PCR required specific conditions, then try these for the screen as well. When using ampicillin as the selectable marker, we have found that colonies with lots of satellites surrounding them tend not to yield products during the PCR screen. If this is the case, try retransforming and always store the plates at 4 °C when you are not screening them.
21. For the plasmid miniprep, we have found that any media richer than LB yields pellets too large for efficient clearing during the miniprep process.
22. It is important to mix your cells when preparing glycerol stocks to ensure the viability of the stock—should you need to go back to it.
23. The volume of each buffer used to isolate the plasmid DNA is 100 μl instead of 150 μl which is recommended in the manufacturer's instruction booklet. In addition, we assemble our clearing plate above the manifold, with the plasmid plate inside the manifold, and apply ~300 mbar pressure. This is contrary to the manufacturer's instructions due to risk of cross-contamination; however, we find this to be more effective and have had no issue with samples missing wells when this level of pressure is applied.

Acknowledgements

We would like to thank all the SGC scientists (past and present) who contributed towards the development of the method. The SGC is a registered charity (number 1097737) that receives funds from the Canadian Institutes for Health Research, Genome Canada, GlaxoSmithKline, Lilly Canada, the Novartis Research Foundation, Pfizer, Takeda, AbbVie, the Canada Foundation for Innovation, the Ontario Ministry of Economic Development and Innovation, and the Wellcome Trust.

References

1. Abdellah Z, Ahmadi A, Ahmed S et al (2004) Finishing the euchromatic sequence of the human genome. Nature 431:931–945
2. Haun RS, Serventi IM, Moss J (1992) Rapid, reliable ligation-independent cloning of PCR products using modified plasmid vectors. Biotechniques 13:515–518
3. Hartley JL, Temple GF, Brasch MA (2000) DNA cloning using in vitro site-specific recombination. Genome Res 10:1788–1795
4. Invitrogen (2010) Gateway® Technology: a universal technology to clone DNA sequences for functional analysis and expression in multiple systems. Invitrogen Life Technologies, Carlsbad
5. Walhout AJ, Temple GF, Brasch MA et al (2000) GATEWAY recombinational cloning: application to the cloning of large numbers of open reading frames or ORFeomes. Methods Enzymol 328:575–592
6. Li MZ, Elledge SJ (2005) MAGIC, an in vivo genetic method for the rapid construction of recombinant DNA molecules. Nat Genet 37:311–319
7. Clontech (2012) In-Fusion® HD cloning kit user manual. Clontech, Mountain View
8. Li MZ, Elledge SJ (2007) Harnessing homologous recombination in vitro to generate recombinant DNA via SLIC. Nat Methods 4:251–256
9. Li MZ, Elledge SJ (2012) SLIC: a method for sequence- and ligation-independent cloning. Methods Mol Biol 852:51–59
10. Hanahan D, Jessee J, Bloom FR et al (1991) Plasmid transformation of Escherichia coli and other bacteria. Methods Enzymol 204:63–113
11. Graslund S, Sagemark J, Berglund H et al (2008) The use of systematic N- and C-terminal deletions to promote production and structural studies of recombinant proteins. Protein Expr Purif 58:210–221
12. Bray JE (2012) Target selection for structural genomics based on combining fold recognition and crystallisation prediction methods: application to the human proteome. J Struct Funct Genomics 13:37–46
13. Savitsky P, Bray J, Cooper CD et al (2010) High-throughput production of human proteins for crystallization: the SGC experience. J Struct Biol 172:3–13

Chapter 5

Medium-Throughput Production of Recombinant Human Proteins: Protein Production in *E. coli*

Nicola A. Burgess-Brown, Pravin Mahajan, Claire Strain-Damerell, Opher Gileadi, and Susanne Gräslund

Abstract

In Chapter 4 we described the SGC process for generating multiple constructs of truncated versions of each protein using LIC. In this chapter we provide a step-by-step procedure of our *E. coli* system for test expressing intracellular (soluble) proteins in a 96-well format that enables us to identify which proteins or truncated versions are expressed in a soluble and stable form suitable for structural studies. In addition, we detail the process for scaling up cultures for large-scale protein purification. This level of production is required to obtain sufficient quantities (i.e., milligram amounts) of protein for further characterization and/or crystallization experiments. Our standard process is purification by immobilized metal affinity chromatography (IMAC) using nickel resin followed by size exclusion chromatography (SEC), with additional procedures arising from the complexity of the protein itself.

Key words *E. coli*, Bacteria, Expression, Recombinant Protein, Purification, Immobilized metal affinity chromatography (IMAC), Size exclusion chromatography (SEC), Gel filtration

1 Introduction

Choosing from which expression system to produce your protein can depend on many different factors such as its size, location within the cell and the requirement for posttranslational modifications (PTMs) [1]. To provide a starting point for researchers, structural genomics groups collectively identified trends and common strategies for producing proteins for structural determination [2]. At the SGC, we preferentially start with *E. coli* for testing and producing human intracellular proteins, specifically a tRNA-enhanced strain of BL21(DE3) which often compensates for codon bias [3, 4]. This low-cost prokaryotic expression system is easy to use, is suitable for increasing throughput, and has a high success rate for many targets, particularly when truncated or mutated versions of the protein are screened [5, 6].

Yu Wai Chen (ed.), *Structural Genomics: General Applications*, Methods in Molecular Biology, vol. 1091, DOI 10.1007/978-1-62703-691-7_5,

In 2010, we showed that 48 % of the human proteins attempted in *E. coli* were successfully purified, and of those, the structures of approximately 40 % were solved by X-ray crystallography [7]. Protein crystallization demands availability of soluble, pure, monodisperse, and homogeneous proteins in sufficient quantities (usually in milligram quantities). Nevertheless, the limited amount of protein obtained from initial small-scale expression testing can provide valuable information on protein solubility, expression level, molecular weight, and PTMs of target proteins. In addition to our standard histidine (His)-tagged vectors, we have engineered a number of other vectors harboring different tags and/or fusion partners (some of which are listed in Chapter 4, table 1) and a variety of *E. coli* host strains [7]. All of these vectors also contain a six or ten His tag enabling the use of IMAC purification for fast and efficient capture of recombinant proteins from cell lysates.

A version of the bacterial methods from expression testing to large-scale protein production has been published previously. The method presented here has been modified, in particular, the changes in the method used to test protein expression in small-scale (1 ml) cultures has provided better correlation with the results of large-scale expression. We found that using *n*-Dodecyl β-D-maltoside (DDM) to lyse the bacterial membranes gave hits most comparable to those from large-scale cultures lysed by sonication or homogenization. The previous method we employed, extracting the protein with BugBuster®, produced many false negative results (unpublished data) and often required purification from a 50 ml culture to distinguish the true positives from the false hits. Since we implemented this change in procedure, our false negative rate has declined substantially. Although we screen for expression in a 96-well format, the methods do not require expensive or specialized equipment and are easily adaptable to lower throughput in individual tubes and flasks. As a consequence, they can be performed in any lab, with minimal equipment, at whatever scale is required.

The methods for large-scale protein expression and purification are also described in this chapter to provide the researcher with a complete process for obtaining quality protein in quantities sufficient for crystallization experiments or developing assays for functional screening. The generic methods described here are routinely used in our laboratory for expression and purification of a large number of proteins. Following the standard IMAC purification, many highly expressed proteins only require one additional step of SEC to yield pure protein, but for difficult to purify proteins, additional steps such as His tag cleavage using TEV protease and rebinding to nickel resin or ion exchange chromatography are often required. Moreover, occasionally variations in the methodology are incorporated to address the need arising from complexity of the proteins, by introducing changes such as buffer type, pH, ionic strength, and use of additives to the buffer in order to stabilize

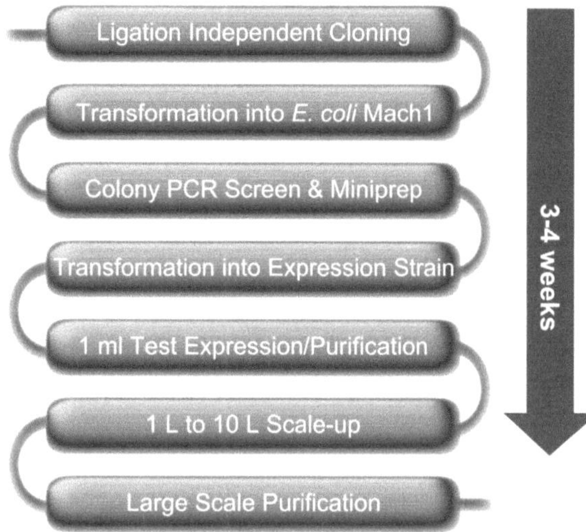

Fig. 1 Overview of the bacterial expression pipeline. The process takes ~3–4 weeks from LIC to scale up

the proteins. The pipeline from cloning to expression testing through to large-scale protein expression and purification is outlined in Fig. 1. The process that we use for screening and producing proteins in the baculovirus expression vector system (BEVS) is described in the subsequent chapter.

2 Materials

Unless otherwise stated, all solutions are prepared using ultrapure water (prepared by purifying deionized water to reach a resistivity of 18 MΩ cm at 25 °C) and analytical grade reagents.

2.1 Transformation and Test Expression

1. BL21(DE3)-R3-pRARE2 *E. coli* strain: Phage resistant derivative of BL21(DE3) isolated in-house containing the pRARE2 plasmid which was extracted from the strain Rosetta2 from Novagen. This strain supplies tRNAs for seven rare codons (AGA, AGG, AUA, CUA, GGA, CCC, and CGG) on a compatible chloramphenicol-resistant plasmid. Chemically competent bacterial cells are prepared in-house as described [8].
2. 60 % (v/v) glycerol: autoclaved to sterilize.
3. 1,000× Antibiotic stocks: Kanamycin (50 mg/ml), filtered through a 0.20 μm syringe filter (Sartorius); Chloramphenicol (34 mg/ml in ethanol), both stored at −20 °C.
4. 1 M IPTG (Isopropyl β-D-1-thiogalactopyranoside): filtered through a 0.20 μm syringe filter (Sartorius) and stored at −20 °C.

5. LB-agar: 22.5 g premixed LB-broth and 13.5 g agar dissolved in 800 ml of water. Volume adjusted to 900 ml and autoclaved on the same day.
6. LB-agar plates: LB-agar melted slowly in a microwave. Once cooled to hand-hot, the appropriate antibiotic is added and swirled vigorously to mix. 10 ml of the molten agar is poured into each 50 mm petri dish and once set, upturned and left open to dry. These can be prepared ahead of time and stored at 4 °C sealed in a plastic bag to prevent over-drying.
7. 1× LB: 22.5 g premixed LB-broth dissolved in 800 ml of water. Volume adjusted to 900 ml and autoclaved on the same day.
8. 2× LB: 45 g premixed LB-broth dissolved in 800 ml of water. Volume adjusted to 900 ml and autoclaved on the same day.
9. SOC medium: 18 g tryptone, 4.5 g yeast extract, 0.45 g NaCl, and 2.25 ml of 1 M KCl dissolved in 800 ml of water. Volume adjusted to 900 ml and autoclaved on the same day. Once cooled, 9 ml of 2 M $MgCl_2$ hexahydrate and 18 ml of 1 M (18 %) glucose are added; both solutions are filtered through a 0.20 μm syringe filter (Sartorius) prior to use (*see* **Note 1**).
10. TB (Terrific Broth) medium: 12 g of tryptone, 24 g of yeast extract, and 4 ml of glycerol dissolved in 800 ml of water. Volume adjusted to 900 ml and autoclaved on the same day. Once cooled to room temperature (RT), volume adjusted to 1 l with 100 ml of a separately autoclaved solution of 0.17 M KH_2PO_4 and 0.72 M K_2HPO_4.
11. Virkon (Appleton Woods).
12. 24-Well tissue culture (TC) plates (Corning).
13. 96-Well PCR plates.
14. 96-Well microtiter plates.
15. 96-Deep-well blocks (Thomson or similar).
16. Disposable sterile spreaders.
17. Disposable sterile inoculation loops, 1 μl.
18. AirOtop porous seals (Thomson).
19. Adhesive tape pads (Qiagen).
20. Disposable Cuvettes (Fisher Scientific).
21. Reagent reservoir for multichannel pipetting (Fisher).
22. Multichannel pipettes and repeat pipettors are used to dispense reagents into a 96-well format.

23. Micro-Express Glas-Col shaker (Glas-Col, Indiana, USA) or alternative that ranges in temperature from 18 to 37 °C and shakes up to 800 rpm.
24. Water bath set at 42 °C.
25. Incubator set at 37 °C.
26. 96-Well block mixer (eppendorf MixMate or similar).
27. A visible light spectrophotometer for measuring OD_{600nm} (optical density) of bacterial cultures (for individual cuvettes).
28. A 96-well plate reader is also useful but not essential.

2.2 Test Purification

The following reagents, consumables, and equipment are required in addition to those listed above:

1. Benzonase (Novagen, HC, 250 units/μl).
2. Protease Inhibitor Cocktail Set VII (Calbiochem).
3. 10 mg/ml Lysozyme (Sigma-Aldrich), prepared fresh.
4. 10 % (w/v) DDM, Sol-grade (Anatrace or Glykon), filtered through a 0.20 μm syringe filter (Sartorius) and stored at –20 °C.
5. 1 M TCEP (Tris (2-Carboxyethyl) phosphine Hydrochloride), filtered through a 0.20 μm syringe filter (Sartorius) and stored at –20 °C.
6. 1 M DTT (Dithiothreitol), filtered through a 0.20 μm syringe filter (Sartorius) and stored as 1 ml aliquots at –20 °C.
7. SeeBlue Plus2 (Invitrogen).
8. InstantBlue™ (Expedeon Protein Solutions).
9. 20× XT Mes running buffer (Bio-Rad).
10. Stock solutions: 1 M HEPES, pH 7.5; 5 M NaCl; 3 M imidazole, pH 8.0; 200 mM MgSO4 (all filtered through a 0.2 μm membrane filter and stored at RT); 50 % (v/v) glycerol (autoclaved and stored at RT).
11. Lysis buffer (1 l): 100 mM HEPES, pH 7.5, 500 mM NaCl, 10 % (v/v) glycerol, and 10 mM imidazole, filtered through a 0.2 μm membrane filter and stored at 4 °C. On the day of purification, Lysozyme (50 μl/ml), Benzonase (0.2 μl/ml), Protease inhibitor cocktail (1 μl/ml), DDM (10 μl/ml), MgSO4 (5 μl/ml), and TCEP (0.5 μl/ml) are added (*see* **Note 2**).
12. Wash buffer (1 l): 20 mM HEPES, pH 7.5, 500 mM NaCl, 10 % (v/v) glycerol, 25 mM imidazole, filtered through a 0.2 μm membrane filter and stored at 4 °C. 0.5 mM TCEP added on the day of purification.

13. Elution buffer (0.1 l): 20 mM HEPES, pH 7.5, 500 mM NaCl, 10 % (v/v) glycerol, 500 mM imidazole, filtered through a 0.2 μm membrane filter and stored at 4 °C. 0.5 mM TCEP added on the day of purification.
14. Affinity buffer: 50 mM HEPES, pH 7.5, 500 mM NaCl, 10 % glycerol, 10 mM imidazole.
15. 50 % Ni-IDA Metal Chelate Resin (Generon) or Ni-NTA-agarose (Qiagen): IMAC resin is generally supplied in 20 % ethanol. To equilibrate, the resin is washed twice in water and then three times in Affinity buffer in a 50 ml tube, by inverting to resuspend the resin and centrifuging at 500 × g for 1 min. After the final wash, the resin is resuspended in Affinity buffer as 50 % slurry and stored at 4 °C when not in use.
16. SB: Stock of NuPAGE LDS sample buffer (Invitrogen) containing DTT (1:4 dilution of 1 M DTT in NuPAGE LDS sample buffer) and stored at −20 °C.
17. MultiScreen® Filter Plates, 1.2 μm (Millipore).
18. MultiScreenHTS Vacuum Manifold (Millipore).
19. Pre-cast 26-Lane SDS-PAGE gradient gels (4–20 %) (Bio-Rad).
20. Protein gel electrophoresis apparatus (Bio-Rad).
21. 96-Well thermocycler with heated lid.
22. All gels are imaged on a Gel Logic 200 Imaging System (Kodak).
23. Centrifuge suitable for 96-deep-well blocks (3,000 × *g*).

2.3 Large-Scale Expression

The following reagents, consumables, and equipment are required in addition to those listed above:

1. Glycerol stocks of transformed expression strain.
2. 2.5 l Ultra Yield baffled flasks (Thomson) or glass flasks.
3. Shaker-incubators with cooling capacity: Innova 44R (New Brunswick) or Multitron (Infors HT).
4. Avanti J-20XP or Avanti J-26XP centrifuge or similar (Beckman Coulter) with a JLA 8.1000 rotor for harvesting large volumes of cells.

2.4 Protein Extraction and Large-Scale Purification

The following reagents, consumables, and equipment are required in addition to those listed above:

1. 5 % (w/v) Polyethyleneimine (PEI): The 50 % solution (Sigma-Aldrich, P3143) is diluted tenfold then adjusted to pH 7.5 with HCl.

2. 2× Lysis buffer: 100 mM HEPES, pH 7.5, 1 M NaCl, 20 % (v/v) glycerol, 20 mM imidazole. Filtered through a 0.2 μm membrane filter and stored at 4 °C. On the day of purification, Benzonase (0.2 μl/ml of cell lysate), 2 μl of Protease inhibitor cocktail (1 μl/ml of cell lysate), and 1 mM TCEP are added.
3. Lysis buffer: 50 mM HEPES, pH 7.5, 500 mM NaCl, 10 % (v/v) glycerol, 10 mM imidazole. Filtered through a 0.2 μm membrane filter and stored at 4 °C. On the day of purification, Benzonase (0.1 μl/ml of cell lysate), Protease inhibitor cocktail (1 μl/ml cell lysate) or Complete EDTA-free protease inhibitor cocktail (1 tablet/25 ml cell lysate), and 0.5 mM TCEP are added.
4. Affinity buffer: 50 mM HEPES buffer, pH 7.5, 500 mM NaCl, 10 % glycerol, 10 mM imidazole. 0.5 mM TCEP added on the day of purification.
5. Wash buffer: 50 mM HEPES, pH 7.5, 500 mM NaCl, 10 % glycerol, 30 mM imidazole. 0.5 mM TCEP added on the day of purification.
6. Elution buffer: 50 mM HEPES, pH 7.5, 500 mM NaCl, 10 % glycerol, 300 mM imidazole. 0.5 mM TCEP added on the day of purification.
7. Size Exclusion Chromatography (SEC) buffer: 20 mM HEPES, pH 7.5, 500 mM NaCl, 5 % glycerol. Filtered through a 0.2 μm membrane filter and stored at 4 °C. 0.5 mM TCEP added on the day of purification.
8. Minisart syringe filters, 0.20, 0.45, and 0.80 μm (Sartorius).
9. Millex®-GV Low Protein Binding Filter, 0.22 μm (Millipore).
10. Amicon Ultra protein concentrators (Millipore).
11. Sonicator (Sonics Vibra-Cell, VCX 750, Sonics & Materials INC) or basic Z model cell disruptor (Constant Systems Ltd.) or EmulsiFlex-C5 high-pressure homogenizer (Avestin).
12. Econo-Columns (Bio-Rad or similar).
13. ÄKTA-Xpress or ÄKTA-Purifier liquid chromatography system (GE).
14. HiTrap 5 ml FF columns (GE) for His-tagged protein purification.
15. Ion exchange chromatography columns such as HiTrap 5 ml Q FF and SP FF (GE).
16. HiLoad Superdex columns (GE) for preparative size exclusion chromatography such as HiLoad 16/60 Superdex S75 or Superdex S200.

17. UV-spectrophotometer for measuring DNA and protein concentration (e.g., The NanoDrop spectrophotometer allows measurements from as low as 1.5 μl volumes).
18. General purpose benchtop centrifuge (Sorvall Legend RT, Kendro).
19. JA-17 rotor for centrifugation of cell lysates.
20. Microcentrifuge.

3 Methods

3.1 Transformation into E. coli BL21(DE3)-R3-pRARE2

1. Prepare four 24-well tissue culture plates containing 1 ml of LB-agar, supplemented with 50 μg/ml kanamycin and 34 μg/ml chloramphenicol and once set allow to dry, inverted at RT.
2. Using a multichannel pipette, add 3 μl of recombinant DNA (see Chapter 4, Subheading 3.8, **step 6**) to a 96-well PCR plate. Place on ice and add 30 μl of chemically competent *E. coli* BL21(DE3)-R3-pRARE2 cells using a repeat pipettor (*see* **Note 3**). Cover with an adhesive tape pad and incubate for 30 min on ice. It is advisable to include a positive control protein (i.e., a protein that has previously shown soluble expression in *E. coli*) in position H12 of the 96-well plate.
3. Heat shock in a water bath at 42 °C for 45 s, return to ice briefly then add 100 μl of SOC (or 2× LB) medium (*see* **Note 4**). Cover with a porous seal and incubate for 1 h at 37 °C.
4. Pipette 30 μl of the transformation mixture onto the agar in the 24-well TC plates according to the format presented (Fig. 2). Gently rock the plates to cover the surface and allow them to dry before incubating at 37 °C inverted overnight (*see* **Note 5**).
5. Inoculate three colonies or a streak of colonies from each well (*see* **Note 6**) into the corresponding well of a 96-deep-well block containing 1 ml of LB (or 2× LB) medium supplemented with 50 μg/ml kanamycin and 34 μg/ml chloramphenicol.
6. Cover the block with porous film and place in the Glas-Col shaker in the afternoon at 37 °C, with shaking at 700 rpm.
7. The following morning, prepare four replicate glycerol stocks. Dispense 30 μl of 60 % (v/v) glycerol into 96-well microtiter plates. Transfer 120 μl of each culture into the corresponding wells of the microtiter plates and mix by pipetting. Seal the plate with an adhesive tape pad and store at −80 °C. Keep the remainder of the overnight culture for setting up the test expression.

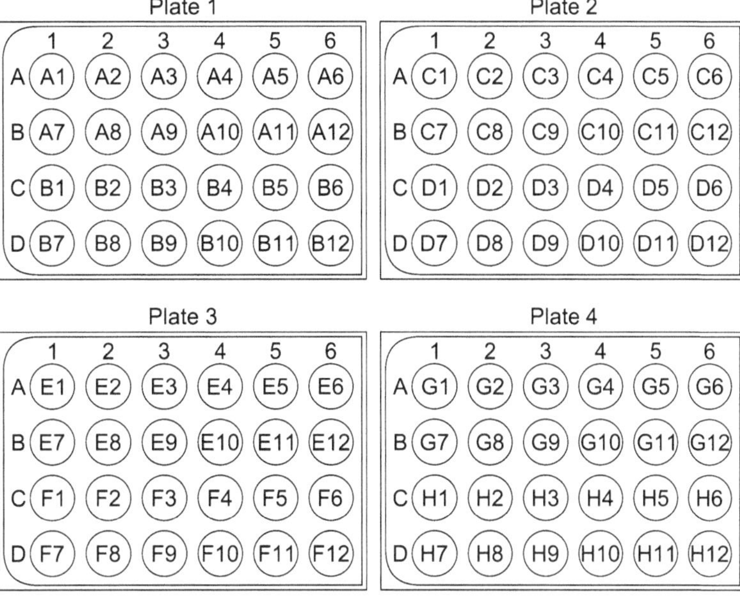

Fig. 2 Format for plating cultures grown in a 96-well block onto four 24-well agar plates. This template can be printed out to scale and placed underneath the 24-well plates when plating

3.2 Test Expression

1. Inoculate 20 μl of the overnight culture (or thawed glycerol stock) into each well of two 96-deep-well blocks containing 1 ml of fresh TB medium, supplemented with 50 μg/ml kanamycin (*see* **Note 7**) and grow to an OD_{600nm} of 2–3 (approximately 5 h) in a Glas-Col shaker set at 37 °C and 700 rpm. Label one block as "OD measurement block" and the other as "test block."
2. Determine the OD measurement for a few wells (at least one appearing visually to have low density and one high density) by diluting 1 in 4 in TB medium and using a visible light spectrophotometer. If the OD_{600nm} is between 2–3, and you have available a 96-well plate reader, dilute aliquots of the test block 1 in 4 (in a total of 200 μl) in a flat-bottomed clear microtiter plate for OD measurement using the plate reader (*see* **Note 8**).
3. Leave the cultures to cool down at RT for 30 min and change the temperature setting of the shaker to 18 °C.
4. Induce expression by adding 0.1 mM IPTG (10 mM stock prepared in TB medium and 10 μl added to the block) and incubating in the Glas-Col shaker overnight at 18 °C and 700 rpm.

3.3 Test Purification

1. Centrifuge the 96-deep-well block at 3,000 × *g* for 20 min, pour off the medium into a waste pot containing 1 % Virkon and tap the block on absorbent paper (*see* **Notes 9** and **10**).
2. Add 200 μl of Lysis buffer (*see* **Note 2**) and resuspend the pellets either using the Glas-Col shaker at 18 °C and 700 rpm or a 96-well block mixer (Eppendorf) at 1,000–2,000 rpm. Use a multichannel pipette to resuspend any remaining solid pellets and store the block at −80 °C for at least 20 min, until all pellets are completely frozen.
3. Thaw the pellets in a shallow water bath (at RT) for approximately 15 min and resuspend in the Glas-Col shaker for 10 min. Remove 3 μl (Total fraction) and pipette into a PCR plate containing 27 μl of water and 10 μl of SB for storage at 4 °C until required.
4. Centrifuge the block at 3,000 × g for 10 min.
5. Meanwhile, add 50 μl of a 50 % slurry of Ni-IDA (or Ni-NTA) resin to each well of a MultiScreen® Filter Plate, 1.2 μm (Millipore).
6. Transfer the clarified lysate (*see* **Note 11**) using a 1 ml capacity multichannel pipette, to the filter plate containing the resin, taking care to avoid transferring any pelleted material (*see* **Note 12**).
7. Place an adhesive tape pad on top and incubate the plate in the Glas-Col shaker at 18 °C for 1 h at 400 rpm (*see* **Note 13**).
8. Assemble the vacuum manifold according to the Manufacturer's instructions and then filter the contents through the plate into waste for approximately 20 s, taking care not to dry out the resin (*see* **Note 14**). Turn off the vacuum.
9. Add 200 μl of Wash buffer and filter quickly. Repeat this step three more times, turning the vacuum off after each step to prevent over-drying, and then place the filter plate on top of a waste block and centrifuge for 2 min at 300 × g to remove all trace of the Wash buffer (*see* **Note 15**).
10. Place a fresh 96-well microtiter plate under the filter plate, add 40 μl of Elution buffer and seal the plate with an adhesive tape pad.
11. Incubate the plate in the Glas-Col shaker for 10–20 min at 400 rpm and 18 °C (or at RT on a shaking platform).
12. Elute the protein by centrifugation at 300 × g for 3 min.
13. Store the eluate (Purified fraction) at 4 °C until required (or −20 °C for long term storage).

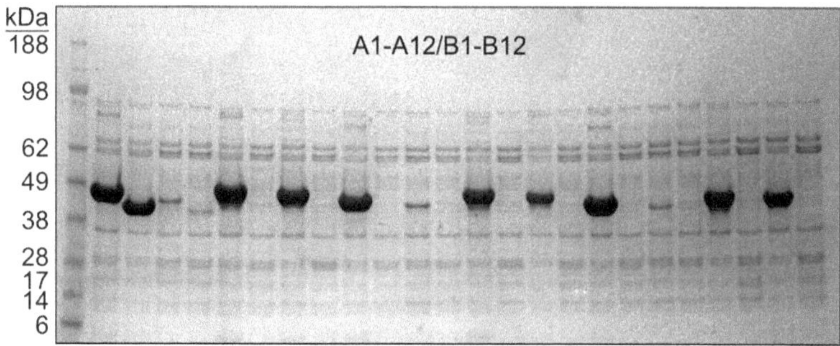

Fig. 3 Image showing the Coomassie SDS-PAGE result of a test purification from *E. coli*. The gel shows a range of high, medium and low expressing proteins of different molecular weights. Note that samples loaded using a multichannel pipette are interleaved

14. Dispense 5 µl of SB in all wells of a 96-well PCR plate, add 15 µl of each Purified fraction, denature by heating at 80 °C for 10 min and load 15 µl samples into each lane of the SDS-PAGE gels using a multichannel pipette, by alternating rows (e.g., A1, B1, A2, B2, *see* **Note 16**). Include a marker in the first lane (e.g., SeeBlue Plus2, Invitrogen).
15. Run the gels at 150 V for approximately 1 h or until the first dye front has reached the bottom of the gel, then stain with InstantBlue™ (Expedeon Protein Solutions) or similar to identify which constructs are positive for soluble expression (*see* Fig. 3).

3.4 Large-Scale Expression

1. After identifying the positive constructs from the test expression and purification, prepare a starter culture by inoculating a loop of the glycerol stock into 10 ml of TB medium containing 50 µg/ml kanamycin and 34 µg/ml chloramphenicol in a 50 ml tube (*see* **Notes 17** and **18**). Grow the starter culture overnight at 37 °C in a shaker incubator.
2. The next morning, inoculate 10 ml of the starter culture into a 2.5 l Ultra Yield or baffled glass flask containing 1 l of sterile TB medium, freshly supplemented with 50 µg/ml kanamycin only (*see* **Note 7**). Cover the flask with a porous seal and incubate at 37 °C, with shaking at 200 rpm (*see* **Note 19**).
3. Monitor OD_{600nm} by taking 1 ml of the sample every hour and continue the incubation at 37 °C until the OD_{600nm} reaches 2.00 ± 1 (*see* **Note 20**).
4. Reduce the temperature of the incubator to 18 °C and after approximately 30 min, induce protein expression by adding IPTG (from a 1 M stock solution) to a final concentration of 0.1 mM (*see* **Note 21**), then continue the incubation overnight at 18 °C.

5. Measure OD_{600nm} by diluting 25 μl of the culture into 1 ml of the TB medium (*see* **Note 22**) and harvest the remaining cells by centrifugation at 6,000 rpm (~9,000 × *g*) for 20 min using a JLA-8.1000 rotor or similar. Pour the supernatant back into the original culture flask and decontaminate using Virkon.
6. Remove traces of the medium from the cell pellet using a 1 ml pipette and transfer the cell pellet to a 50 ml tube. Record the wet-weight of the cells (generally 12–30 g from 1 l of culture) and store the pellets at −80 °C until required for purification. The cell pellets can be stored at −80 °C for many months (*see* **Note 23**).

3.5 Protein Extraction

All the following steps of protein extraction and purification are performed at 4 °C or on ice. Prechill all buffers and centrifuges.

1. If protein purification is performed straight after harvesting the cells, transfer the cell pellets to ice or if the cells were frozen, thaw the pellets in a water bath set at 37 °C for as long as required to thaw, then immediately transfer to ice.
2. Resuspend the cells in 1 volume of 2× Lysis buffer (1 ml/g wet-weight) and mix thoroughly using a glass rod or serological pipette. Add 2–3 more volumes of 1× Lysis buffer until the sample is manageably fluid with no cell lumps.
3. Prechill the cell disruptor and lyse the cells resuspended in step 2 above. Refer to the manufacturer's instructions of the instrument that is used (e.g., for the basic Z model cell disruptor, two to three rounds at ~15,000 psi are sufficient for cell lysis). Recover the lysate in the disruptor by flushing it with Lysis buffer (20–40 ml). Save 10 μl of the lysate which represents the Total fraction (*see* **Note 24**).
4 Add PEI to the cell lysate to a final concentration of 0.15 % and mix thoroughly by inverting the tube several times or using a pipette. At this stage the lysates turn milky (*see* **Note 25**).
5 Transfer the lysates to centrifuge tubes, balance the tubes pairwise and centrifuge at 17,000 rpm (~39,000 × *g*) in a JA-17 rotor (or similar) for at least 30 min at 4 °C.
6 Transfer the clear supernatant into a clean tube taking care to avoid transferring any pelleted material. This clarified supernatant represents the Soluble fraction (*see* **Note 26**).

3.6 Large-Scale Protein Purification

The purification scheme described here for His-tagged proteins is generic and applied to a diverse set of proteins; however, it may not be applicable to every individual protein. Other buffer compositions may be substituted to address issues such as protein instability and requirements of final applications. Careful optimization of the buffer composition with respect to the buffering system, pH, salt concentrations and additives is particularly critical for difficult to purify proteins (*see* **Note 27**). The protein purification scheme described here is a two-step procedure (1) IMAC and (2) SEC. Manual IMAC provides the flexibility to use a specific volume of resin to the amount of lysate and collection of several elutions with gradual increase in imidazole concentration. Automated protein purification systems allow rapid purification of target proteins while using multiple chromatography steps with minimal intervention. An important point to mention when working with large culture volumes is the problems associated with applying large volumes of lysate to small IMAC columns which can result in reduced protein binding capacity due to depletion of nickel ions from the column [9].

1. To perform manual IMAC, prepare the Ni-IDA (or Ni-NTA) resin as described in Subheading 2.2, **step 15**.
2. Add the resin to the clarified cell lysate in a 50 ml tube. Depending on the estimated protein expression level, add 0.5–2 ml of the 50 % (w/v) resin to the clarified lysate obtained per liter of culture and rotate the tubes gently for 30 min to 1 h at 4 °C.
3. Centrifuge at 500×g for 10 min, remove, and save the supernatant in a fresh tube which represents the Unbound fraction, taking care not to disturb the resin while removing the supernatant.
4. Resuspend the resin in 2–3 column volumes (CV) of Affinity buffer and transfer to an empty chromatography column (such as an Econo-Column, Bio-Rad). Alternatively, prepare a proportionate resin bed in an empty column, apply the clarified cell lysate, and collect the Unbound fraction by gravity flow through the column.
5. Wash the resin in the column with 10 CV of Affinity buffer and save the flow through for SDS-PAGE analysis.
6. Wash the resin with 20 CV of Wash buffer, again saving the flow through for gel analysis.
7. Elute the bound protein in fractions of at least five elutions of 2 CV of Elution buffer, generally a total of 10–15 CV. Analyze 15 μl of each elution by SDS-PAGE (see Fig. 4a) prior to proceeding to the next step (*see* **Note 28**).
8. To prepare the sample for SEC, pool the fractions and concentrate using an Amicon Ultra protein concentrator (Millipore) according to the Manufacturer's instructions. Transfer the concentrated sample into a 50 ml tube and centrifuge at 4,000×*g* for 10 min or filter through a 0.22 μm low protein binding filter (Millipore) to remove aggregates and particulates before loading onto the SEC column.

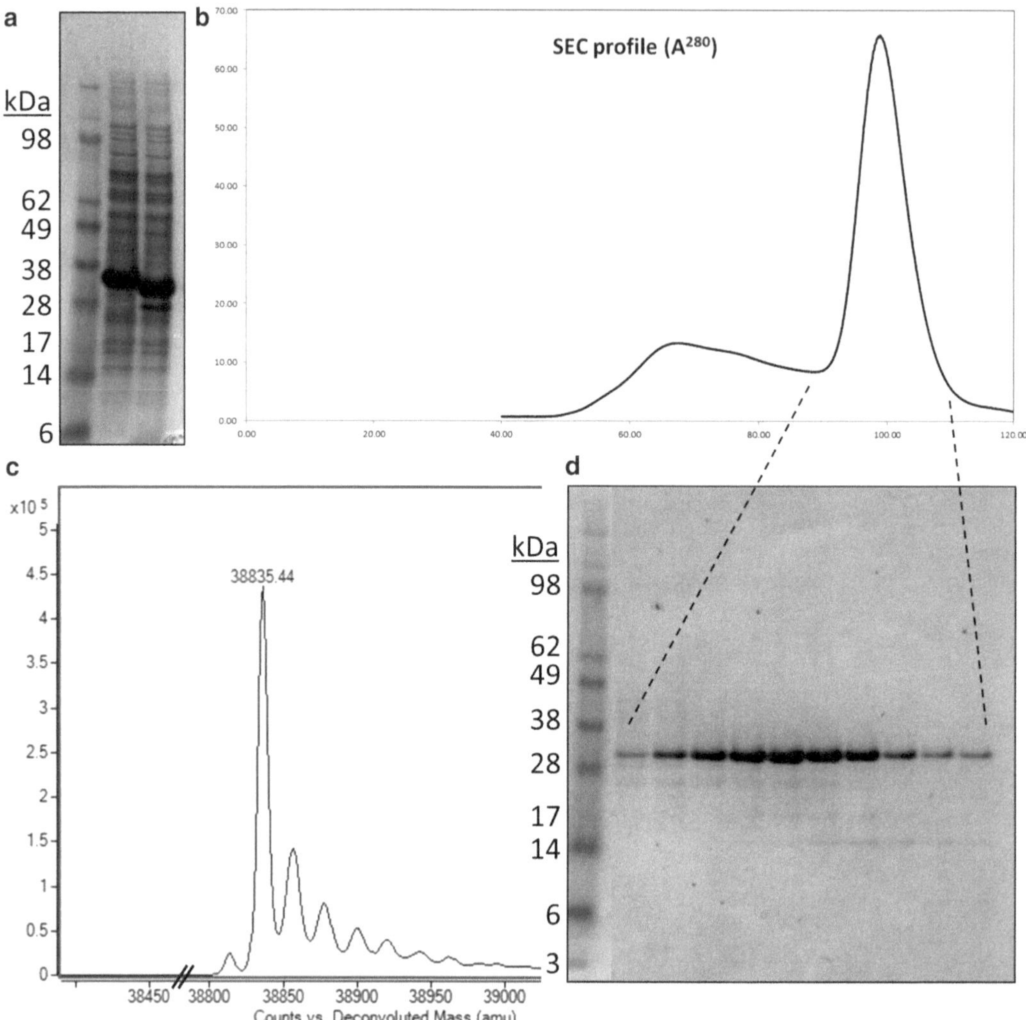

Fig. 4 Image showing quality assurance measures in protein purification. (**a**) The initial IMAC elutions are analyzed by SDS-PAGE to determine approximate size and yield. In the example shown, gel A *lane 3* shows the product of TEV-mediated cleavage of the His-tag following IMAC purification. (**b**) The cleaved protein is then purified by SEC. (**d**) The resulting fractions from SEC are assessed for purity by SDS-PAGE before pooling and concentrating the protein. (**c**) The identity of the purified protein is then confirmed by intact mass spectrometry (MS) analysis. Note that in the example shown the expected mass of the protein is 38.8 kDa, as confirmed by MS analysis. The size discrepancy shown in *inserts* **a** and **d** is due to the inaccuracy of size determination by SDS-PAGE

9. To perform SEC, follow the method from step 16 onwards; however, the protein sample will have to be injected onto the SEC column manually (*see* **Note 29**).

10. To perform automated IMAC and SEC using an ÄKTA-Xpress chromatography system, prepare the system in a cold cabinet or cold room in advance by employing the desired number of IMAC columns (e.g., HisTrap FF crude) and a SEC column

(e.g., HiLoad 16/60 Superdex 75 prep grade or HiLoad 16/60 Superdex 200 prep grade).

11. Prepare the HisTrap columns by washing first with 10 CV of water and then by equilibrating with 10 CV of Affinity buffer at 0.8 ml/min flow rate.
12. Prepare the SEC column by washing first with 2 CV of water (inlet A5) and then with 2 CV of SEC buffer (inlet A4).
13. Set up an IMAC and SEC purification method. Change the default parameters as described in the notes (*see* **Note 30**). Steps 14–17 are performed automatically on the ÄKTA-Xpress.
14. Apply the clarified and filtered cell lysate to the pre-equilibrated IMAC column at 0.8 ml/min flow rate.
15. Wash the IMAC column with 5–10 CV of Affinity buffer using inlet A1 until the A_{280nm} stabilizes. Wash with 10 CV of Wash buffer using inlet B1. Elute with 5 CV of Elution buffer using inlet A3. The eluted peak is automatically identified by detection of an increased A_{280nm} and is collected into the reinjection loop.
16. The eluted peak is automatically injected onto the SEC column at a flow rate of 1.2 ml/min, followed by running 1.2 CV of SEC buffer at the same flow rate using inlet A4.
17. Collect 2 ml fractions based on the A_{280nm} peak (see Fig. 4b) into a 96-deep-well block.
18. Analyze the fractions by SDS-PAGE for purity and homogeneity (see Fig. 4d). Avoid high molecular weight aggregates and pool peaks corresponding to different oligomeric forms, e.g., monomers, dimers, separately (*see* **Note 31**).
19. If the purified protein is to be used at a later date, concentrate the protein using an Amicon Ultra protein concentrator (Millipore) and aliquot in small volumes, flash-freeze in liquid nitrogen, and store at −80 °C until needed. To prevent damage to the protein during freezing and thawing add glycerol, if not already included in the buffer (*see* **Note 32**).
20. If required purity is not achieved in this two-step purification scheme, additional steps such tag removal followed by IMAC purification or ion exchange chromatography may be included to obtain pure and homogeneous protein (*see* **Note 33**).

3.7 Quality Assurance

In addition to SDS-PAGE and SEC, if available, mass spectrometry analysis of every purified protein is highly recommended (*see* Fig. 4c). This confirms the molecular weight of the protein, with mass discrepancies indicating mutations or cloning artifacts and potential PTMs. The protein is loaded into a small C3 HPLC column for de-salting and eluted onto an in-line electrospray ionization time-of-flight analyser (Agilent). Any discrepancy needs to be explained, either by sequencing the DNA or by enzymatic removal of suspected modifications or by MS/MS analysis of proteolytic fragments.

4 Notes

1. It is advisable to prepare small batches (10–100 ml) of SOC medium as this can become contaminated very quickly.
2. On the day of purification, only prepare the required amount of buffer for the number of samples to be purified, e.g., for one 96-well plate you will need about 25 ml of Lysis buffer, 70 ml of Wash buffer, and 5 ml of Elution buffer. The stock buffers can be stored at 4 °C for at least 1 month.
3. Be careful not to splash the cells up the sides of the wells whilst using the repeat pipettor and also check that the cells are at the bottom of the well before continuing. This step can also be done using a single channel pipette but will take more time.
4. The SOC medium can be added using a multichannel pipette with the medium in a reagent reservoir.
5. The transformation can be performed for individual clones. In this case, plate 80 μl of the transformation mixture onto a 50 mm petri dish and spread with a sterile spreader.
6. Multiple colonies are selected at this stage in order to account for clone to clone variation in expression levels of the protein.
7. We recommend not adding chloramphenicol at this stage as the pRARE2 plasmid is not lost during expression; however, its addition may significantly slow down the bacterial growth.
8. Using the 96-well plate reader to determine the OD_{600nm} of the cultures across the whole 96-well block indicates any inconsistencies with growth for particular targets, or constructs, and can therefore be used to identify proteins which failed to express because of a lack of proper growth.
9. The cell pellets can also be stored at −80 °C for 1–2 weeks if necessary.
10. It is useful to set up two 96-well plates of test proteins in parallel to provide a balance for the centrifugation steps; however, a balance block can be used containing water instead.

11. At this point, you can remove 15 μl of clarified lysate (as the Soluble fraction) and mix with 5 μl of SB in a PCR plate before transferring to the filter plate.

12. Take care to avoid transferring insoluble material to the resin as it may block the filter plate in subsequent steps. To avoid disturbing the Insoluble fraction tilt the plate and drive the tips down the side of the wells at an angle. Stop just above the pellet, on most plates there is a ridge just off the bottom—feel for this with the tips. Gently pipette up the supernatant and then transfer to the new plate. Do not go back into the wells as this will resuspend the pellets, if this happens then re-spin the sample and try again.

13. Alternatively, incubate at RT on a shaking platform. It is advisable to place the filter plate on top of a 96-well microtiter plate to avoid any drips coming through onto the shaker.

14. This step can also be done using centrifugation (200 × g for 1 min).

15. Removing all trace of Wash buffer is essential to ensure that the subsequent elution step does not become diluted with Wash buffer.

16. As standard, we only run the Purified fractions on gels to identify which proteins are expressed, soluble, and purified. We will only analyze the Total and Soluble fractions if we want to determine whether or not a protein has been expressed but is insoluble or if the control protein has failed to purify.

17. Alternatively, retransform the expression plasmid into BL21(DE3)-R3-pRARE2 as described in Subheading 3.1, except plate 80 μl of the transformation mixture onto a 50 mm petri dish and spread with a sterile spreader.

18. One 10 ml starter culture is required per L of culture scaled up. If you are planning to scale up more than 1 l, prepare starter cultures proportionately. We generally find that 1 l scale is sufficient to obtain milligram quantities of highly expressed proteins, i.e., those having large visible bands on Coomassie SDS-PAGE after test purification (see Fig. 3). If the bands are weak, you may need to scale up to 3 l of culture or more.

19. The flasks can be autoclaved with the media in them but do not use porous seals during autoclaving, instead cover the flasks with a piece of aluminum foil and use porous seals only during cell growth. The bacterial growth is an important determinant for protein expression and is mainly affected by aeration, stirring, and temperature. Efficient aeration in shaker flasks can be achieved by optimizing the ratio of culture

volume to the total capacity of flask and shaking speed. The wide mouth design of the 2.5 l Ultra Yield flasks with straight walls and baffles at the bottom of the flasks facilitate good oxygenation for culture volumes up to 1 l. Conventional baffled Erlenmeyer flasks provide comparable aeration but with lower culture to vessel ratios (typically 1:4).

20. Using a 5 ml serological pipette, remove 1 ml of sample and measure OD_{600nm}. OD measurements above 0.5 are not linear, dilute the culture if it is at higher OD before measurement and use the corrected value to obtain the precise OD. When cells are grown in TB medium, induction at an OD_{600nm} value of 1–3, followed by overnight growth at reduced temperature is optimum for protein expression. However, this may need to be optimized for individual proteins.

21. A concentration of 0.1 mM IPTG is sufficient for most strains; however, others such as pLysS may require higher concentrations in the range of 1–2 mM for efficient induction. We find that the optimum temperature of induction for the majority of the human proteins that we express in *E. coli* is 18–25 °C. It may be beneficial to test a number of temperatures (ranging from 15 to 37 °C) at the test expression stage for specific proteins.

22. At this stage you can remove a 5 ml sample and harvest the cell pellet by centrifugation in a 15 ml tube to perform a test purification which will confirm if the scale up expression has been successful, before proceeding to large-scale purification.

23. If the cell pellets are not used for protein purification immediately, they can be frozen directly or after resuspension in a small volume of Lysis buffer at −80 °C. If the cells are frozen after resuspension in buffer, thawing may result in a very viscous solution because of cell lysis and release of nucleic acids. Viscosity can be reduced by the addition of Benzonase nuclease to the cell lysate at a concentration of 25–50 U/ml. The addition of protease inhibitors is important when freezing pellets in Lysis buffer to reduce protein degradation. However, it is advisable to test your protein by purification in small scale first to determine how sensitive it is to degradation. Some proteins require purification immediately from cell harvesting to prevent them from proteolytic degradation.

24. Although many methods are available to lyse cells, high-pressure cell disruption is a very efficient way of lysing large volumes of cell suspensions. For smaller volumes (<100 ml) sonication can be used effectively. However, both methods can cause localized heating which can result in protein precipitation or denaturation; therefore, it is important to keep samples on ice at all times and prechill the cell disruptor.

Cell disruption by sonication can also help in reducing viscosity by shearing nucleic acids. Use of detergents should be avoided for cell lysis if its presence will interfere with the downstream applications such as protein crystallization. To lyse your cells by sonication, transfer the cell suspension to a 50 ml conical tube or a beaker depending on the volume and place the container on ice. Sonicate the cell suspension using 10–15 bursts of 10 s on, 10 s off. Generally, an amplitude of 35 % using a 750 W Sonics Vibra-Cell sonicator is sufficient for lysis of a 50 ml cell suspension. The sonication time may need to be adjusted depending on the volume of the cell suspension. Avoid excessive foaming and heating of the suspension by keeping the cell suspension on ice at all times.

25. Polyethyleneimine (PEI) is a highly positively charged polymer at neutral pH and can be used to remove negatively charged nucleic acids from cell lysates by precipitation in the presence of high salt. At lower ionic strength, nucleic acid binding proteins may remain bound to the nucleic acid and the use of PEI may precipitate proteins of interest along with the nucleic acid. Therefore, it is crucial to maintain a high salt concentration (>0.5 M NaCl) during this step. Alternatively, a pre-equilibrated anion exchange column such as DEAE-cellulose (DE52) may be used prior to IMAC purification.

26. If the supernatant is still turbid after centrifugation or if the pellet dislodges, add additional PEI to a final concentration of 0.05 % and repeat the centrifugation step to obtain clear supernatant. If the lysate is still turbid, before proceeding to the IMAC step, filter the supernatant using a 0.80 μm syringe filter (Sartorius) first, followed by a 0.45 μm syringe filter (Sartorius) to remove large particulates and cell debris which can delay the binding, washing and elution steps in the chromatography procedure.

27. Phosphate and HEPES buffers with 0.5 M NaCl concentration work equally well for the IMAC; HEPES is preferred, as divalent ions (e.g., Mg2+, Ca2+, or Zn2+) are included to avoid precipitation. If the purified protein is to be used for crystallization, care must be taken to exchange the buffer from phosphate to HEPES during later stages of purification (such as SEC) because phosphates may form salt crystals with many of the crystallization solutions. A commonly observed problem in IMAC is co-purification of intrinsic proteins from host cells due to affinity of exposed histidines or metal binding moieties towards immobilized metal ions. Success of the technique depends on buffer composition, pH, and ionic strength. The binding of His-tagged proteins to the resin is optimal at physiological pH; therefore, it is important to keep the Lysis buffer

pH close to 7.5–8.0. Higher salt concentration (>0.5 M NaCl) is also responsible for avoiding nonspecific binding of proteins to IMAC resin. Salt concentration also plays an important role in protein stability in solution; therefore, it is crucial that the ionic strength of the buffers should not be reduced too far as this may promote protein precipitation. The presence of 5–10 % glycerol is useful to promote protein stability; however, it may inhibit protein crystallization in some cases.

28. It is important to collect the eluates in fractions and analyze them on SDS-PAGE before pooling them together. Pooling the fractions together before SDS-PAGE analysis can result in mixing of the purified sample with other contaminated fractions or dilution of concentrated fractions. Protein purified through IMAC may be pure enough for some functional studies but it is rarely pure enough for crystallization. Many host proteins bind nonspecifically during affinity chromatography which can be separated from the target protein by introducing a size exclusion chromatography step. This step also gives important information on oligomeric state of the protein and is useful in separating any contaminant proteins as well as aggregates.

29. To obtain high resolution separation on the SEC column, load a maximum volume of 5 ml. It may be necessary to concentrate your protein before applying to the SEC column to reduce the volume, and remember to filter the sample before loading to remove any precipitated protein.

30. If using an ÄKTA-Xpress system for purification, the detection parameters should be changed to accommodate varying protein loads. We recommend using the default parameters with the following changes:

    Affinity Peak Collection

    Start: Watch level greater than 20 mAU, slope greater than 25 mAU/min.

    Stop: Peak max factor 0.5, watch level less than 20 mAU, watch stable plateau for 0.5 min.

    Delta plateau 5 mAU/min.

    Gel Filtration Peak Collection

    Elution volume before fractionation: 0.3 CV

    Elution volume with fractionation: 0.8 CV

    Peak fractionation algorithm: level_OR_slope

    Start level 10 mAU, start slope 5 mAU/min

    Peak max factor 0.5, minimum peak width 0.5 min

    Stop level 10 mAU, slope 5 mAU/min

31. Care must be taken while pooling protein fractions. Pay special attention to the concentration of the target protein and the level of contaminant proteins on the gel, analyze the SEC spectra and compare with molecular weight standards (which have been separated on the same column). Eliminate aggregated proteins (eluted in the void volume of SEC) and pool together fractions corresponding to monomer or oligomer peaks separately. Pool fractions from well-formed and symmetrical peaks and avoid mixing fractions from long tails which may represent some heterogeneity.

32. Protein aggregation can occur at any stage of the expression/purification procedure, but is very common during the process of concentration. This becomes clearly apparent when attempting to concentrate by ultrafiltration as protein aggregates rapidly block the filter and it becomes impossible to further concentrate the protein. Therefore, it is important to test a small volume of protein for its ability to concentrate before committing to concentrate the whole protein sample. Measure the protein concentration using a NanoDrop or similar before starting to concentrate. Choose an appropriate protein concentrator with molecular weight cutoff size that is two times smaller than the protein molecular weight. Transfer 200–500 μl of the sample to a concentrator that fits into a 1.5 ml microcentrifuge tube. Centrifuge according to the manufacturer's instructions at 4–15 °C. Check the volume every 10–15 min and more regularly when the volume is low. The sample should concentrate quite rapidly to a protein concentration of at least 5–10 mg/ml. If the process is stuck with no apparent reduction in volume, it is likely that the protein is aggregating. The aggregates can be detected by analytical SEC or light scattering. If the protein aggregates easily, change in buffer pH, NaCl concentration or use of additives should be considered. Once the concentration conditions are established, the remaining protein can be concentrated using those parameters.

33. Impurity can be a result of co-purification of contaminant proteins. To improve the purity of such samples, additional chromatographic steps can be employed. An effective general purification step is removal of the tag by cleavage with TEV protease followed by rebinding to Ni-IDA resin which is an efficient way to remove contaminants. An overnight digestion with TEV protease at 4 °C removes the His tag (*see* Fig. 4a), the cleaved protein is then applied to Ni-IDA resin, which will isolate the cleaved His tag as well as other contaminant proteins by their affinity for the beads and the target protein is collected in the flow through. In order for this protocol to work, the protein solution must not contain more than 25 mM

imidazole; this can be achieved by SEC (before or after cleavage), or by performing the TEV cleavage during dialysis of the protein. Further purification can be achieved using ion exchange and other chromatographic methods that need to be specifically tailored for each protein.

Acknowledgements

We would like to thank all the SGC scientists (past and present) who contributed towards the development of the methods. The SGC is a registered charity (number 1097737) that receives funds from the Canadian Institutes for Health Research, Genome Canada, GlaxoSmithKline, Lilly Canada, the Novartis Research Foundation, Pfizer, Takeda, AbbVie, the Canada Foundation for Innovation, the Ontario Ministry of Economic Development and Innovation, and the Wellcome Trust.

References

1. Sorensen HP (2010) Towards universal systems for recombinant gene expression. Microb Cell Fact 9:27
2. Graslund S, Nordlund P, Weigelt J et al (2008) Protein production and purification. Nat Methods 5:135–146
3. Burgess-Brown NA, Sharma S, Sobott F et al (2008) Codon optimization can improve expression of human genes in Escherichia coli: a multigene study. Protein Expr Purif 59:94–102
4. Tegel H, Tourle S, Ottosson J et al (2010) Increased levels of recombinant human proteins with the *Escherichia coli* strain Rosetta (DE3). Protein Expr Purif 69:159–167
5. Cornvik T, Dahlroth SL, Magnusdottir A et al (2005) Colony filtration blot: a new screening method for soluble protein expression in *Escherichia coli*. Nat Methods 2:507–509
6. Cornvik T, Dahlroth SL, Magnusdottir A et al (2006) An efficient and generic strategy for producing soluble human proteins and domains in *E. coli* by screening construct libraries. Proteins 65:266–273
7. Savitsky P, Bray J, Cooper CD et al (2010) High-throughput production of human proteins for crystallization: the SGC experience. J Struct Biol 172:3–13
8. Hanahan D, Jessee J, Bloom FR (1991) Plasmid transformation of Escherichia coli and other bacteria. Methods Enzymol 204:63–113
9. Magnusdottir A, Johansson I, Dahlgren LG et al (2009) Enabling IMAC purification of low abundance recombinant proteins from *E. coli* lysates. Nat Methods 6:477–478

Chapter 6

Medium-Throughput Production of Recombinant Human Proteins: Protein Production in Insect Cells

Pravin Mahajan, Claire Strain-Damerell, Opher Gileadi, and Nicola A. Burgess-Brown

Abstract

This chapter describes the step-by-step methods employed by the Structural Genomics Consortium (SGC) for screening and producing proteins in the baculovirus expression vector system (BEVS). This eukaryotic expression system was selected and a screening process established in 2007 as a measure to tackle the more challenging kinase, RNA–DNA processing and integral membrane protein families on our target list. Here, we discuss our platform for identifying soluble proteins from 3 ml of insect cell culture and describe the procedures involved in producing protein from liter-scale cultures. Although not discussed in this chapter, the same process can also be applied to integral membrane proteins (IMPs) with slight adaptations to the purification procedure.

Key words Insect cells, Baculovirus, BEVS, Expression, Recombinant Protein, Purification, IMAC, SEC, Gel filtration

1 Introduction

Availability of a pure protein is essential for obtaining information on protein structure and function. Heterologous protein production in *Escherichia coli* has remained the preferred system for many research laboratories as it is low-cost, fast, and easy to handle. However, there is no guarantee that *E. coli* cells will produce eukaryotic proteins in a soluble and biologically active form because of a number of limitations such as codon bias, lack of posttranslational modifications (PTMs), or disulfide bond formation. Exploring other protein expression hosts such as mammalian cells, yeast, and insect cells is often required if *E. coli* fails to produce soluble protein after attempting different strains, solubility enhancing tags, etc. Among the alternatives available, the baculovirus expression vector system (BEVS) is increasingly becoming popular for expression of recombinant proteins as it is non-pathogenic to humans [1], capable of

Yu Wai Chen (ed.), *Structural Genomics: General Applications*, Methods in Molecular Biology, vol. 1091, DOI 10.1007/978-1-62703-691-7_6, © Springer Science+Business Media, LLC 2014

producing high levels of soluble proteins with PTMs similar to those observed in mammalian cells and easily scalable in suspension culture [2]. This system is also proving popular for the production of large protein complexes, production of virus like particles, gene delivery, viral vector vaccines, expression of proteins in mammalian cells, and display of proteins and peptides on the baculovirus envelope [3]. Baculoviruses are double-stranded DNA viruses [4] most of which infect insects of the order Lepidoptera [5]. The most widely used baculovirus used as a BEVS is *Autographa californica* multiple nuclear polyhedrosis virus (AcMNPV). Two major genes that express in the very late phase of baculovirus infection of insects are p10 and polyhedrin which are strong expressers but dispensable for viral replication. This discovery has allowed exploitation of the p10 and polyhedrin promoters to be used for driving recombinant protein expression in BEVS; the polyhedrin promoter in particular has been described as a workhorse promoter of BEVS [6]. The most common insect cell lines utilized as hosts of BEVS are Sf9 and Sf21, derived from pupal ovarian tissue of the fall army worm, *Spodoptera frugiperda* [7] and High Five cells (BTI-Tn-5B1-4) derived from ovarian cells of the cabbage looper, *Trichoplusia ni* [8].

Since the first use of baculoviruses for protein expression in 1983 [9], the system has gone through numerous technological advances that have allowed it to be widely accessible. Various baculovirus expression systems are commercially available to produce baculoviruses, most notably Bac-to-Bac® (Invitrogen), flashBAC (Oxford Expression Technologies), BaculoDirect™ (Invitrogen), BacVector®-3000 (Novagen), BacPAK (Clontech), Bac-n-Blue™ (Invitrogen), etc. About 5 years ago, it became evident in our laboratory that the bacterial expression system was unable to cope with more challenging proteins on our target list such as many protein kinases, RNA-DNA processing proteins and integral membrane proteins (IMPs). To address this issue, we established an efficient process based on the Bac-to-Bac® system [10] for screening multiple versions of each protein in insect cells to identify those that were amenable to purification and crystallization. The 96-well cloning procedure is described in detail in Chapter 4. In this chapter we continue the methodologies for expression screening and scaling up expression of proteins in suspension culture. To describe our series of standardized protocols for protein production in insect cells, this chapter is broadly divided into the following stages: (1) transposition, bacmid production, and PCR screen, (2) growth and maintenance of insect cell lines in adherent and suspension culture, (3) transfection into Sf9 cells, baculovirus generation and small-scale test expression/purification, and (4) large-scale protein expression and purification. The screening process has been miniaturized to 24-well format. The steps involved in the pipeline from cloning to large-scale expression are outlined in Fig. 1.

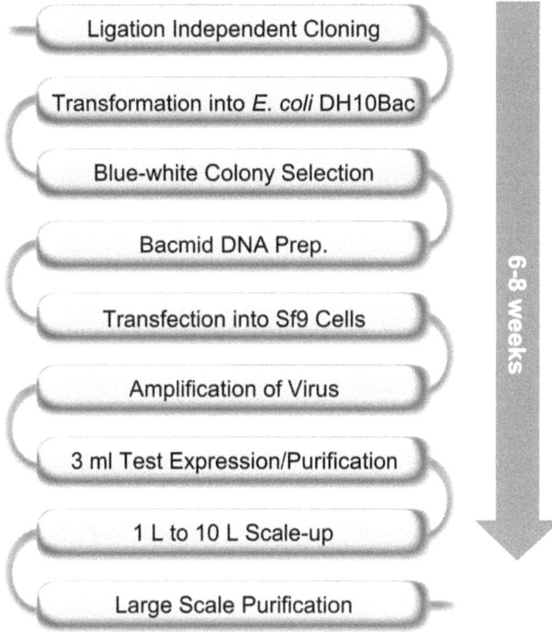

Fig. 1 Overview of the baculovirus expression process. The process takes ~6–8 weeks from LIC to scale up

2 Materials

Unless otherwise stated, all solutions are prepared using ultrapure water (prepared by purifying deionized water to reach a resistivity of 18 MΩ cm at 25 °C) and analytical grade reagents.

2.1 Transposition and Bacmid Preparation

1. *E. coli* DH10Bac (Invitrogen). Chemically competent bacterial cells are prepared in house as described [11].
2. Primers: Primers are supplied by either MWG-Biotech or Sigma-Aldrich and are HPSF purified at 0.01 μmol scale or DST purified at 0.025 μmol scale, respectively. Primer stocks are either supplied at or diluted (in 10 mM Tris–HCl buffer, pH 8.0) to 100 μM and stored at −20 °C.
3. BIOTAQ™ Red DNA polymerase (1 U/μl, Bioline).
4. Molecular biology grade water (Thermo Scientific HyClone).
5. 2 mM dNTP solution: 2 mM dATP, 2 mM dTTP, 2 mM dGTP, and 2 mM dCTP (prepared from 100 mM dNTP set, Invitrogen) diluted in molecular biology grade water and stored at −20 °C.
6. 50× TAE buffer (1 l): 242 g Tris base, 57.1 ml glacial acetic acid and 100 ml of 0.5 M EDTA, pH 8.0, pH adjusted to 8.5. Filtered through a 0.2 μm membrane filter and used as a 1× solution.

7. 96-Well 1.5 % TAE-agarose gels: 3 g agarose powder (Invitrogen), 200 ml of 1× TAE buffer, and 8 μl of SYBR-safe DNA gel stain (Invitrogen), cast in a Sub-cell Model 96 (Bio-Rad or similar) gel cast.
8. DNA ladder: For the bacmid screen, the 1 kb Plus DNA ladder (Invitrogen) prepared in 1× BlueJuice™ (Invitrogen) diluted in molecular biology grade water.
9. TE (Tris-EDTA) Buffer (50 ml): 10 mM Tris–HCl, pH 8.0 and 1 mM EDTA filtered through a 0.20 μm syringe filter (Sartorius) to sterilize and stored at RT.
10. 60 % (v/v) Glycerol autoclaved to sterilize.
11. 70 % (v/v) Ethanol.
12. 1,000× Antibiotic stocks: Kanamycin (50 mg/ml); Tetracycline (10 mg/ml in ethanol); Gentamycin (7 mg/ml), stored at −20 °C. All stocks prepared in water are filtered through a 0.20 μm syringe filter (Sartorius).
13. 1,000× Selection reagents: Blue-gal (Glycosynth, 100 mg/ml in DMSO); IPTG (40 mg/ml) stored at −20 °C. All stocks prepared in water are filtered through a 0.20 μm syringe filter (Sartorius).
14. LB-agar: 22.5 g premixed LB-broth and 13.5 g agar dissolved in 800 ml of water. Volume adjusted to 900 ml and autoclaved on the same day.
15. Recombinant bacmid selection plates: LB-agar melted slowly in a microwave. Once cooled to hand-hot, the appropriate antibiotic/reagent is added and swirled vigorously to mix. 10 ml of the molten agar is poured into each 50 mm petri dish and once set, upturned and left open to dry. These can be prepared ahead of time and stored at 4 °C in the dark sealed in a plastic bag to prevent over-drying.
16. 2× LB: 45 g premixed LB-broth dissolved in 800 ml of water. Volume adjusted to 900 ml and autoclaved on the same day.
17. Virkon (Appleton Woods).
18. Montage Plasmid Miniprep$_{\text{HTS}}$ 96 Kit (Millipore, *see* **Note 6**).
19. 50 mm Petri dishes.
20. 96-Well PCR plates.
21. 96-Well microtiter plates that can hold up to 200 μl of sample (Sterilin Ltd. UK or similar).
22. 24-Well blocks (Microplate Devices Uniplate®, GE Healthcare, or similar).
23. 96-Deep-well blocks (Thomson or similar).
24. Adhesive tape pads (Qiagen).

25. Adhesive PCR seals (ABgene).
26. AirOtop porous seals (Thomson).
27. Silicone 96 Square Well AxyMat (Axygen).
28. Disposable sterile spreaders (Fisher).
29. Disposable sterile inoculation loops (1 μl).
30. Reagent reservoirs for multichannel pipetting (Fisher).
31. Minisart syringe filters, 0.20 μm (Sartorius).
32. Membrane filters, 0.2 μm and unit.
33. Multichannel pipettes and repeat pipettors are used to dispense reagents into a 96-well format.
34. 96-Well PCR thermocycler with heated lid.
35. 96-Well gel cast and tank (Subcell Model 96 Bio-Rad or similar).
36. All gels are imaged on a Gel Logic 200 Imaging System (Kodak).
37. A UV-spectrophotometer for measuring DNA and protein concentration (e.g., The NanoDrop spectrophotometer allows measurements from as low as 1.5 μl volumes).
38. Scanlaf Mars recirculating class II biological safety cabinet (or similar laminar airflow (LAF) workstation).
39. Micro-Express Glas-Col shaker (Glas-Col, IN, USA) or alternative that ranges in temperature from 18 to 37 °C and shakes up to 800 rpm.
40. 96-Well block mixer (Eppendorf MixMate or similar).
41. Water bath set at 42 °C.
42. Incubator set at 37 °C.
43. Centrifuge suitable for 96-deep-well blocks (3,000 × *g*).

2.2 Test Expression

The following reagents, consumables, and equipment are required in addition to those listed above:

1. Cell lines: Sf9 insect cells, SFM adapted (Invitrogen). High Five cells, SFM adapted (Invitrogen).
2. Media: Sf-900™ II SFM (1×) (Invitrogen). Unsupplemented Grace's Insect Medium (1×) (Invitrogen).
3. Reagents: Fetal bovine serum (FBS), insect cell culture tested (Invitrogen). Cellfectin reagent (Invitrogen). Pen/Strep (used at 50 U penicillin and 50 μg streptomycin per ml medium) (Gibco). 0.4 % Trypan Blue Stain (Invitrogen).
4. Cryo-vials.
5. 24-Well tissue culture (TC) plates (Corning).
6. 250, 500, and 1,000 ml flasks with vented cap (Corning).

7. Inverted light microscope (Axiovert 25, CarlZeiss).
8. Hemocytometer, improved Neubauer (VWR International).
9. Shaker-incubators with cooling capacity: Multitron (Infors HT).

2.3 Test Purification

The following reagents, consumables, and equipment are required in addition to those listed above:

1. Benzonase (Novagen, HC, 250 U/μl).
2. Protease Inhibitor Cocktail Set III (Calbiochem).
3. 1 M TCEP (Tris (2-Carboxyethyl) phosphine Hydrochloride), stored as 1 ml aliquots at −20 °C.
4. 1 M DTT (Dithiothreitol), stored as 1 ml aliquots at −20 °C.
5. SeeBlue Plus2 (Invitrogen).
6. InstantBlue™ (Expedeon Protein Solutions).
7. XT Mes running buffer (20×) (Bio-Rad).
8. PBS: Five tablets of PBS (Sigma) dissolved in 1 l of water, filtered through a 0.2 μm membrane filter and stored at 4 °C.
9. Stock solutions: 1 M HEPES, pH 7.5; 5 M NaCl; 3 M imidazole, pH 8.0; 200 mM $MgSO_4$ (all filtered through a 0.2 μm membrane filter and stored at RT); 50 % (v/v) glycerol (autoclaved and stored at RT).
10. Lysis buffer (1 l): 50 mM HEPES, pH 7.5, 300 mM NaCl, 5 % (v/v) glycerol, and 10 mM imidazole in advance, filtered through a 0.2 μm membrane filter and stored at 4 °C. On the day of purification, Benzonase (0.2 μl/ml), Protease inhibitor cocktail (1 μl/ml), and TCEP (0.5 μl/ml) are added.
11. Wash buffer (1 l): 50 mM HEPES, pH 7.5, 300 mM NaCl, 5 % (v/v) glycerol, 30 mM imidazole in advance, filtered through a 0.2 μm membrane filter and stored at 4 °C. 0.5 mM TCEP added on the day of purification.
12. Elution buffer (0.1 l): 50 mM HEPES, pH 7.5, 300 mM NaCl, 5 % (v/v) glycerol, 500 mM imidazole in advance, filtered through a 0.2 μm membrane filter and stored at 4 °C. 0.5 mM TCEP added on the day of purification.
13. 50 % Ni-IDA Metal Chelate Resin (Generon) (or Ni-NTA-agarose (Qiagen): The IMAC resins are generally supplied in 20 % ethanol. To equilibrate, the resin is washed twice in water and then three times in Affinity buffer (*see* Subheading 2.5, **step 4**) in a 50 ml tube, by inverting to resuspend the resin and centrifuging at 500×*g* for 1 min. After the final wash, the resin is resuspended in Affinity buffer as 50 % slurry and stored at 4 °C when not in use.

14. SB: Stock of NuPAGE LDS sample buffer (Invitrogen) containing DTT (1:4 dilution of 1 M DTT in NuPAGE LDS sample buffer) and stored at −20 °C.
15. 96-Well filter plates (Thomson).
16. Pre-cast 26-Lane SDS-PAGE gradient gels (4-20 %) (Bio-Rad).
17. Protein gel electrophoresis apparatus (Bio-Rad).
18. Vibra-Cell Sonicator with 24-well probe (Sonics®).
19. General purpose benchtop centrifuge (Sorvall Legend RT, Kendro).

2.4 Large-Scale Expression

The following reagents, consumables, and equipment are required in addition to those listed above:

1. Media: Insect-XPRESS serum-free and protein-free medium (Lonza).
2. Non-baffled Erlenmeyer flasks: glass or polycarbonate in various sizes 250 ml, 500 ml, and 1 l and glass flasks of 3 l capacity for large-scale expression.
3. Cell freezing container: Mr. Frosty (Nalgene).
4. Avanti J-20XP or Avanti J-26XP centrifuge or similar (Beckman Coulter) with a JLA 8.1000 rotor for harvesting large volumes of cells.

2.5 Protein Extraction and Large-Scale Purification

The following reagents, consumables, and equipment are required in addition to those listed above; however, the majority of materials required for these procedures are the same as in Chapter 5, Subheading 2.4:

1. Complete EDTA-free protease inhibitor (Roche).
2. 2× Lysis buffer: 100 mM HEPES buffer, pH 7.5, 1 M NaCl, 20 % (v/v) glycerol, 20 mM imidazole. Filtered through a 0.2 μm membrane filter and stored at 4 °C. On the day of purification, Benzonase (0.2 μl/ml of cell lysate), Protease inhibitor cocktail (2 μl/ml of cell lysate) or Complete EDTA-free protease inhibitor cocktail (1 tablet/25 ml of cell lysate), and 1 mM TCEP are added.
3. Lysis buffer: 50 mM HEPES, pH 7.5, 500 mM NaCl, 10 % (v/v) glycerol, 10 mM imidazole. Filtered through a 0.2 μm membrane filter and stored at 4 °C. On the day of purification, Benzonase (0.1 μl/ml of cell lysate), Protease inhibitor cocktail (1 μl/ml of cell lysate) or Complete EDTA-free protease inhibitor cocktail (1 tablet/50 ml of cell lysate), and 0.5 mM TCEP are added.
4. Affinity buffer: 50 mM HEPES, pH 7.5, 500 mM NaCl, 10 % glycerol, 10 mM imidazole. Filtered through a 0.2 μm membrane filter and stored at 4 °C. 0.5 mM TCEP added on the day of purification.

5. Wash buffer: 50 mM HEPES, pH 7.5, 500 mM NaCl, 10 % glycerol, 30 mM imidazole. Filtered through a 0.2 μm membrane filter and stored at 4 °C. 0.5 mM TCEP added on the day of purification.
6. Elution buffer: 50 mM HEPES, pH 7.5, 500 mM NaCl, 10 % glycerol, 300 mM imidazole. Filtered through a 0.2 μm membrane filter and stored at 4 °C. 0.5 mM TCEP added on the day of purification.
7. Size Exclusion Chromatography (SEC) buffer: 20 mM HEPES, pH 7.5, 500 mM NaCl, 5 % glycerol. Filtered through a 0.2 μm membrane filter and stored at 4 °C. 0.5 mM TCEP added on the day of purification.
8. Minisart syringe filters, 0.20, 0.45, and 0.80 μm (Sartorius).
9. Amicon Ultra protein concentrators (Millipore).
10. Sonicator (Sonics Vibra-Cell, VCX 750, Sonics & Materials INC) or basic Z model cell disruptor (Constant Systems Ltd).
11. ÄKTA-Xpress or ÄKTA-Purifier liquid chromatography system (GE).
12. HiTrap 5 ml FF columns (GE) for His-tagged protein purification.
13. Ion exchange chromatography columns such as HiTrap 5 ml Q FF and SP FF (GE).
14. HiLoad Superdex columns (GE) for preparative size exclusion chromatography such as HiLoad 16/60 Superdex S75 or Superdex S200).
15. JA-25.50 rotor for centrifugation of cell lysates.

3 Methods

3.1 Transposition in *E. coli* DH10Bac

The transposition process is outlined in Fig. 2.

1. Prepare at least 100 petri dishes (50 mm) containing approximately 10 ml of LB-agar, supplemented with 50 μg/ml kanamycin, 7 μg/ml gentamycin, 10 μg/ml tetracycline, 40 μg/ml IPTG and 100 μg/ml Blue-gal (*see* **Note 1**) and once set allow to dry, inverted at RT.
2. Using a multichannel pipette, add 3 μl of recombinant DNA (see Chapter 4, Subheading 3.8, **step 6**) to a 96-well PCR plate.
3. On ice, add 20 μl of chemically competent *E. coli* DH10Bac cells using a repeat pipettor (*see* **Note 2**), cover with an adhesive tape pad and incubate for 30 min. It is advisable to include a positive control (i.e., a construct that has previously shown soluble protein expression in BEVS) in position H12 of the 96-well plate.

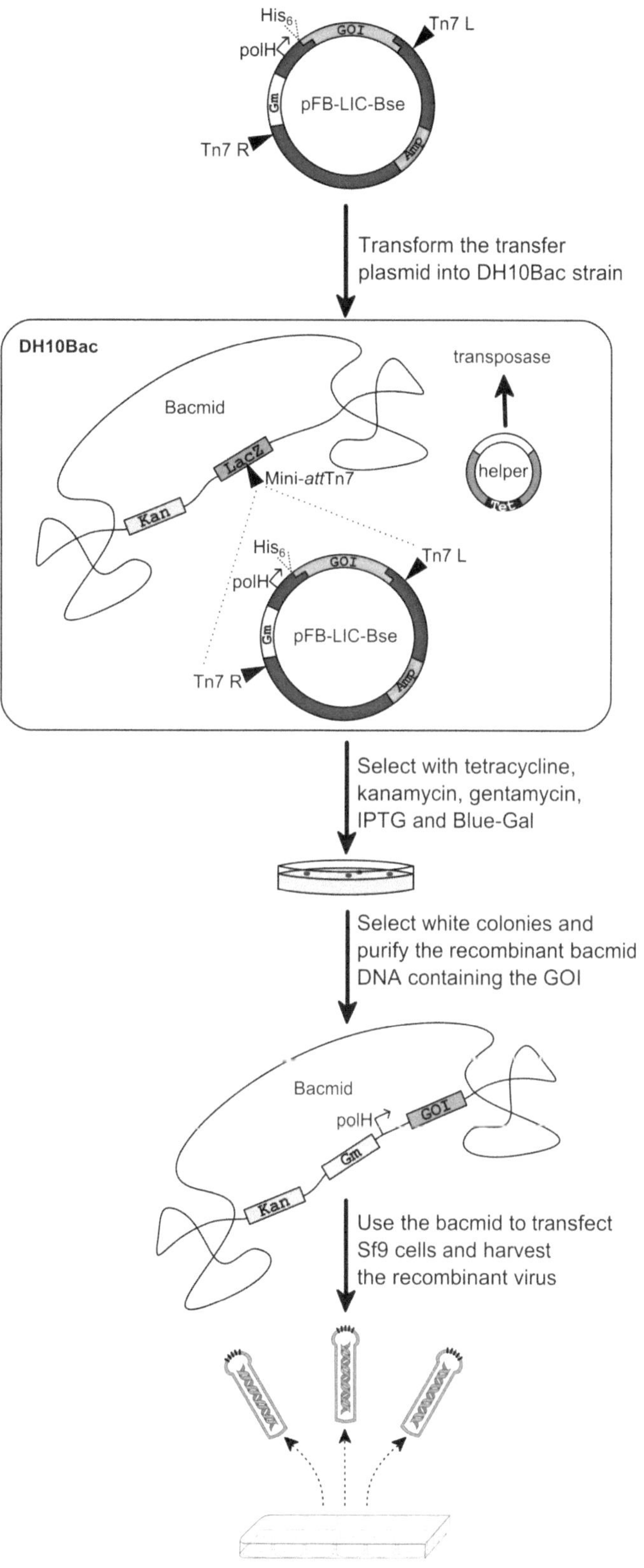

Fig. 2 Diagram describing the transposition process. The construct DNA is transformed into the DH10Bac *E. coli* strain, which contains both bacmid DNA and a helper plasmid. The transposase, expressed from the helper plasmid, will facilitate transfer of the transposable element including the gene of interest (GOI) into the bacmid. The recombinant bacmid DNA can then be purified and used directly to transfect Sf9 insect cells

4. In the meantime, add 900 μl of pre-warmed 2× LB medium to each well of a 96-deep-well block using a reagent reservoir.
5. Heat shock the cells in the PCR plate for 45 s in a 42 °C water bath and return briefly to ice.
6. Transfer the bacterial suspension into the pre-warmed medium block (**step 4**), cover with a porous seal and incubate in a Glas-Col shaker (or equivalent) at 37 °C with shaking at 700 rpm for 5 h.
7. Dilute the culture (10 μl into 90 μl) into LB (or 2× LB) medium in a 96-well microtiter plate and spread 50 μl onto the recombinant bacmid selection plates (*see* **Note 3**).
8. Incubate the plates at 37 °C for 48 h, covered with foil (*see* **Note 4**).
9. White colonies contain the recombinant bacmid DNA and the blue ones do not (see Fig. 2). To ensure that the colonies are white, divide a selective plate into six or eight sectors using a marker pen and label with the well position (e.g., A1). Pick single colonies, streak to dilution using a sterile loop and incubate at 37 °C overnight.

3.2 Bacmid Production

1. Using a multichannel pipette and a reagent reservoir, dispense 50 μl of 2× LB medium into each well of a 96-well PCR plate.
2. Inoculate the recombinant white colonies (isolated from the restreaked plates) into the corresponding wells, then transfer 20 μl of this cell suspension into two 96-deep-well blocks, each containing 1 ml of 2× LB medium per well, supplemented with 50 μg/ml kanamycin, 7 μg/ml gentamycin and 10 μg/ml tetracycline (*see* **Note 5**).
3. Cover with a porous seal and incubate at 37 °C overnight at 700 rpm in a Glas-Col shaker.
4. The following morning, prepare one or two 96-well microtiter plates containing 120 μl of culture and 30 μl of 60 % (v/v) glycerol and store at −80 °C.
5. Centrifuge the deep-well blocks at 3,000 × *g* for 30 min. Decant the supernatant into a suitable container for Virkon decontamination. Invert the blocks and tap gently on absorbent paper.
6. Add 250 μl of the Solution 1 from the 96-well miniprep kit (Millipore) (*see* **Note 6**) to each well of one block using a multichannel pipette.
7. Seal the block with a silicone sealing mat (*see* **Note** 7) and mix in the Glas-Col incubator for 2 min at 700 rpm or a 96-well block mixer (Eppendorf) at 1,000–2,000 rpm. If necessary, resuspend using a multichannel pipette.
8. Transfer the suspension to the corresponding wells of the second block. Seal and repeat the mixing process.

9. Add 250 μl of Solution 2 (Millipore) to each well, seal with a silicone sealing mat, invert gently five times and incubate at RT for 5–10 min.
10. Add 300 μl of Solution 3 (Millipore), seal with a silicone sealing mat and mix gently but thoroughly by inverting five times.
11. Place the sample on ice for 20 min then centrifuge at 3,000 × *g* for 30 min at 4 °C.
12. Transfer the clear supernatant to a fresh 96-deep-well block, cover with an adhesive tape pad and centrifuge again at 3,000 × *g* for 30 min at 4 °C (*see* **Note 8**).
13. In another fresh 96-deep-well block, dispense 0.8 ml of isopropanol into each well and add 0.8 ml of the clarified supernatant to the corresponding wells (*see* **Note 9**).
14. Using a 1 ml capacity multichannel pipette, gently mix up and down, cover with an adhesive tape pad and then incubate on ice for 30 min (*see* **Note 10**).
15. Centrifuge at 3,000 × *g* for 30 min at 4 °C.
16. Spray the outside of the 96-deep-well block with 70 % (v/v) ethanol (*see* **Note 11**) and inside the LAF workstation, remove the cover from the block and discard the supernatant by decanting into a suitable container and blotting on absorbent paper.
17. Add 500 μl of 70 % (v/v) ethanol to each well and tap the block gently to wash the pellets. Cover with an adhesive tape pad and then centrifuge at 3,000 × *g* for 30 min at 4 °C.
18. Inside the LAF workstation, open the block and discard the supernatant by decanting. Tap the block very gently on absorbent paper to remove the ethanol. Allow the block to dry inside the hood for approximately 2 h (*see* **Note 12**).
19. Inside the LAF workstation, add 50 μl of sterile TE buffer, cover with an adhesive tape pad and allow to stand for about 1 h. Very gently resuspend the bacmid DNA using a multichannel pipette (*see* **Note 13**) and transfer to a 96-well microtiter plate. Remove a couple of microliters of DNA from a few wells to measure the concentration using a UV-spectrophotometer and also pipette 1 μl of each DNA into a PCR plate for the bacmid PCR screen as described in Subheading 3.3, then seal with a fresh adhesive tape pad.
20. Store bacmid DNA at 4 °C until the test purification is complete, then store at −20 °C.

3.3 Bacmid PCR Screen

1. Prepare a 10 μM primer stock (50 μl each of the 100 μM forward and reverse primers added to 400 μl of molecular biology grade water) of the bacmid screening primers (Table 1). Store at −20 °C.

Table 1
Primers used to confirm correct insertions at the bacmid PCR screen stage

Primer name	Primer sequence
Fbac-1	TATTCATACCGTCCCACCA
M13-rev	CAGGAAACAGCTATGAC

2. Dilute the bacmid DNA 1 in 50 in molecular biology grade water in a 96-well PCR plate (*see* **Note 14**).
3. Set up a PCR master mix as follows: 200 μl of 10× NH4 Reaction Buffer (Bioline), 60 μl of 50 mM MgCl2, 60 μl of 100 % DMSO, 1.24 ml of molecular biology grade water, 200 μl of 2 mM dNTPs, 200 μl of 10 μM bacmid screening primers (**step 1**), and 40 μl of BIOTAQ™ Red DNA polymerase (1 U/μl, Bioline). Using a repeat pipettor or a multichannel pipette, pipette 20 μl into each well of a 96-well PCR plate.
4. Transfer 1 μl of the diluted bacmid (**step 2**) to the PCR plate (**step 3**) and mix well.
5. Seal the PCR reaction plate with an adhesive PCR seal and set a thermocycler with the following conditions making sure that the block is up to 95 °C before placing your sample plate in the instrument:

   95 °C, 5 min.

   (95 °C, 45 s; 50 °C, 45 s; 72 °C, 5 min) ×25 cycles.

   72 °C, 7 min.

   15 °C hold.
6. Whilst the PCR cycle is running, prepare a 96-well 1.5 % TAE-agarose gel.
7. Using a multichannel pipette, load 10 μl of the PCR reaction mixtures directly onto the gel. Note that the spacing of the wells means that samples will be interleaved (*see* Chapter 4, Fig. 4). Load 6 μl of 1 kb Plus DNA ladder and run the gel at 150 V for 1 h.
8. Confirm the sizing of the products and repeat the screen for any constructs that do not produce a band of the correct size in the first screen (*see* **Note 14**). The size of the PCR fragments should be 700 bp larger than the cloned insert.

3.4 Growth and Maintenance of Insect Cell Lines

Insect cell lines can be maintained in adherent culture as well as in suspension culture. Their ability to grow in suspension at high densities allows expression of recombinant proteins in large scale; however, their ability to grow in monolayers can be utilized for the initial stage of transfection to generate baculoviruses. The most widely used insect cell lines for BEVS-based protein expression are Sf9, Sf21, and High Five, all of which are adaptable to serum-free, protein-free medium. We routinely use Sf9 cells for all the steps from transfection to large-scale protein expression simply because of their robustness and ease in manipulation; however, occasionally High Five cells are used for large-scale expression of proteins. Use of Sf9 cells for all steps in routine protocols ensures that uniform parameters are applied to a number of protein targets initially and if needed other cell lines can be tested later on to improve protein expression. Insect cell culture methods are described previously [6, 12, 13] in detail. Some important points when working with insect cells are mentioned in **Note 15**.

3.5 Reviving Sf9 Cell Line from Frozen Stock

Sf9 cells can be revived straight into suspension culture without first reviving them into adherent culture, provided there are sufficient cryo-vials of cells available in liquid nitrogen. Alternatively revive cells into adherent culture using T-flasks, then transfer to suspension culture at a density of 1×10^6 cells/ml. Cells can be kept in suspension culture for 6–8 weeks, after which time a new stock should be revived as older cells may show a decline in protein expression. There are different commercial formulations of serum-free insect cell media available; however, we use Sf-900 II SFM (Invitrogen) mainly for initial revival of cells, transfection, expression testing, and virus amplification and Insect-XPRESS (Lonza) mainly for large-scale protein expression. Sf9 cells adapt quickly from one medium to another. All the cell culture steps described below are performed in aseptic conditions inside a LAF workstation.

1. Bring Sf-900 II medium to RT and pipette 15 ml of the medium into a T-75 flask.
2. Remove a cryo-vial containing the cells from liquid nitrogen and thaw rapidly in a 37 °C water bath (~1.5 min), making sure not to leave the cells at 37 °C after they have thawed (*see* **Note 16**).
3. Decontaminate the outside of the vial by wiping with 70 % (v/v) ethanol.
4. Transfer the thawed cells immediately into the T-75 flask containing the medium.
5. Gently mix the cell suspension and transfer the flask to a 27 °C incubator.
6. Allow the cells to attach for approximately 30 min then observe them under the microscope to confirm attachment and good health.
7. Remove the medium and add 15 ml of fresh medium.
8. Continue to incubate the cells at 27 °C until the flask is confluent and change the medium if needed.

9. When the flask is confluent, dislodge the cells by streaming medium over the monolayer using a 10 ml pipette and by tapping the flask gently against your palm.
10. Split the cells 1:3 or 1:4 into T-75 flasks and add fresh medium to the final volume of 15 ml in each flask.

3.6 Suspension Culture of Sf9 Cells in Shake Flask

Cells previously cultured in an anchorage-dependent manner need complete adaptation to suspension culture. The cells can be grown in suspension using either shake flasks or spinner flasks; however, our method of choice is the former. The use of simple shake flasks makes the process of protein expression in insect cells easily scalable from 10 ml to more than 10 l volume and does not require specialized equipment, which would be needed for spinner flasks and bioreactors.

1. After growing a sufficient number of cells in 3–4 confluent T-75 flasks, dislodge the cells as described in Subheading 3.5, **step 9** above and determine the viable cell count using Trypan Blue Stain (Invitrogen) and a hemocytometer (*see* **Note 17**).
2. Seed the cells (1×10^6 cells/ml) into a 500 ml non-baffled polycarbonate or glass flask in Sf-900 II medium.
3. Incubate the flask at 27 °C with shaking set at 90–105 rpm (*see* **Note 18**).
4. When the cells reach a density of 4×10^6 cells/ml, dilute them back to 1×10^6 cells/ml and expand the cell volume depending on requirement of the cells (*see* **Note 19**).

3.7 Cell Freezing

Once the cells start doubling regularly after revival, it is advisable to freeze down the low passage number cells in several cryo-vials.

1. Prepare freezing medium containing 72.5 % (v/v) Sf-900 II medium, 20 % (v/v) FBS and 7.5 % (v/v) DMSO and store at 4 °C.
2. Label sterile cryo-vials with the name of the cell line, date of freezing and any other relevant information and store the vials at 4 °C until ready to use.
3. Take a small suspension of cells from a shake flask and count viable cells using a hemocytometer. Alternatively cells from adherent cultures can be used for freezing.
4. Take the required volume of cell suspension for 1×10^7 cells per vial.
5. Centrifuge the cells at $500 \times g$ for 10 min and discard the supernatant.
6. Resuspend the cells in the freezing medium (prepared in **step 1**) so that after resuspension the cell density is ~1×10^7 cells/ml.
7. Quickly aliquot 1 ml of the cell suspension into the cryo-vials (prepared in **step 2**).

8. Place the vials into a suitable freezing container (e.g., Mr. Frosty, Nalgene) and transfer the container to a −80 °C freezer overnight (*see* **Note 20**).
9. The following day transfer the vials to liquid nitrogen storage.

3.8 Decontamination and Cleaning of Shake Flasks

It is extremely important to clean the shake flasks properly so that they can be reused without affecting the cell health or growth. Any residual disinfectant or scum of dead cells can adversely affect the cells and protein expression.

1. Pour off any spent media into a waste container and add ~10 g of Virkon powder per liter of the spent media.
2. Completely fill the empty culture flask with 1 % (w/v) Virkon and leave for at least 20 min (but no longer than 30 min). Make sure that every surface of the flask that has come in contact with virus is covered with the diluted Virkon (*see* **Note 21**).
3. Use a laboratory brush to remove any scum of dead cells from the internal surface of the flask.
4. Discard the waste, fill the flask with water and scrub again with the brush to make sure there is no visible cell debris or dead cell scum remaining inside the flask.
5. Repeat the rinsing and brushing procedure until the flask looks visibly clean.
6. If available, wash the flasks using a washer-disinfector (such as a G 7883, Miele, or similar) according to the manufacturer's instructions.
7. Dry the flasks in a drying cabinet set at 50–60 °C, cover with two layers of aluminum foil and autoclave.

3.9 Transfection into Sf9 Cells

1. Prepare ~100 ml of Sf9 cells one day in advance by diluting the cell count to 1×10^6 cells/ml in Sf-900 II medium.
2. The next day dilute the mid-log phase Sf9 cells to 2×10^5 cells/ml in Sf900-II medium.
3. Label four 24-well TC plates with "plate 1" to "plate 4" to cover your 96 samples (*see* Fig. 3 for how to transfer samples between 96-well and 24-well blocks or plates).
4. Using a 1 ml 12-channel multichannel pipette (with 6 tips spaced two apart), dispense 1 ml of diluted cells (**step 2**) into each well of four 24-well TC plates. Include controls: one for Cellfectin-only and the other for untreated cells (*see* **Note 22**). Incubate the plates at 27 °C for 1 h to allow cell attachment (*see* **Note 23**).
5. Bring Unsupplemented Grace's insect medium (serum free) and Cellfectin to RT. Using a multichannel pipette, add 50 μl of

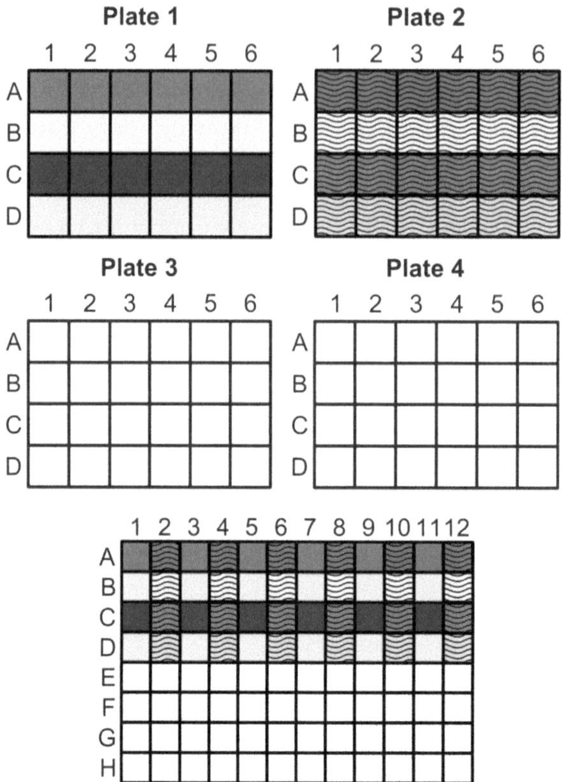

Fig. 3 The format for transferring samples between 24 and 96-well blocks

Unsupplemented Grace's insect medium into a sterile flat-bottomed 96-well microtiter plate (that can hold up to 200 μl of sample, e.g., Sterilin Ltd. UK). Transfer 5 μl of recombinant bacmid DNA (concentration should be 1–3 μg/μl) into each well and mix by shaking the plate gently or pipetting (*see* **Note 13**).

6. Mix Cellfectin thoroughly by tapping the tube gently. In a 15 ml tube, combine 5 ml of Unsupplemented Grace's insect medium and 0.3 ml of Cellfectin, then add 50 μl to each well of the 96-well microtiter plate containing the recombinant bacmid DNA (**step 5**). Do not add Cellfectin mixture to the cell control well, just an equivalent amount of medium. Mix the plates by gently tapping.
7. Cover the microtiter plate with an adhesive tape pad and incubate the mixture inside the LAF workstation for 45 min then dilute the solution by adding 100 μl of Unsupplemented Grace's insect medium (serum free) (*see* **Note 24**).
8. Remove the 24-well TC plates containing the cells (**step 4**) from the incubator and aspirate the medium using a multichannel pipette (*see* **Note 25**). Immediately overlay the cells with the 0.2 ml of Cellfectin-DNA complexes (**step 5**) using a 12-channel multichannel pipette (with six tips spaced two

apart) following the layout from Fig. 3. Do not leave the plate open for too long, the cells will dry out from the center and this will lead to cell death.

9. Add a further 0.2 ml of Unsupplemented Grace's insect medium (Serum free) to each well and incubate the cells for 5 h at 27 °C.
10. Remove the transfection mixture and add 0.8 ml of Sf900-II insect medium containing 2 % (v/v) FBS and antibiotics 0.1 % (v/v) Pen/Strep to each well (*see* **Note 26**). Incubate the cells at 27 °C for 72–96 h.
11. Signs of infection should be seen in the transfected cells 4–5 days post transfection, by comparing with the control cells under an inverted microscope. Confluent growth of cells will be seen in control wells, whereas areas of clearing will be prominent in wells with infected cells. Infected cells are usually larger and deformed or elongated compared to uninfected cells.
12. Harvest the viruses when the cells are well infected (this may take up to 96 h or more) by transferring the liquid contents from the 24-well TC plate into a sterile 96-deep-well block (*see* Fig. 3 for layout) and centrifuging at 1,500 × *g* for 20 min at RT. Collect the clear supernatant (<700 μl) in another sterile 96-deep-well block. This is the P0 baculovirus (BV) stock, which is stored at 4 °C, protected from light.

3.10 Test Expression

1. Using a 1 ml multichannel pipette, dispense 3 ml of Sf9 cells (in Sf900-II medium, containing 2 % (v/v) FBS, at a density of 2×10^6 cells/ml) into each well of four 24-deep-well blocks.
2. Following the layout shown in Fig. 3, infect the cells with 120 μl of P0 BV stock (*see* **Note 27**) and incubate at 27 °C, with shaking at 450 rpm in a Glas Col shaker for 66–72 h (i.e., set up late on day 1 and harvest early on day 3).
3. Pellet the cells by centrifugation at 1,500 × g for 20 min and harvest the supernatant by pipetting into a two 96-deep-well blocks in the LAF workstation according to the layout shown in Fig. 3. Store as P1 BV stock at 4 °C in the dark.
4. Wash the cell pellets once with 1 ml of ice cold PBS, spin as above and discard the supernatant.
5. Resuspend in 1 ml of Lysis buffer, supplemented with protease inhibitors, and store at −80 °C for test purification at a later date (or preferably purify directly).

3.11 Test Purification

1. If frozen, thaw pellets in a water bath at RT, then sonicate on ice for 3 min (5 s on, 10 s off with 35 % amplitude on a 750 W sonicator) using a 24-head probe (check that the probe is level and all tips are in the liquid; after sonication check for clearing).

2. Remove 15 μl of the total cell lysate into a PCR plate as the Total fraction and store at 4 °C.
3. Transfer the remaining sample into a 96-deep-well block according to the layout shown in Fig. 3 and centrifuge at 3,000 × *g* for 30 min at 4 °C.
4. Remove the clarified supernatant to a fresh 96-deep-well block using a multichannel pipette, taking care to avoid transferring any pelleted material (*see* **Note 28**).
5. Add 100 μl of 50 % Ni-NTA (or Ni-IDA) slurry to each well using a multichannel pipette with cut tips, mixing well before each row (*see* **Note 29**).
6. Seal the block with a silicone mat and place another 96-deep-well block on top, tape together, and incubate at 18 °C on their side in any shaker, for 1 h, with shaking at 90 rpm (*see* **Note 30**).
7. Centrifuge the block for 30 s at 200 × *g* to remove the liquid from the lid and load the mixture on to a 96-well filter plate (Thomson) placed on top of a 96-deep-well waste collection block.
8. Allow the liquid to drip through the filter plate or centrifuge at 200 × *g* for 1 min.
9. Add 800 μl of Wash buffer to the resin block to wash out the remaining resin and then transfer to the corresponding wells of the filter plate. Allow the buffer to flow-through or centrifuge briefly at 200 × *g*. Pour off the buffer from the waste block after this and all subsequent washing steps.
10. Add 800 μl of Wash buffer and allow the buffer to flow through or centrifuge briefly at 200 × *g*.
11. Repeat the wash step a further three times and after the final wash, spin the plate for 2 min at 300 × *g* to remove any residual Wash buffer. Pour off Wash buffer from the waste block and spin for a further 1 min to remove all trace of Wash buffer (*see* **Note 31**).
12. Place the filter plate on top of a fresh 200 μl V-bottomed 96-well microtiter plate and add 50 μl of Elution buffer to each filter well.
13. Incubate at RT with shaking for 20 min, then centrifuge for 3 min at 300 × *g* to collect the elution (Purified fraction).
14. In a 96-well PCR plate, mix 15 μl of each Purified fraction with 5 μl of 4× sample buffer, containing DTT. Heat denature at 80 °C for 10 min.
15. Prepare four SDS-PAGE pre-cast gels by rinsing with water, adding 1× XT MES buffer and rinsing the wells. Rinse the packaging and save for use as a staining tray.

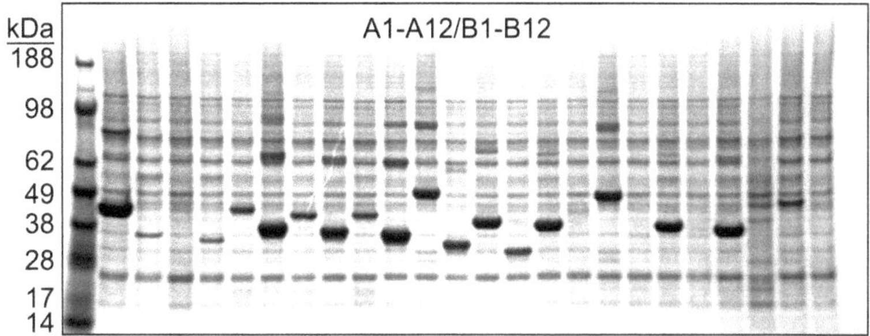

Fig. 4 Image showing the SDS-PAGE result of a test purification from insect cells. The gel shows a range of high, medium, and low expressions of various proteins of different molecular weights. Note that samples loaded using a multichannel pipette will be interleaved (e.g., A1, B1, A2, B2)

16. Using a multichannel pipette, load 15 μl of your samples onto the gels, note that samples will be interleaved (e.g., A1, B1, A2, B2). Also load 5 μl of the SeeBlue Plus2 (Invitrogen) protein ladder in one lane of the gel.
17. Run the gel at 150 V for at least 1 h, or as long as required for the dye-front to reach the bottom of the gel.
18. Break open the cast and carefully remove the gel into the rinsed packaging from **step 14**. Add a cap full of InstantBlue (Expedeon Protein Solutions) and stain for ~1 h with shaking at RT.
19. Discard the stain and wash twice with water, taking care not to tear the gel. Leave in water with shaking to destain for as long as required.
20. Confirm the sizing of your products against the protein ladder (*see* **Note 32** and Fig. 4).

3.12 Virus Amplification

The volumes of P0 (0.70 ml) and P1 (3 ml) viruses generated as described in Subheadings 3.9 and 3.10 respectively are low in volume and insufficient to be used for large-scale expression experiments. Therefore, it is necessary to amplify the virus in a larger volume, typically to the scale of 50–100 ml. The virus can be stored at 4 °C for months but it is advisable to re-amplify the virus, if stored at 4 °C for a longer period of time. For virus amplification, insect cells are generally infected with low Multiplicity of Infection (MOI—number of virus particles per cell) to avoid generating non-infectious particles in the virus stocks. Use a healthy log phase culture of Sf9 cells with more than 95 % viability. All of our virus stocks are made in Sf-900 II, but other media formulations may work equally well.

1. Take a sterile 250 or 500 ml flask and seed 50 ml of suspension-adapted Sf9 cells (2×10^6 cells/ml) in Sf-900 II medium.

2. Add FBS to the final concentration of 2 % (*see* **Note 33**).
3. Add 100 μl of the P1 BV stock to the cells and gently swirl the flask.
4. Transfer the flask to a 27 °C shaking incubator with shaking speed set at 100 rpm and incubate the flask for 72 h.
5. At 72 h post-infection take a small aliquot of cells and observe under the microscope for signs of infection (*see* **Note 34**) and absence of any form of microbial contamination.
6. Transfer the cells to a 50 ml tube and centrifuge at 900 × g for 20 min.
7. Collect the supernatant into a fresh 50 ml tube and store at 4 °C. This represents P2 BV stock.
8. The cell pellet generated in the process of virus amplification can be utilized for protein purification using IMAC. Protein purified from this pellet can be used for any intended application. Moreover, this purification validates the ability of the virus stock to express protein.

3.13 Large-Scale Expression

This protocol is successfully applied for the expression of a broad range of proteins but for some proteins the expression time point and MOI can be highly specific and will require optimization (*see* **Note 35**).

1. Seed log phase Sf9 cells to the density of 1×10^6 cells/ml in Insect-XPRESS or Sf-900 II medium. Keep the volume of culture to 1 l in a 3 l capacity flask. If more than 1 l scale-up is needed, use multiple 3 l flasks with 1 l culture volume in each (*see* **Note 36**).
2. Incubate flasks at 27 °C with shaking set at 100 rpm and allow the cells to grow for 24 h.
3. The next day, check the cell density using a hemocytometer and also check cell health and for any signs of contamination. Cells should go through one doubling cycle in 24 h and the cell count should be ~2×10^6 cells/ml.
4. Add 1.5–3.0 ml of P2 BV stock per liter of culture, swirl the flask gently and transfer the culture to a 27 °C shaker incubator set at 100 rpm (*see* **Note 37**).
5. Incubate the flask for 66–72 h.
6. Take a small sample of the infected culture and look under the microscope for signs of infection but not lysis of the cells and also look for absence of any bacterial, yeast or fungal contamination (*see* **Note 38**).
7. Take out 3 ml of the culture and centrifuge separately (at 900 × *g* for 20 min) from the remaining culture for expression testing (*see* **Note 39**).

8. Without waiting for results from **step** 7 above, transfer the remaining cells to 1 l centrifuge pots, balance pairwise and centrifuge at 900 × *g* for 20 min using JLA 8.1000 rotor on Avanti J-20XP or Avanti J-26XP centrifuge (*see* **Note 40**).
9. Pour the supernatant into a waste container for decontamination using Virkon (this can be done in the culture flask).
10. Resuspend the cell pellet obtained from 1 l of the culture in 25–30 ml of PBS by swirling and pipetting gently and transfer to 50 ml tubes.
11. Balance the tubes pairwise and centrifuge at 900 × *g* for 20 min using a benchtop centrifuge.
12. Discard the PBS in Virkon and purify the protein from the cell pellet as described in Subheading 3.14 or freeze the cell pellets at −80 °C for purification at a later date.

3.14 Protein Extraction

All of the following steps of protein extraction and purification are performed at 4 °C or on ice. Prechill the buffers and centrifuges.

1. If protein purification is performed straight after harvesting the cells, transfer the cell pellets to ice or if the cells were frozen, thaw the pellets in a water bath set at RT or 37 °C. Do not leave pellets in the water bath for any longer than is required to thaw them and transfer onto ice immediately once thawed.
2. Resuspend the cells in one volume of ice cold 2× Lysis buffer (1 ml/g wet-weight of cells) using a pipette and add additional Lysis buffer until the suspension is homogeneous.
3. Place the cell suspension container on ice. Set the amplitude to 35 % on a 750 W Sonics Vibra-Cell sonicator and sonicate with 10–15 bursts of 10 s on, 10 s off (*see* **Note 41**). Save 10 μl of the lysate which represents the Total fraction.
4. Transfer the lysates to centrifuge tubes, balance the tubes pairwise and centrifuge at 21,000 rpm using a JA-25.50 rotor (~53,000 × *g*) for at least 30 min at 4 °C.
5. Transfer the clear supernatant into a clean tube taking care to avoid transferring any pelleted material. This clarified supernatant represents the soluble fraction.

3.15 Large-Scale Protein Purification

The protein purification scheme for insect cells is similar to protein purification from *E. coli* as described in Chapter 5, Subheading 3.6. However, we recommend paying particular attention to the following points while purifying proteins from insect cells:

1. The buffer compositions described here work for a diverse set of proteins but the buffers can be substituted to address issues such as protein instability and requirements of final applications. Careful optimization of the buffer composition with respect to the buffering system, pH, salt concentrations and additives is particularly critical for difficult to purify proteins.

2. In comparison to *E. coli* cell lysates, insect cell lysates are denser because of higher background protein concentration. This can result in clogging of pre-packed IMAC columns, therefore we recommend doing manual IMAC using the gravity-flow procedure for purification of proteins from insect cells.
3. Often intrinsic proteins from insect cells co-purify due to the affinity of exposed histidines or metal binding moieties of endogenous proteins towards the immobilized metal ions. Therefore it is often the case that IMAC followed by SEC is not enough to obtain very pure protein from insect cells, which necessitates inclusion of additional purification steps such as ion exchange chromatography or tag cleavage and rebinding to IMAC.

3.16 Quality Assurance

Follow the guidelines as recommended in Chapter 5, Subheading 3.7.

4 Notes

1. X-gal does not produce sufficiently dark blue non-recombinant colonies in our hands therefore we use Blue-gal instead. The plates can be stored for up to 1 month at 4 °C, covered with foil to prevent exposure to light.
2. Be careful not to splash the cells against the sides of the wells whilst using the repeat pipettor and also check that the liquid is at the bottom of the well before continuing. This step can also be done using a single channel pipette but will take more time.
3. When there are no colonies, plate 50 μl of undiluted culture instead.
4. This step can be performed at RT on the bench over the weekend if necessary.
5. One 96-well block should provide sufficient bacmid DNA for transfection. However, we find it useful to set up two blocks to provide a balance for the centrifugation step.
6. We only use the reagents from the Montage Plasmid Miniprep$_{\text{HTS}}$ 96 Kit (Millipore) for purifying the recombinant bacmid DNA, not the filter plates. The reagents can also be purchased from Millipore individually.
7. Covering the block with an adhesive tape pad or alternative will result in leaking and cross-contamination of wells. Make sure the silicone sealing mats are suitable for either round or square 96-deep-well blocks, depending on which 96-well blocks you use.
8. This second centrifugation step is important to remove as much of the insoluble pelleted material as possible in order to obtain clean bacmid DNA at the end of the prep.

9. It is recommended not to remove all of the supernatant to avoid transferring insoluble material.
10. Incubation can also be done overnight at 4 °C, and will result in a higher yield of bacmid DNA, but is not necessary.
11. If you have more than one block, be careful not to remove the marker labels when using 70 % (v/v) ethanol.
12. Do not allow the pellets to dry out completely.
13. The bacmid DNA is very fragile so mix gently and do not over-pipette.
14. High concentrations of bacmid DNA will inhibit the bacmid PCR screen so we dilute the bacmid prior to addition. Where the yields of bacmid are low it may be necessary to use a lower dilution instead.
15. All cell culture steps must be performed under aseptic conditions in a LAF workstation, making sure that sterility is maintained throughout the procedures. To keep the cultures free from contamination by bacteria, yeast, fungi and viruses, it is crucially important to keep the benches, LAF workstation and incubators clean. Use 70 % (v/v) ethanol to wipe the LAF workstation before and after use, also wipe the outside of media bottles, pipettors, flasks, and other containers with 70 % (v/v) ethanol before transferring them into the LAF workstation. Wear clean lab coats and gloves and wash hands before and after working with cell culture. Any spillage inside the LAF workstation, incubators, etc. should also be cleaned immediately with 70 % (v/v) ethanol or MicroSol. Use separate media bottles for general cell culture maintenance and for virus work. We recommend adding penicillin and streptomycin to the final concentration of 50 U/ml and 50 μg/ml respectively to the cell culture media to prevent bacterial contamination during culture growth.
16. Always wear protective clothing (lab coat, gloves, and safety specs) when thawing vials containing frozen cells as they sometimes explode on contact with the water.
17. The % cell viability is calculated by counting the number of viable cells and also the number of total cells on the hemocytometer grid. Viable cells do not take up Trypan Blue Stain (Invitrogen); whereas, non-viable cells take up the stain and appear blue under the microscope. To determine cell viability, mix 0.1 ml of Trypan Blue Stain (Invitrogen) with 1 ml of cell suspension and load a hemocytometer. Count the number of blue-stained cells and also the total number of cells and then calculate the number of viable cells per ml and correct for the dilution factor. Cell viability should be at least 95 % for a healthy log phase culture before it can be used for transfection, virus amplification or protein expression.

18. For better aeration of the cells, it is important to keep the culture volume between 25 and 35 % of the total volume capacity of shake flask and shaking between 90 and 105 rpm. Cells form clumps initially but should start growing in single cell suspension within a week or so.
19. Cells can be transferred gradually to 1 l and then 3 l flasks, keeping the culture volume between 25 and 35 % of the total volume capacity of shake flask. Ideally do not allow the cell density to exceed 5×10^6 cells/ml or fall below 0.7×10^6 cells/ml. Cell growth may slow down if diluted to the density of less than 0.7×10^6 cells/ml.
20. If a freezing container is not available, vials can be transferred to a −20 °C freezer for 2–3 h followed by transfer to −80 °C overnight.
21. It is not necessary to keep Virkon solution in flasks for more than 20 min. Leaving Virkon for longer may make it difficult to remove the traces from flasks. Glass flasks are easier to clean than the polycarbonate flasks. Polycarbonate flasks for suspension culture are meant to be disposable but they can be reused several times if cleaned properly after Virkon treatment. We have noticed that if Virkon is left in polycarbonate flasks for an extended time, the plastic starts leaching and the flasks become unusable. Other disinfectants may be used but if using Virkon for decontaminating cell culture glassware, special attention should be paid to remove any residues of Virkon from the flasks before autoclaving.
22. The Cellfectin and cell-only controls are important for determining the success of the transfection as they allow the user to distinguish cytotoxic effects and uninfected cells from infected cells.
23. Cell attachment can be observed using an inverted microscope by focusing through the sample; the cells should be visible in one plane of view once successfully attached.
24. Adding serum to the transfection will inhibit the process.
25. Pipette the medium off gently and avoid touching the bottom of the plate so as to not disturb the cells. Keep the tips on one edge of the wells and tilt the plate slightly to ease aspiration. As soon as you have removed the medium, replace it with the lipid-DNA complex mixtures to prevent the cells drying out. When doing this step it is beneficial to have the tips already in place on the multichannel pipette ready to add your transfection mixture. Also have a waste container to pipette the media into and note that you can reuse your 1 ml tips that you aspirate your medium off with.
26. This step again needs to be done quickly without disturbing the cells. It is beneficial to use two 1 ml multichannel pipettes. Have 96 1 ml tips ready for aspirating off the transfection mixture and

have another set of tips ready for adding the medium. Tilt the plate as in **Note 25** to aspirate off the mixture and then add the medium slowly by pipetting it gently down the side of the well.

27. For some targets it may be necessary to use the P1 virus to infect for test expression. However, we have found that there is little difference in the yields when expressing from P1 rather than P0. We therefore use P0 virus, which shortens the expression process by at least 3 days.
28. To avoid disturbing the Insoluble fraction, tilt the plate and drive the tips down the side of the wells at an angle. Stop just above the pellet, on most plates there is a ridge just off the bottom—feel for this with the tips. Gently pipette up the supernatant and then transfer to the new plate. Do not go back into the wells as this will resuspend the pellets, if this happens then re-spin the sample and try again.
29. The resin tends to clump and settles quickly. We recommend using 200 μl tips with ~5 mm cut from the ends to prevent clogging the tips and ensure even loading. Also, continually mix the resin by pipetting up and down in addition to shaking the reservoir from side to side to prevent settling.
30. When the silicone matting seal is pressed down firmly and held in place with another deep-well block the block will not leak when placed on its side. If you prefer, you can incubate the plate upright but the resin tends not to mix as well when done this way, we would therefore recommend keeping the samples in a 24-well format for this step as this provides greater surface area for binding.
31. Removing all trace of Wash buffer is essential to ensure that the subsequent elution step does not become diluted with Wash buffer.
32. It is beneficial to grade the expression level of your proteins to more easily identify ones that you may wish to scale up. At this point we also recommend confirming the targets using quality control steps such as intact mass determination (if quantities are sufficient) or by in-gel tryptic digest MSMS analysis.
33. Baculovirus stability is known to improve in the presence of FBS. As Sf-900 II is a serum-free medium, addition of FBS to the final concentration of 2 % is recommended to stabilize the virus and maintain its infectivity when it is stored at 4 °C.
34. Signs of baculovirus infection: baculovirus infected insect cells look swollen, nuclei appear to fill the cells and the cells do not show any clumps when compared to a healthy cell control. If the cells are in very late phase of infection, they will start to lyse.

35. Availability of healthy viable cells is very important for successful scale-up of a broad range of targets. Culture conditions such as temperature, pH, dissolved oxygen, osmolality, and nutrient composition of the culture medium can influence the infection of the insect cells. In addition, factors such as cell line, expression time point, MOI and cell density at the time of infection can have significant effects on protein expression in insect cells. This protocol is generically applied to a large number of proteins; however, occasionally for some proteins, optimization at protein expression level is necessary to improve the results. Optimization experiments should be performed on a small scale initially and can be later applied to large-scale expressions. The following conditions could be tested for expression optimization: range of MOI, two harvesting time points (48 and 72 h), two cell lines (Sf9 and High Five), or different cell densities (2×10^6 cells ml and 4×10^6 cells/ml). It should be noted that baculoviruses are lytic viruses for insect cells and will eventually lyse the cells if left long enough after infection. This also means that a harvesting time of 48 or 72 h is also determined by the volume of virus added. The cells can be infected with low MOI (0.05–0.3 pfu/cell) and harvested at 72 h or they can be infected with a high MOI (>1 pfu/cell) and harvested at 48 h. Cells infected with high MOI and harvested at 72 h may show significant lysis.
36. Before diluting the cells, check for the health of the cells and absence of any signs of infection or contamination under a microscope. If less than 1 l scale-up is enough, smaller flasks should be used. However, remember to use a culture volume of only 25–35 % of the total volume capacity of the flask.
37. The amount of virus added is determined by the titer of virus stock. We do not routinely measure viral titers but various methods for baculovirus titration have been developed based on cell viability, plaque formation, antibody-based assays, etc. [14] For the 72 h expression time point, we recommend an MOI of 0.05–0.3 pfu/cell. If the titer of virus stock is 1×10^8 pfu/ml and 2 ml of virus is added to 1 l of the cells (total of 2×10^9 cells), that would be an MOI of 0.1. Addition of more virus can affect the expression and can also cause cell lysis.
38. It should be noted that good signs of infection are desirable but more than 10 % lysis of cells can be detrimental to protein purification.
39. This small volume of cells can be used for expression testing before committing to purify a large batch of cells. This can give a quick estimate of protein expression levels or any failure of the batch to express the protein of interest. To purify the protein from 3 ml of culture, follow the protocol as described in Subheading 3.11.

40. Sf9 cells become very fragile after infection and can rupture if centrifuged at very high speed resulting in loss of protein in the medium itself. We recommend harvesting the cells by centrifugation at 900 × *g* for 20 min and handling cell pellets gently.
41. Sonication time may need to be adjusted depending on volume of the cell suspension. Avoid excessive foaming and heating of the suspension by adjusting the instrument settings and keeping the cell suspension on ice all the time to reduce the potential for protein precipitation or denaturation. Cell disruption by sonication can also help in reducing viscosity by shearing nucleic acids.

Acknowledgements

We would like to thank all the SGC scientists (past and present) who contributed towards the development of the methods. The SGC is a registered charity (number 1097737) that receives funds from the Canadian Institutes for Health Research, Genome Canada, GlaxoSmithKline, Lilly Canada, the Novartis Research Foundation, Pfizer, Takeda, AbbVie, the Canada Foundation for Innovation, the Ontario Ministry of Economic Development and Innovation, and the Wellcome Trust.

References

1. Ignoffo CM (1975) Entomopathogens as insecticides. Environ Lett 8:23–40
2. Invitrogen (2002) Guide to Baculovirus expression vector systems (BEVS) and insect cell culture techniques. Invitrogen Life Technologies, Carlsbad
3. Kost TA, Condreay JP, Jarvis DL (2005) Baculovirus as versatile vectors for protein expression in insect and mammalian cells. Nat Biotechnol 23:567–575
4. Summers MD, Anderson DL (1972) Granulosis virus deoxyribonucleic acid: a closed, double-stranded molecule. J Virol 9:710–713
5. Matthews REF (1982) Classification and nomenclature of viruses. Intervirology 17:1–181
6. O'Reilly D ML, Luckow V (1992) Baculovirus expression vectors: a laboratory manual
7. Vaughn JL, Goodwin RH, Tompkins GJ et al (1977) The establishment of two cell lines from the insect *Spodoptera frugiperda* (Lepidoptera; Noctuidae). In Vitro 13:213–217
8. Granados RR, Guoxun L, Derksen ACG et al (1994) A new cell line from *Trichoplusia ni* (BTI-Tn-5B1-4) susceptible to *Trichoplusia ni* single enveloped nuclear polyhedrosis virus. J Invertebr Pathol 64:260–266
9. Smith GE, Summers MD, Fraser MJ (1983) Production of human beta interferon in insect cells infected with a baculovirus expression vector. Mol Cell Biol 3:2156–2165
10. Invitrogen (2010) Bac-to-Bac® Baculovirus expression system. Invitrogen Life Technologies, Carlsbad
11. Hanahan D, Jessee J, Bloom FR (1991) Plasmid transformation of *Escherichia coli* and other bacteria. Methods Enzymol 204:63–113
12. Shrestha B, Smee C, Gileadi O (2008) Baculovirus expression vector system: an emerging host for high-throughput eukaryotic protein expression. Methods Mol Biol 439:269
13. Invitrogen (2010) Growth and maintenance of insect cell lines. Invitrogen Life Technologies, Carlsbad
14. Roldao A, Oliveira R, Carrondo MJ et al (2009) Error assessment in recombinant baculovirus titration: evaluation of different methods. J Virol Methods 159:69–80

Chapter 7

OmniBac: Universal Multigene Transfer Plasmids for Baculovirus Expression Vector Systems

Deepak B. Thimiri Govinda Raj, Lakshmi S. Vijayachandran, and Imre Berger

Abstract

Current baculovirus expression vector systems (BEVS) rely on either using homologous recombination or site specific transposition (Tn7 transposition) to obtain recombinant baculovirus. Each approach has its own merits. To date, the choice of transfer plasmids limited expression of target proteins to only one of the two types of BEVS. Here we describe OmniBac, comprising novel universal multigene transfer plasmids that can access all BEVS currently in use for protein production in the community. Detailed protocols are presented for integrating OmniBac plasmids into baculoviral genomes used for heterologous protein production in insect cells.

Key words BEVS, Bac-to-Bac, FlashBac, BacVector series, MultiBac, OmniBac, Tn7 transposition, Cre-LoxP fusion, Homologous recombination, Co-expression, Multiprotein complexes

1 Introduction

A number of recombinant baculovirus expression systems (BEVS) are in use for the production of recombinant proteins and their complexes in insect cells. Examples include Invitrogen's *Bac-to-Bac*® system [1], the FlashBAC system from Oxford Expression Technologies (OET) [2–4], the BacVector series from Novagen [5], and the MultiBac [6] and SweetBac [7] systems developed by academic research groups. The baculovirus genomes that are utilized by these systems are accessed by using plasmids called transfer plasmids. Two methods are predominantly used to react the transfer plasmids with the baculovirus genomes to insert the recombinant genes, relying either on homologous recombination [2–5] or, alternatively, on site specific transposition [1, 6, 7], Currently, the choice of transfer plasmid decides the entry method used to integrate heterologous genes into the viral genome, which in turn dictates the baculovirus genome to be used for insect cell infection

Yu Wai Chen (ed.), *Structural Genomics: General Applications*, Methods in Molecular Biology, vol. 1091, DOI 10.1007/978-1-62703-691-7_7, © Springer Science+Business Media, LLC 2014

and protein production. A range of baculoviruses is available to the community relying on either one or the other system, each with its own merit. However, transfer plasmids that could access all commonly used baculoviruses, by either transposition or homologous recombination, respectively, was lacking to date. Such universal transfer plasmids, however, would be desirable as they would provide flexibility to switch baculoviruses without having to reclone the genes of interest into a different transfer plasmid. This is particularly relevant for multiprotein complexes with many subunits and therefore many encoding genes, where recloning could become a significant burden.

We created new "OmniBac" transfer plasmids that contain both the functionalities required for Tn7 transposition (Tn7R and Tn7L DNA sequences) and also DNA elements required for homologous recombination (Orf1629 and lef2/603 sequences) [8]. The OmniBac transfer plasmids are fully compatible with the multigene construct generation methods of our previous MultiBac system [6, 9]. Moreover, our acceptor–donor tandem recombineering (TR) approach [10, 11] can be likewise used with the OmniBac plasmids, which act then as acceptors, to put together multigene expression constructs.

Here we present protocols for recombinant baculovirus generation by using our novel OmniBac transfer plasmids. Two protocols are detailed, one each for homologous recombination and for the BAC/Tn7 entry approach (Fig. 1).

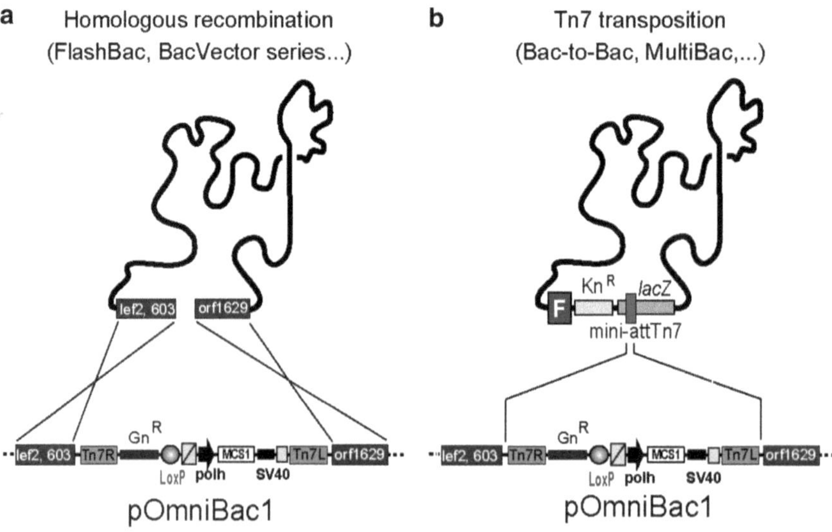

Fig. 1 OmniBac transfer plasmids can universally access available baculovirus genomes using either homologous recombination (**a**) or Tn7 transposition (**b**) to generate recombinant baculoviruses for protein expression in insect cells

2 Materials

All solutions should be prepared using ultrapure and sterilized water (Millipore Milli-Q system or equivalent; conductivity of 18.2 MΩ cm at 25 °C) and analytical grade reagents. Prepared solutions are sterilized wherever it is necessary and stored at room temperature (unless it is indicated otherwise). It is recommended to diligently follow all biosafety rules and regulations when performing the protocol and while disposing waste materials.

2.1 Reagents Required for Integration by Tn7 Transposition

1. pOmniBac1 or pOmniBac2 plasmid (*see* **Note 1**).
2. DH10Bac, DH10MultiBac, DH10EMBacY bacterial cell strains.
3. Buffers from QIAprep Spin Miniprep kit (Qiagen, cat. no. 27104) (or similar, or self-made buffer solutions).

2.2 Reagents Required for Integration by Homologous Recombination

1. pOmniBac1 or pOmniBac2 plasmids (*see* **Note 1**) (500 ng of transfer plasmid DNA).
2. 100 ng of purified baculoviral DNA (Novagen, OET systems).

2.3 Reagents and Equipment Required for Both Approaches

1. Sterile Erlenmeyer flasks (100 ml).
2. Tabletop Centrifuge.
3. Fluorescence spectrophotometer and cuvettes.
4. Sonicator.
5. YFP standard.
6. Lysis buffer (e.g., PBS).
7. 6X protein gel loading dye (125 mM Bis/Tris-Cl, pH 6.8, 20 % (v/v) glycerol, 4 % (w/v) sodium dodecyl sulfate, 10 % (v/v) β-mercaptoethanol, 0.4 mg/ml bromophenol blue).
8. 95 °C heating block.
9. Incubator set at 27 °C.
10. 6-well tissue culture plate.
11. Sterile pipettes, tips, and sterile hood.
12. Inverted phase-contrast microscope.
13. Sf21 insect cells (or Sf9, others).
14. Insect cell serum free media (Invitrogen or Hyclone).
15. Transfection reagent such as Fugene (Roche), Gene Juice (Novagen), Lipofectin (Invitrogen), etc.

3 Methods

Heterologous genes of interests (GOI) are cloned into the expression cassettes of the OmniBac plasmids and, if multiprotein expression is planned, also into the expression cassettes of Donor plasmids which are identical to the Donors supplied with the original MultiBac system (Fig. 2a) [11]. Gene insertion can be performed by the user's method of choice (restriction enzymes and ligase, ligation-independent cloning methods, PCR-based methods, others). Our preferred method of choice is sequence and ligation independent cloning (SLIC) [12]. Several expression cassettes can be placed on each of the plasmids, by taking advantage of the multiplication module comprising a homing endonuclease (HE) site and a BstXI that flank the expression cassettes [9, 13], OmniBac plasmids containing one or more expression cassettes are transformed into common *E. coli* cloning strains, and positive clones identified by restriction mapping and by sequencing using standard protocols.

Expression of several genes may be desired. Donor plasmid, each containing one or several expression cassettes, can be combinatorial assembled with the OmniBac acceptor plasmid by Cre recombinase (Fig. 2b) [10, 14, 15]. Donors and OmniBac plasmids are mixed with the Donors slightly in excess, and incubated

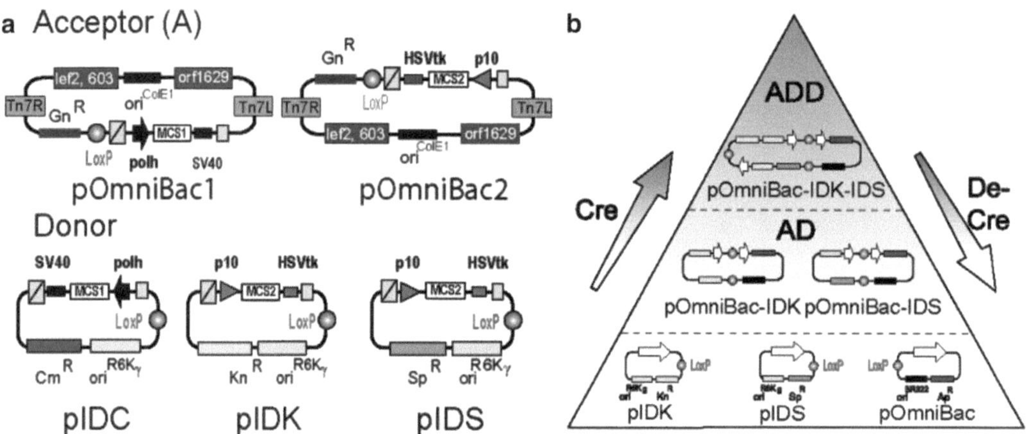

Fig. 2 The OmniBac system. (**a**) Acceptors (pOmniBac1, pOmniBac2) and Donors (pIDC, pIDK, pIDS) are shown in a schematic fashion. Origins of replication (ColE1 and R6Kγ) are indicated. Plasmids contain expression cassettes controlled by late baculoviral promoters (polh or p10) as well as eukaryotic polyadenylation signals (from SV40 or HSVtk). Homing endonuclease sites and matching BstXI sites (*blue squares*) flanking the expression cassettes are shown. *Ap* stands for ampicillin, *Cm* for chloramphenicol, *Kn* for kanamycin, *Gn* for gentamycin, *Sp* for spectinomycin resistance markers [8]. (**b**) Combinatorial assembly of acceptor–donor fusions using Cre-recombinase. The Cre reaction is an equilibrium reaction resulting in acceptor–donor fusions that are characterized by unique resistance marker combinations which can be used for selection [10]. *A* stands for Acceptor, *D* for Donor, *AD* and *ADD* denote Acceptor–Donor fusions by the Cre-LoxP reaction at the equilibrium of assembly (marked as Cre) and disassembly (marked as DeCre)

with the Cre enzyme following published protocols [9]. After transformation, the desired fusion constructs are selected based on the combination of resistance markers, and validated by restriction mapping and DNA sequencing. The integration of OmniBac-based transfer plasmids into the baculovirus of choice using Tn7 transposition or homologous recombination follows the same protocols (below) as for the OmniBac plasmids alone (Subheading 3.1 or Subheading 3.2, respectively).

3.1 Integration of OmniBac Plasmid into Baculovirus Genome by Homologous Recombination in Insect Cells (FlashBac, BacVector Series, etc.) and Production of Recombinant Virus

1. Prepare transfection dilution mixture containing the transfection reagent and insect cell media as recommended by the manufacturer's protocol. (BacVector series, Novagen, or FlashBac, OET)
2. Prepare 0.5–0.8 × 106 Sf21 cells per well in 2 ml serum free medium in the 6-well culture plate.
3. Incubate at 27 °C for 1 h for the cells to adhere on the surface of the plate.
4. In the meantime, prepare the co-transfection mix containing 100 ng of BacVector or Flashbac DNA (5 μl), 500 ng of recombinant OmniBac plasmid (5 μl), lipofectin or alternative transfection reagent (5 μl), and 1 ml of serum free, antibiotic free medium (*see* **Note 2**).
5. Incubate at room temperature for 15–30 min to generate Liposome–DNA complexes.
6. Remove the culture medium from the 6-well plates without disrupting the monolayer (*see* **Note 3**).
7. Add immediately 1 ml of transfection mixture (liposome–DNA complexes) drop-wise such that monolayer is not affected. Incubate in dark for minimum 5 h or overnight at 27 °C
8. After first incubation period, add further 1 ml of serum free insect cell medium into each well. Continue the second incubation period of 4–5 days at 27 °C in dark.
9. After the second incubation period, collect the the supernatant from the wells which is *V0 virus* (*see* **Note 4**). Proceed for virus amplification (*see* **step 9** in Subheading 3.2).

3.2 Integration of OmniBac Plasmid into Baculovirus Genome by Tn7 Transposition (Bac-to-Bac, Multibac, Others) and Production of Recombinant Virus

1. Transform chemical competent DH10Bac, DH10MultiBac, DH10EMBacY, etc. cells by mixing 10–15 μl of the annealing reaction with 200 μl of cell suspension on ice.
2. Incubate for 30 min on ice, heat-shock at 42 °C for 45–60 s, incubate on ice for 2 min, add 600 μl of LB Broth, and incubate in a 37 °C shaker for overnight.
3. Streak out 150 μl on plates containing the antibiotics, IPTG (1 mM) and Bluogal or X-Gal at standard concentration. Use dilution series (1:1, 1:10, 1:100, 1:1,000), this results usually in optimal separation of colonies on one of the plates.

4. Pick 2–3 white clones for each construct, start mini-cultures overnight, and (optionally) restreak. Proceed for bacmid preparation for insect-cell infection.
5. Isolate baculoviral DNA using solution I, II, III of the QIAprep Spin Miniprep kit by following the Qiagen manual. Precipitate the resultant supernatant (900 μl) with isopropanol (700 μl) and wash the pellet twice with 200 μl 70 % EtOH. Dry the pellet and resuspend in 20 μl sterilized ddH_2O. Further add 200 μl of sterilized medium.
6. Seed 1×10^6 Sf21 cells in duplicates in 6-well plates and incubate for 15–30 min at 27 °C.
7. Prepare transfection reagent solution of 100 μl media with 10 μl transfection reagents (Fugene reagent from Roche). Add this mixture (100 μl) to the volume containing dissolved MultiBac baculoviral DNA supplemented with 200 μl media (**step 4**).
8. Add half of the above transfection mixtures to each of the two wells marked for the construct to be tested.
9. Incubate for 60 h at 27 °C. Then, collect the supernatant from the well which is *V0 virus*. Add 3 ml of fresh media to each well and proceed for protein expression test on the samples.
10. Prepare Erlenmeyer flasks containing 25 ml Sf21 cell suspension at a density of $0.5–1 \times 10^6$ cells/ml. Make sure cells divide properly in the flasks used (test once or twice the doubling rate, which should be around 18–20 h for most insect cells at 27 °C).
11. Infect 25 ml Sf21 cell culture with 3 ml of V0 virus.
12. Monitor cell growth by withdrawing aliquots (24 h intervals) and counting cells. If cells double, dilute culture in fresh flask and cell density must maintain at $1–1.5 \times 10^6$ cells/ml (*see* **Note 5**).
13. Identify the time when cells stop doubling (day of proliferation arrest, *dpa*). Aliquot 1×10^6 cells (i.e., 0.9 ml culture if cell count is at 1.1×10^6/ml). Pellet the cells at high speed for 1 min (i.e., "dpa" probe). Take also probes (1×10^6 cells) every 12 or 24 h after dpa (dpa +12, dpa +24, etc.).
14. After taking probes after dpa +48/60 h, Sf21 cell suspension is transferred to a sterile 50 ml Falcon tube, centrifuged gently (100–150 rcf, 3 min) and the supernatant is retained in a fresh sterile 50 ml Falcon tube (this is *V1 virus*). The Sf21 pellet is then gently resuspended in fresh media (50 ml) and placed back into the shaker flask. Continue to withdraw probes (1×10^6 cells) until dpa +96 h, or, if EMBacY virus is used, until the YFP expression reaches a plateau (*see* **Note 6**).
15. Harvest cell pellet and store the pellet at −20 °C.

16. Analyze dpa; dpa +12, dpa +24 h probe from protein expression by SDS-PAGE or Western blot analysis. Herein, sonicate and resuspend the pellets in appropriate amount of lysis buffer, mix with Protein gel loading buffer, and run on SDS-PAGE Gel. Perform western blot analysis if necessary.

4 Notes

1. OmniBac plasmid sequence information can be downloaded from the link: http://www.embl.fr/multibac/multiexpression _technologies/.
2. For the control, omit baculoviral DNA from the transfection mix.
3. Make sure that insect cell monolayer is not dry out during the step. Include the washing step with serum-free medium, if cells were maintained in serum-supplemented medium and repeat the step twice.
4. V0 virus is the seed stock of recombinant virus which is used for virus amplification.
5. If couple of doubling are observed (indicative of very week V0 virus), the culture needs to be diluted and split using a fresh flask or suspension discarded. If cells continue to double after 4–5 days, it is recommended to repeat the bacmid preparation and transfection reaction.
6. Centrifuge the resuspended pellet at maximum speed in a tabletop centrifuge at room temperature for 3 min. Transfer the supernatant into fresh eppendorf tube and measure YFP fluo rescence (excitation: 488 nm, emission max: ~520 nm) using spectrofluorometer and having YFP standard as a control.

Acknowledgements

We thank all members of the Berger laboratory for helpful discussions. This work was supported by the Centre National de Recherche Scientifique (CNRS) through a PEPS discovery grant (to IB), and by the European Commission (EC) Framework Program 7 (PCUBE, BioSTRUCT-X and ComplexINC, to IB). DBTGR is recipient of an EC/EMBL CoFund EIPOD fellowship.

Competing Financial Interest Statement
The authors declare competing financial interests. IB is the author on patents or patent applications detailing reagents and parts of the methods here described.

References

1. Luckow VA, Lee SC, Barry GF et al (1993) Efficient generation of infectious recombinant baculoviruses by site-specific transposon-mediated insertion of foreign genes into a baculovirus genome propagated in Escherichia coli. J Virol 67:4566–4579
2. Hitchman RB, Possee RD, Crombie AT et al (2010) Genetic modification of a baculovirus vector for increased expression in insect cells. Cell Biol Toxicol 26:57–68. doi:10.1007/s10565-009-9133-y
3. Hitchman RB, Possee RD, King LA (2009) Baculovirus expression systems for recombinant protein production in insect cells. Recent Pat Biotechnol 3:46–54
4. Possee RD, Hitchman RB, Richards KS et al (2008) Generation of baculovirus vectors for the high-throughput production of proteins in insect cells. Biotechnol Bioeng 101:1115–1122. doi:10.1002/bit.22002
5. Kitts PA, Possee RD (1993) A method for producing recombinant baculovirus expression vectors at high frequency. Biotechniques 14:810–817
6. Berger I, Fitzgerald DJ, Richmond TJ (2004) Baculovirus expression system for heterologous multiprotein complexes. Nat Biotechnol 22:1583–1587. doi:10.1038/nbt1036
7. Palmberger D, Wilson IB, Berger I et al (2012) SweetBac: a new approach for the production of mammalianised glycoproteins in insect cells. PLoS One 7:e34226. doi:10.1371/journal.pone.0034226
8. Vijayachandran LS, Thimiri Govinda Raj DB, Edelweiss E et al (2013) Gene gymnastics: synthetic biology for baculovirus expression vector system engineering. Bioengineered 4:279–287
9. Fitzgerald DJ, Berger P, Schaffitzel C et al (2006) Protein complex expression by using multigene baculoviral vectors. Nat Methods 3:1021–1032. doi:10.1038/nmeth983
10. Bieniossek C, Nie Y, Frey D et al (2009) Automated unrestricted multigene recombineering for multiprotein complex production. Nat Methods 6:447–450. doi:10.1038/nmeth.1326
11. Bieniossek C, Imasaki T, Takagi Y et al (2012) MultiBac: expanding the research toolbox for multiprotein complexes. Trends Biochem Sci 37:49–57. doi:10.1016/j.tibs.2011.10.005
12. Haffke M, Viola C, Nie Y et al (2013) Tandem recombineering by SLIC cloning and Cre-LoxP fusion to generate multigene expression constructs for protein complex Research. Methods Mol Biol 1073:131–140
13. Li MZ, Elledge SJ (2007) Harnessing homologous recombination in vitro to generate recombinant DNA via SLIC. Nat Methods 4:251–256. doi:10.1038/nmeth1010
14. Trowitzsch S, Bieniossek C, Nie Y et al (2010) New baculovirus expression tools for recombinant protein complex production. J Struct Biol 172:45–54. doi:10.1016/j.jsb.2010.02.010
15. Vijayachandran LS, Viola C, Garzoni F et al (2011) Robots, pipelines, polyproteins: enabling multiprotein expression in prokaryotic and eukaryotic cells. J Struct Biol 175:198–208. doi:10.1016/j.jsb.2011.03.007

Chapter 8

Multiprotein Complex Production in Insect Cells by Using Polyproteins

Yan Nie, Itxaso Bellon-Echeverria, Simon Trowitzsch, Christoph Bieniossek, and Imre Berger

Abstract

A powerful approach utilizing polyproteins for balancing stoichiometry of recombinant multiprotein complexes overproduced in baculovirus expression vector systems (BEVS) is described. This procedure has been implemented here in the MultiBac system but can also be directly adapted to all commonly used BEVS. The protocol details the design principles of polyprotein-expressing DNA constructs, the generation of composite baculovirus for polyprotein production, and the expression and in vivo processing of polyproteins in baculovirus infected insect cells.

Key words Polyprotein, Multiprotein complexes, Subunit stoichiometry, Baculovirus-insect cell expression system, Tobacco etch virus (TEV) NIa protease, In vivo proteolysis, Multigene delivery, MultiBac system, Cre recombinase, Tn7 transposition

1 Introduction

Multiprotein complexes catalyze essential cellular activities. Studying their structure and function is an emerging focus of biological research. In most cases, recombinant production is required for obtaining sufficient amounts of homogenous material for detailed analyses [1]. MultiBac is an advanced expression system that has been designed for overexpressing multiprotein complexes in insect cells infected by a single composite multigene baculovirus. MultiBac has enabled production of many challenging multiprotein complexes, setting the stage to unlock their mechanism [2–4]. A bottleneck which can be encountered when many heterologous proteins are co-produced from individual expression cassettes derives from imbalanced expression levels of the individual proteins prohibiting proper complex assembly [3, 5]. Some subunits may be expressed stronger, others weaker, and occasionally one subunit is expressed at such a low level that it becomes detrimental

Yu Wai Chen (ed.), *Structural Genomics: General Applications*, Methods in Molecular Biology, vol. 1091, DOI 10.1007/978-1-62703-691-7_8, © Springer Science+Business Media, LLC 2014

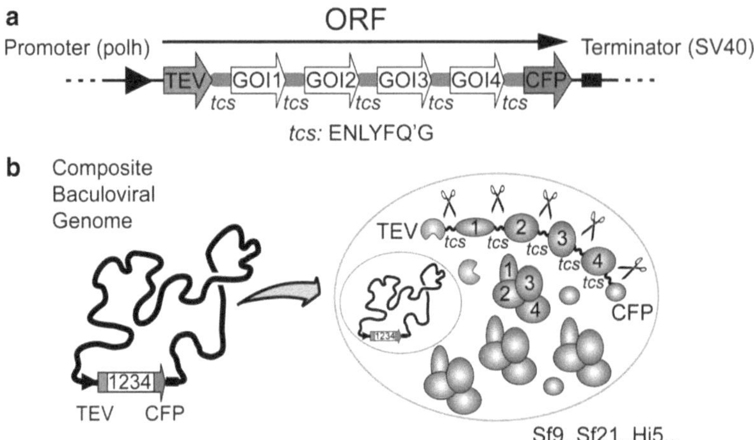

Fig. 1 Design and in vivo processing of polyproteins. (**a**) Genes of interest (GOIs) are assembled into a single open reading frame (ORF), giving rise to a polyprotein. In this polyprotein, individual genes of interest are spaced apart by cleavage sites (tcs) for tobacco etch virus (TEV) NIa protease which is also encoded by the ORF. The C-terminal CFP serves to monitor heterologous expression level. (**b**) Composite MultiBac baculoviral genome DNA containing the polyprotein expression cassette (*left*) is used to transfect cultured insect cells for multiprotein complex production (*right*)

to overall yield and a complex containing all desired subunits cannot be obtained.

In order to balance expression levels and achieve properly assembled complexes with correct subunit stoichiometry, we have implemented a novel strategy based on polyproteins that are processed in vivo into individual subunits by a highly specific protease [3, 5] (Fig. 1). This approach derives from the strategy used by certain viruses such as Coronavirus to realize their proteome [6]. To facilitate polyprotein production with the MultiBac BEVS, new transfer plasmids have been created that rely on Tn7 transposon-mediated gene integration into the MultiBac baculovirus (Fig. 2). Other baculoviruses that are in use rely on homologous recombination for composite baculovirus generation (flashBAC from OET, BacVector series from Novagen, others). These baculoviruses can likewise be accessed for polyprotein expression by using the pOmni-PBac plasmid [7].

All polyprotein transfer plasmids contain the same expression cassette which encompasses a very late viral promoter (polyhedrin) followed by the gene encoding for NIa protease from tobacco etch virus (TEV) for subunit liberation, a short oligonucleotide sequence presenting a BstEII and an RsrII restriction endonuclease site, and finally a gene encoding for cyan fluorescent protein (CFP) for direct read-out of polyprotein expression. A TEV NIa protease cleavage site (tcs) is placed upstream of the CFP encoding gene (Fig. 2a). Heterologous genes of interest (GOIs) can be inserted into this polyprotein expression cassette by using the

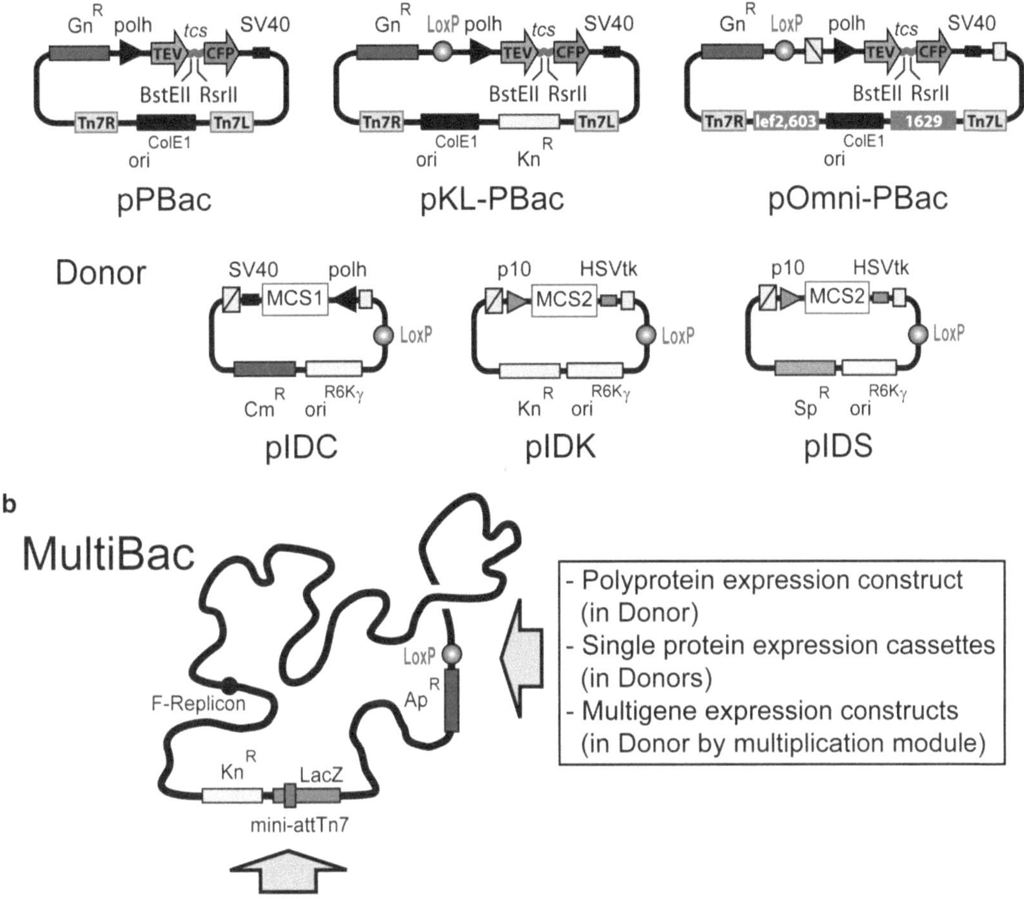

Fig. 2 Integration of polyprotein expression cassettes into MultiBac baculovirus. (**a**) MultiBac Acceptor vectors pPBac, pKL-PBac, and pOmni-PBac, tailored for polyprotein production, are shown schematically (*top*). They contain a regular ColE1 origin of replication and a polyprotein expression cassette, which encodes an N-terminal TEV protease and a C-terminal CFP spaced by a TEV cleavage site (tcs). BstEII and RsrII sites are used for inserting the polyprotein encoding ORF of interest. Donor vectors pIDC, pIDK, and pIDS contain a conditional origin of replication derived from the R6Kγ phage [13]. The multiplication module flanking the expression cassettes contain a homing endonuclease site and a complementary BstXI site (*boxes in light blue*). Polh and p10 are baculoviral very late promoters; SV40 and HSVtk are polyadenylation signals. MCS1 and MCS2 stand for multiple cloning sites. Tn7L and Tn7R are specific DNA sequences for Tn7 transposition; the lef2/603 and Ori1629 homology regions are shown as *gray boxes*. LoxP sites are shown as *red balls*. *Cm* stands for chloramphenicol, *Gn* for gentamicin, *Kn* for kanamycin, *Sp* for spectinomycin. (**b**) Besides polyprotein expression cassettes, single protein and multigene expression cassettes can also be integrated into the Tn7 attachment site (mini-attTn7) harbored by the LacZ (lacZα) gene, or the LoxP site of the MultiBac baculovirus. *Ap* stands for ampicillin. The F-Replicon is a single-copy bacterial origin of replication. For reagents contact: iberger@embl.fr

BstEII and RsrII sites and restriction–ligation cloning (*see* **Note 1**). Transfer plasmids are then used to integrate the resulting polyprotein expression cassette into baculovirus genomes of choice to generate composite baculovirus for protein expression (Fig. 2b). With this strategy, a number of complexes have been successfully produced with balanced subunit expression levels, including a ~700 kDa physiological core complex of human general transcription factor TFIID [4] (Fig. 3).

Multiprotein complexes can be expressed from a single polyprotein, or alternatively, from several polyproteins that are co-expressed, or a combination of single protein expression cassettes and one or several polyproteins, depending on the complex of choice. We recommend combining a maximum of four to five genes (in addition to the genes encoding for TEV protease and the fluorescent protein) into a single open reading frame (ORF). Otherwise, any later work to modify the genes of interest may become complicated.

We observed that while it is sufficient to provide one TEV NIa protease gene in a co-expression experiment using several polyproteins, it appears that "tagging" all polyproteins with TEV NIa protease at the N-terminus balances overexpression levels between polyproteins (IB, unpublished data).

2 Materials

We strongly recommend carrying out the design of polyproteins in silico using a DNA cloning software of choice (i.e., VectorNTI, ApE, others). Gene synthesis may be preferred for generating the individual genes of interest, in which internal BstEII and RsrII sites must be eliminated. If synthetic genes are used in conjunction with other plasmids of the MultiBac system, we recommend to further eliminate also any restriction sites that are part of the so-called multiplication modules (AvrII, ClaI, SpeI, BstZ17I, NruI, PmeI, BstXI) in the MultiBac plasmids [2, 3, 9]. Thereby, maximum flexibility of gene assembly is achieved for co-expressing proteins.

The modular concept of the MultiBac system allows transferring expression cassettes between various plasmids [2, 9]. This option can be used if several polyproteins are to be co-expressed, for example by inserting a polyprotein expression cassette into a Donor and accessing the LoxP site present on the MultiBac baculoviral backbone (Fig. 2b). Alternatively, Acceptor–Donor fusions can be generated by Cre-LoxP reaction of Donors of choice with pKL-PBac or pOmni-PBac, following published protocols [9, 10]. When co-expressing several polyproteins, we recommend using different fluorescent markers (CFP, YFP, mCherry, others) to monitor polyprotein expression instead of tagging each polyprotein with the same fluorescent protein.

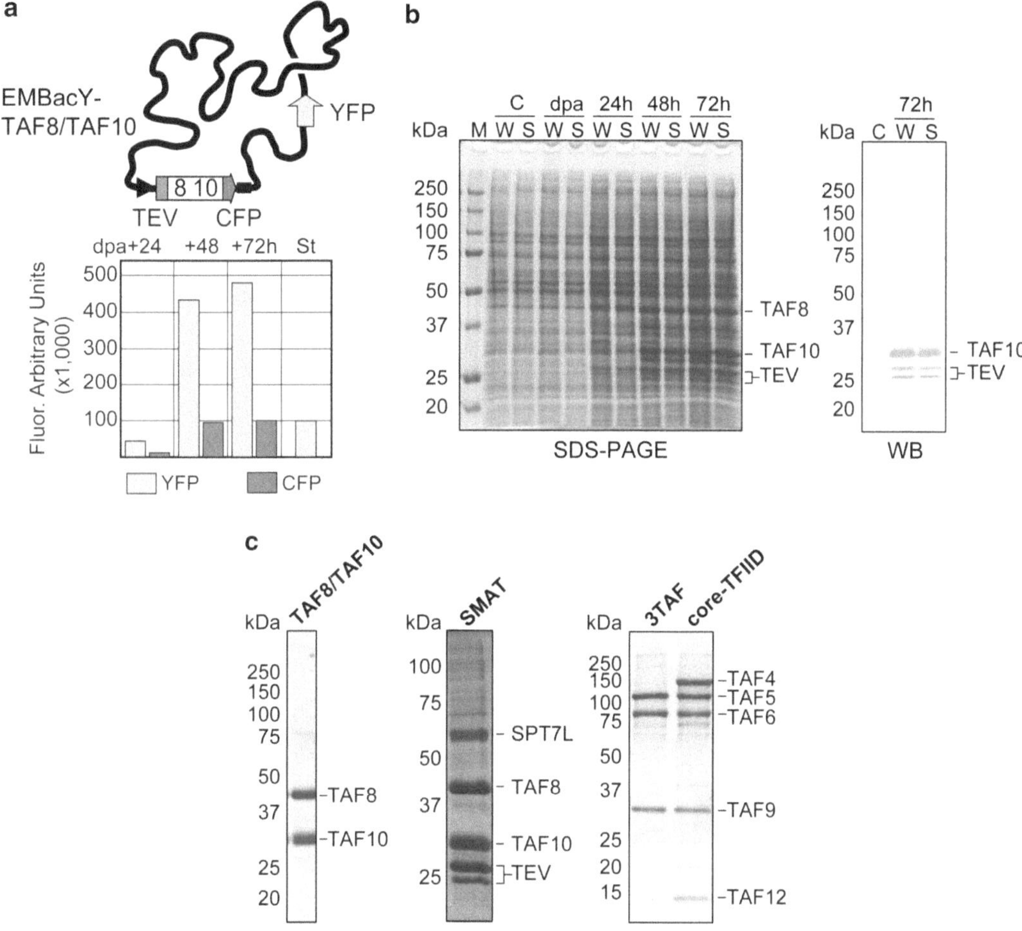

Fig. 3 Multiprotein complexes produced from polyproteins. (**a**) The TAF8/TAF10 dimer (inserted into the Tn7 attachment site) was co-expressed as a polyprotein with the yellow fluorescent protein (YFP) inserted into the viral LoxP site from a composite baculovirus (EMBacY-TAF8/TAF10). YFP and CFP expression per one million cells were tracked for evaluating the viral infection and polyprotein production. St stands for a defined fluorescence standard (used to calibrate for 100,000 arbitrary units), dpa stands for day of proliferation arrest in the infected culture [5]. (**b**) SDS-PAGE (*left*) shows balanced expressions of TAF8 and TAF10. Complete proteolysis of the TAF8/TAF10 polyprotein was confirmed by Western blot (*right*) using antibody specific for the hexa histidine-tags of TAF10 and TEV protease (*doublet*). *M* stands for molecular weight marker, *C* stands for cell control (uninfected insect cells), *W* stands for whole cell extract, *S* stands for supernatant. (**c**) Sections from SDS-PAGE are shown for TAF8/TAF10 dimer from size exclusion chromatography purification, SMAT complex from IMAC batch purification [5], 3TAF and core-TFIID complexes from size exclusion chromatography purification [4]

All reagents are prepared using ultrapure water (Millipore Milli-Q system or equivalent; conductivity of 18.2 MΩ cm at 25 °C) and analytical grade reagents. Buffers, antibiotics, and enzymes are stored at −20 °C.

2.1 Materials for Inserting Polyprotein Constructs into Transfer Vectors via Restriction–Ligation Cloning

1. Restriction endonucleases BstEII and RsrII and reaction buffers (New England Biolabs, NEB).
2. T4 DNA ligase and buffer (NEB).
3. Gel extraction kit (i.e., Qiagen, Germany).
4. Plasmid purification kit (i.e., Qiagen, Germany).
5. Regular *E. coli* competent cells (TOP10, HB101, or comparable).
6. *E. coli* competent cells containing pir gene (if Donor plasmids are used, *see* **Note 2**).
7. Antibiotics: chloramphenicol, gentamicin, kanamycin, spectinomycin (for concentrations *see* ref. 8).
8. Agar for pouring plates.
9. Media (LB, TB, SOC) for growing minicultures.

2.2 Materials for Integrating Polyprotein Expression Cassettes into Baculovirus Genome

1. *E. coli* competent cells (DH10MultiBac, DH10EMBacY, DH10MultiBacCre) (*see* **Note 3**).
2. Antibiotics chloramphenicol, gentamicin, kanamycin, spectinomycin, tetracycline (for concentrations *see* ref. 8).
3. Bluo-Gal or X-Gal.
4. IPTG.
5. Agar for pouring plates.
6. Media (LB, TB, SOC) for growing minicultures.

3 Methods

The genes encoding for the polyproteins are designed in silico, and then inserted into the transfer plasmid of choice. Once designed, polyprotein encoding genes can be created by a variety of means including DNA synthesis, restriction–ligation cloning, or sequence and ligation independent cloning (SLIC) [10, 11] or other methods, according to the preferences of the user. We recommend custom DNA synthesis to facilitate polyprotein construction.

3.1 Polyprotein In Silico *Design*

1. Group genes into polyproteins based on a set of chosen criteria (such as putative interaction partners, physiological (sub)assemblies, subunits with the same copy number within a complex).
2. Decide on the number of polyproteins that should be co-expressed (we recommend not to catenate more than four to five genes in addition to the protease and fluorescent marker encoding genes in each polyprotein ORF).
3. Decide on placement of tags. Note that cleavage sites other than TEV protease cleavage sites have to be used if tags are to

be removed at a later stage by a specific protease (i.e., PreScission protease, thrombin, enterokinase, others).

4. Remove stop codons from individual genes, except for the last gene of interest if the option to monitor polyprotein expression via the plasmid-encoded fluorescent marker protein is not desired. If fluorescence read-out is desired, delete stop codons of all genes that are to be inserted.
5. Decide on TEV protease cleavage site containing linker in between individual protein entities in the polyprotein. In particular if long unstructured tails are already predicted for example at the C-terminus of a given protein, we recommend adjoining the TEV NIa protease cleavage site (typically ENLYFQ'G) directly. The glycine residue replaces the starting methionine of the following protein.
6. Generate the DNA sequence. Add BstEII site to 5′ end and RsrII site to 3′ end.
7. Create complete polyprotein expression construct in silico, predict translation, verify reading frame through the TEV NIa protease and the fluorescent marker.
8. Decide on DNA assembly strategy (SLIC, restriction–ligation, PCR assembly, others).
9. Create all DNA sequences in silico and validate by simulating the reading frame.

3.2 Preparation of Transfer Plasmid DNA

1. Choose from pOmni-PBac, pPBac, or pKL-PBac to generate the polyprotein expressing construct for expression with the baculovirus of choice (*see* **Note 4**). All polyprotein expression cassettes have the same design with BstEII and RsrII sites for DNA insertion between the gene encoding for TEV NIa protease and the gene encoding for CFP. pKL-PBac contains a LoxP site for integrating Donor plasmids with further genes of interest; pOmni-PBac contains elements for homologous recombination in addition to elements for Tn7 transposition.
2. Digest several micrograms transfer plasmid by BstEII and RsrII enzymes according to manufacturers' recommendation. Sequential digestion is recommended as BstEII cuts optimally at 60 °C, while RsrII prefers 37 °C.
3. Analyze the digestions by agarose gel electrophoresis to confirm that the digestions are complete.
4. Purify digested plasmid by using commercial gel extraction kits (for example Qiagen gel extraction kit). It is recommended to elute the extracted DNA in the minimal volume defined by the manufacturer. Determine the concentration of the extracted DNA spectrophotometrically (e.g., Thermo Scientific NanoDrop 2000). Store in frozen aliquots.

3.3 Inserting Polyprotein Expression Cassettes into BstEII/RsrII Digested Transfer Vectors

1. Digest several micrograms of the DNA (generated by DNA synthesis, SLIC, PCR assembly, or other methods of choice) encoding for the desired polyprotein with BstEII and RsrII enzymes according to the manufacturers' recommendation. Sequential digestion is recommended as BstEII cuts optimally at 60 °C, while RsrII prefers 37 °C.
2. Purify digested insert DNA by using a commercial gel extraction kit. It is recommended to elute the extracted DNA in the minimal volume defined by the manufacturer. Determine the concentration of the extracted DNA spectrophotometrically.
3. Set up ligation reactions by mixing purified insert and vector (*see* Subheading 3.2) in 10–20 μL reaction volume with T4 DNA ligase and specific buffer according to the recommendations from the supplier. Perform ligation reactions at 25 °C overnight. It is recommended to analyze the ligation reaction by agarose gel electrophoresis to evaluate the ligation efficiency.
4. Transform regular *E. coli* competent cells (*see* **Note 2**) with ligation reaction by following standard transformation procedures. Incubate the transformation reaction in a 37 °C shaker for 1–2 h and plate on agar plates in a dilution series to ensure optimal colony separation.
5. Pick colonies, grow minicultures, and purify plasmids.
6. Indentify positive clones by restriction digestion and DNA sequencing of the insert.

3.4 Integrating Polyprotein Expression Constructs into the MultiBac Baculoviral Genome via Tn7 Transposition

1. Transform corresponding *E. coli* competent cells (DH10MultiBac or DH10EMBacY) with transfer plasmid by following standard transformation procedures. Incubate the transformation reaction in a 37 °C shaker overnight (*see* **Note 5**).
2. Plate the transformation reaction on agar plates containing antibiotics as described [9], IPTG (1 mM) and Bluo-Gal (or X-Gal) in a dilution series to ensure optimal colony separation. Incubate at 37 °C until blue and white colonies are well distinguishable.
3. Restreak four to eight white colonies to unambiguously confirm that they are positive (white). It is recommended to restreak also a blue colony as negative control.
4. Inoculate four confirmed white colonies in 2 mL aliquots of LB medium supplemented with corresponding antibiotics. After overnight incubation, use two to four of the cell cultures for bacmid purification, transfection, viral amplification, and multiprotein complex overexpression [8].

3.5 Integrating Polyprotein Expression Cassettes into MultiBac Baculoviral Genome via In Vivo Cre-LoxP Reaction

1. Place polyprotein expression cassette into Donor plasmid of choice by SLIC, restriction–ligation, PCR assembly, or other methods of choice. Validate resulting constructs by restriction mapping.
2. Transform DH10MultiBac^{Cre} electro-competent cells (these contain Cre recombinase expressed from a separate plasmid [9]) with this polyprotein expressing Donor plasmid by following standard electroporation procedures. Incubate the transformation reaction in a 37 °C shaker overnight.
3. Plate the transformation reaction on agar plates containing corresponding antibiotics, IPTG (1 mM) and Bluo-Gal (or X-Gal) in a dilution series to ensure optimal colony separation. Incubate at 37 °C until blue color of the colonies is clearly observed.
4. Restreak four to eight blue colonies on the same type of agar plates to confirm they are positive.
5. Inoculate four confirmed blue colonies in 2 mL aliquots of LB medium supplemented with corresponding antibiotics. After overnight incubation, use all four cell cultures (*see* **Note 6**) for bacmid purification, transfection, viral amplification, and multiprotein complex overexpression following published protocols [8].

4 Notes

1. The BstEII enzyme has the asymmetric restriction site G^GTNAC_C, the RsrII restriction enzyme has the asymmetric restriction site CG^GWC CG. In both cases the central base can have different contexts. When constructing the ORF encoding for the polyprotein, the sites have to be chosen such as to be compatible with the transfer plasmids.
2. Donors and their derivatives can only be propagated in cells that express the *pir* gene (such as BW23473, BW23474, or PIR1 and PIR2, Invitrogen) due to the conditional origin present on these plasmids [13]. In contrast, Acceptors and their derivatives contain regular ColE1 origin of replication and can be propagated in regular *E. coli* strains (TOP10, HB101, or comparable) [3, 12].
3. The generation of DH10MultiBac^{Cre} cells by expressing Cre recombinase is detailed in ref. 9.
4. Plasmids pPBac and pKL-PBac rely on Tn7 transposition and a baculovirus genome in form of a bacterial artificial chromosome (bacmid) for composite baculovirus generation (i.e., Bac-to-Bac system from Invitrogen, MultiBac). Plasmid

pOmni-PBac, in contrast, is a universal transfer plasmid that can access baculoviruses by both Tn7 transposition and homologous recombination [7].

5. Besides polyprotein expression constructs, the Tn7 attachment site (mini-attTn7) and the LoxP site can also be used for integrating single protein and multigene expression constructs (Fig. 2b).
6. It is recommended to check at least four blue colonies since the integration efficiency of in vivo Cre-LoxP reaction is generally lower than Tn7 transposition.

Acknowledgements

We thank all the members of the Berger laboratory for helpful discussions. YN was a fellow of the Marie-Curie training network Chromatin Plasticity and the Boehringer Ingelheim Foundation (BIF, Germany). IBE is a Hoffmann-La Roche postdoctoral fellow. ST was supported by a European Commission (EC) Marie Curie IEF postdoctoral fellowship. IB acknowledges support from the Swiss National Science Foundation (SNSF), the Agence Nationale de la Recherche (ANR), the Centre National de la Recherche Scientifique (CNRS), the EMBL and the European Commission (EC) Framework Programme (FP7) projects INSTRUCT, PCUBE, BioStruct-X, and ComplexINC.

Competing Financial Interest Statement

The authors declare competing financial interests. IB is the author on patents and patent applications containing parts of the methods and reagents here described.

References

1. Nie Y, Viola C, Bieniossek C et al (2009) Getting a grip on complexes. Curr Genomics 10:558–572
2. Berger I, Fitzgerald DJ, Richmond TJ (2004) Baculovirus expression system for heterologous multiprotein complexes. Nat Biotechnol 22:1583–1587
3. Bieniossek C, Imasaki T, Takagi Y et al (2012) MultiBac: expanding the research toolbox for multiprotein complexes. Trends Biochem Sci 37:49–57
4. Bieniossek C, Papai G, Schaffitzel C et al (2013) The architecture of human general transcription factor TFIID core complex. Nature 493:699–702
5. Vijayachandran LS, Viola C, Garzoni F et al (2011) Robots, pipelines, polyproteins: enabling multiprotein expression in prokaryotic and eukaryotic cells. J Struct Biol 175:198–208
6. Gorbalenya AE, Enjuanes L, Ziebuhr J et al (2006) Nidovirales: evolving the largest RNA virus genome. Virus Res 117:17–37
7. Vijayachandran LS, Thimiri Govinda Raj DB, Edelweiss E et al (2013) Gene gymnastics: synthetic biology for baculovirus expression vector system engineering. Bioengineered 4: 279–287
8. Bieniossek C, Richmond TJ, Berger I (2008) MultiBac: multigene baculovirus-based eukary-

otic protein complex production. In: Coligan JE, Dunn BM, Speicher DW, Wingfield PT (eds) Current protocols in protein science. Wiley, Hoboken, NJ, USA, Unit 5.20

9. Fitzgerald DJ, Berger P, Schaffitzel C et al (2006) Protein complex expression by using multigene baculoviral vectors. Nat Methods 3:1021–1032
10. Bieniossek C, Nie Y, Frey D et al (2009) Automated unrestricted multigene recombineering for multiprotein complex production. Nat Methods 6:447–450
11. Li MZ, Elledge SJ (2007) Harnessing homologous recombination in vitro to generate recombinant DNA via SLIC. Nat Methods 4:251–256
12. Trowitzsch S, Bieniossek C, Nie Y et al (2010) New baculovirus expression tools for recombinant protein complex production. J Struct Biol 172:45–54
13. Metcalf WW, Jiang W, Wanner BL (1994) Use of the rep technique for allele replacement to construct new Escherichia coli hosts for maintenance of R6Kλ origin plasmids at different copy numbers. Gene 138:1–7

Chapter 9

Expression Screening in Mammalian Suspension Cells

Susan D. Chapple and Michael R. Dyson

Abstract

Proteins naturally expressed in eukaryotic organisms often require host chaperones, binding partners, and posttranslational modifications for correct folding. Ideally the heterologous expression system chosen should be as similar to the natural host as possible. For example, mammalian proteins should be expressed in mammalian expression systems. However this does not guarantee a protein will be expressed in a sufficient high yield for structural or biochemical studies or antibody generation. Often a screening process is undertaken in which many variants including truncations, point mutations, investigation of orthologues, fusion to peptide or protein tags at the N- or C-terminus, the co-expression of binding partners, and even culture conditions are varied to identify the optimal expression conditions. This requires multi-parallel expression screening in mammalian cells similar to that already described for *E. coli* expression. Here we describe in detail a multi-parallel method to express proteins in mammalian suspension cells by transient transfection in 24-well blocks.

Key words Expression screening, HEK293 cells, CHO cells, Transient transfection, Mammalian cell culture, Interaction assays, Antibodies

1 Introduction

Expression of human and mammalian proteins in *E. coli* often results in a poor soluble expression yield [1]. Expression in eukaryotic systems such as yeast or insect cells can aid expression. However the most authentic chaperones, binding partners, and posttranslational modifications for mammalian proteins will be found in mammalian expression systems. There are several reasons why one may wish to perform a multi-parallel expression experiment. Firstly it is common to express single or tandem domains of multi-domain containing proteins to both improve expression and to study their function. Unfortunately domain boundaries are not accurately predicted within the current protein databases [2] and so often several truncations are performed at the DNA level either by rational or combinatorial [3] design followed by expression screening. Secondly individual expression domains can be stabilized and their

Yu Wai Chen (ed.), *Structural Genomics: General Applications*, Methods in Molecular Biology, vol. 1091, DOI 10.1007/978-1-62703-691-7_9, © Springer Science+Business Media, LLC 2014

yield improved by fusion at the N- or C-terminus with peptide or protein tags [4]. Each protein target is different and so it is likely that several fusion partners would need to be investigated. Thirdly, it is well known that protein orthologues and mutations can display different solubility and crystallization properties and so one may wish to investigate a panel of point mutations and orthologues. Lastly some proteins are only stable in the presence of their natural binding partners and so one may wish to investigate co-expression with candidate binding partners [5]. The variables described here soon multiply and a thorough investigation requires the use of a plate based mammalian expression screen.

The optimization of expression parameters is not the only reason an investigator may wish to perform a multi-parallel expression experiment. They may also, for example, need to express a panel of receptor ectodomains for interaction studies [6] or functional screening [7]. Also panels of recombinant antibodies can be expressed for screening in proteomic applications [8, 9] or to aid therapeutic antibody lead isolation and optimization projects [10].

Screening expression in suspension adapted HEK293 or CHO cells allows the convenience of fast scale-up of any hits discovered in a small-scale expression screen [11, 12]. Here we describe a method for transfection of HEK293F cells in 24-well blocks and a dot-blot screen to identify secreted expression screen hits. The dot blot screen could be replaced by a standard western blot procedure or ELISA. Also the methods are transferable to suspension Chinese Hamster Ovary (CHO) suspension cells.

2 Materials

All chemicals were from Sigma unless stated otherwise.

2.1 HEK293F Cell Maintenance

1. For maintenance of cells in Erlenmeyer flasks a humidified CO2 shake incubator is required with a 25 mm orbital throw such as the Infors Multitron.
2. Vented sterile Erlenmeyer flasks (Corning).
3. HEK293F cells and Freestyle media (Life Technologies).
4. A hemocytometer for cell counting.

2.2 HEK293F Cell 24-Well Block Transfection

1. For maintenance of cells in 24-well blocks a humidified CO2 plate shaker incubator is required with a 3 mm orbital throw such as the Infors Multitron plate shaker incubator.
2. Sterile 24-well blocks were from Qiagen (Fig. 1).
3. Linear 25 kDa polyethylenimine (PEI) was from Polysciences Inc. This was prepared at a concentration of 1 mg/ml in MilliQ water. Solubilization was achieved by first adding

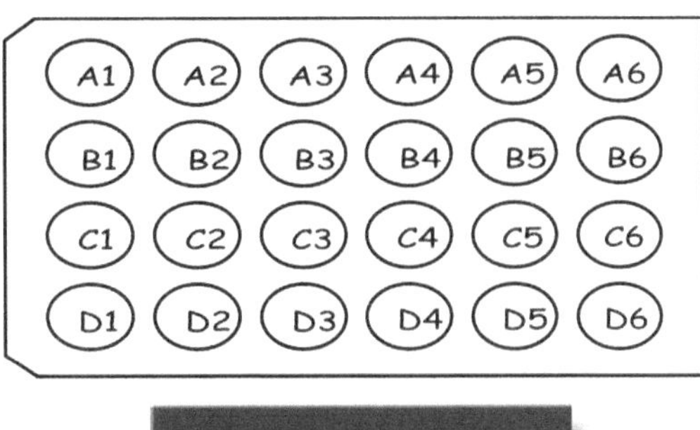

Fig. 1 24-well block for HEK293F cell culturing and transfection

concentrated HCl to a stirred PEI solution until the pH was <2.0 and stirring continued for 2–3 h. The pH was then adjusted to 7.0 using concentrated NaOH. The PEI solution was finally filter-sterilized by passage through a 0.22 μm membrane and 1 ml aliquots stored at −20 °C. Each batch of PEI should be tested for transfection efficiency using a GFP reporter plasmid [11] and different DNA to PEI ratios (e.g., 1:1 and 1:2 [12]).

2.3 Expression Screening by Dot Blot

1. Phosphate-buffered saline (PBS).
2. 8 M urea.
3. Blocking solution (e.g., 3 % milk/PBS/TWEEN).
4. Nitrocellulose from Schleicher and Schuell.
5. Whatman 3MM filter paper.
6. Dot Blot apparatus from Schleicher and Schuell (Manifold I system dot blot apparatus).

3 Methods

3.1 HEK293F Cell Maintenance

1. When the cell density reaches 1–4 × 106 cells/ml passage the cells (*see* **Note 1**).
2. Centrifuge cells for 4 min at 1,500 × *g* (Sorvall Legend centrifuge) at room temperature in a 50 ml Falcon tube.
3. Resuspend the cells in fresh medium (i.e., 1/4 the original culture volume) and pipette to break up any cell clumps.
4. Count viable cells by trypan blue exclusion using a 1:5 dilution (e.g., 200 μl cells: 100 μl trypan blue:700 μl medium).

5. Seed the required culture volume with 2.5×105 cells/ml using Freestyle medium (*see* **Note 2**).
6. Label flask with name, cell line name, passage number, date, seeding density.
7. Incubate at 37 °C, 5 % CO2, 60 % humidity, 125 rpm.
8. The cells will require splitting again 3–4 days later.

3.2 HEK293F Cell 24-Well Block Transfection

1. Split 200 ml of HEK293F cells at 2.5×105 cells/ml in a 1 L sterile vented erlenmeyer flask for each 24-well block (i.e., for 96-well plate 4×200 ml flasks are required), 48 h before the transfection.
2. On the day of transfection, add 400 μl of serum-free media (SFM), warmed to room-temperature (*see* **Note 3**) to the wells of the 24-well block followed by 4 μg of plasmid DNA (*see* **Note 4**).
3. Add 5 μl PEI to the walls of each well with a repeater pipettor or a multichannel pipette with a Varispan to allow pipetting into the 6-well row of the 24-well block. The PEI is placed approximately 0.5–1 cm from the meniscus of the SFM.
4. Vortex the 24-well block for 10 s on plate vortexer. Incubate for 10 min at room temp (*see* **Note 5**).
5. Add Pluronic F68 reagent into each 1 L vented Erlenmeyer flask, now containing 1×106 cells/ml (*see* **Note 6**) to a final concentration of 0.1 % (*see* **Note 7**).
6. HEK293F cells are added (4 ml) to each well of the 24-well block containing the DNA–PEI complex. Cover with an air-pore plate sealer.
7. Incubate the 24-well block at 37 °C, 5 % CO2, 75 % humidity, 400 rpm in a plate shake incubator with a 3 mm orbital throw. Check after 1 h that the cells are still in complete suspension.
8. Harvest after 5 days transfection (*see* **Note 8**), by centrifugation at 2,500×*g* for 5 min, and analyze the supernatant (secreted proteins) or cell lysate (intracellular proteins) by western blot or by dot blot.

3.3 Expression Screening by Dot Blot

1. 8 M urea was added to cleared culture supernatants (or purified proteins) to give a final concentration of 5 M urea (i.e., 125 μl 8 M urea added to 75 μl culture supernatant (*see* **Note 9**)).
2. Incubate the culture supernatant–urea mix for 1 h at room temperature.
3. Set up dot blot apparatus during this time: Pre-soak Whatman 3MM filter paper (2–3 sheets) and nitrocellulose membrane in PBS.
4. Arrange Minifold I apparatus according to the Schleicher–Schuell protocol. In summary: place the middle unit (96 wells with small holes) on top of base collection unit according to the line up pins.

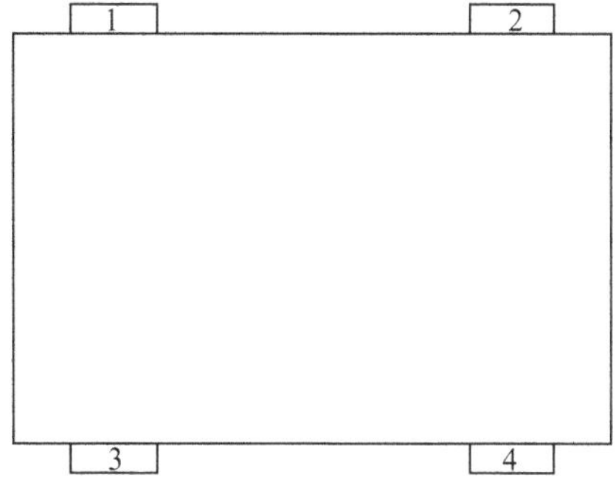

Fig. 2 Dot blot apparatus depicting clip numbering

5. Place 2–3 sheets of PBS-soaked filter paper onto the unit, followed by the membrane.
6. Place the top unit (96-well plate with dispensing holes) in place over the filter paper and membrane using the line up pins.
7. Secure the whole dot blot apparatus in place using the four clips on the side (N.B make sure that they are fixed in place using clip 1 followed by clip 4 then clip 2 followed by clip 3 and NOT clips 1 + 2 followed by clips 3 + 4 as illustrated in Fig. 2).
8. When ready to load the samples: connect dot blot unit to vacuum source and turn on for a few seconds to clear the excess PBS from the wells.
9. Switch vacuum off then load all samples to be analyzed (can use multichannel pipette).
10. Switch on vacuum and allow samples to move onto the membrane (this should take approx 10–20 s). If there are small air bubbles trapping sample in a well, gently tap the apparatus on the bench to move the air bubbles out the way and allow the sample to move onto the membrane.
11. Once finished, remove membrane and place in blocking solution.
12. Probe with antibody as detailed in standard western blot protocols.
13. Finally: rinse dot blot apparatus in water to prevent anything clogging up the apparatus and allow to air dry on bench.

4 Notes

1. Work at all times with good aseptic technique within a functioning tissue culture hood. Pre-warm the culture media in hood for approx 1 h prior to use. Always clean (using ethanol spray) the inside of hood and any equipment to be used prior to use in the hood. Infection of mammalian cell cultures with bacteria or yeast results in poor expression yield and can be major cause for delay.

2. HEK293F cells can be split as low as 1×10^5 cells/ml. It is important not to allow the cells to over-grow ($\geq 3 \times 10^6$ cells/ml) as dead cells can accumulate resulting in a less healthy cell population. Maintaining cells in a good state is essential for high transfection efficiency and thus expression yield.

3. The serum-free media is the media the cells are normally propagated with, minus the addition of serum. For example for HEK293F cells, this would be Freestyle medium (Life Technologies).

4. The plasmid DNA to be used for transfection must be of sufficient purity to allow for an efficient transfection. The DNA should be prepared according to the NAPPA protocol [13], a standard midi- or maxi-prep method involving an isopropanol precipitation, or a commercially available transfection quality plasmid DNA kit from suppliers such as Qiagen or Macherey-Nagel. The OD_{260nm}:OD_{280nm} ratio should be between 1.8 and 1.9. This ensures low protein and endotoxin contamination.

5. 10 minutes is the minimum time to allow for formation of the DNA–PEI complex. Up to 30 min still allows for efficient transfection, but from 30 min to 1 h transfection efficiency gradually decreases due to the formation of higher order DNA–PEI aggregates.

6. The cells should be as close to mid-logarithmic phase as possible (for HEK293 cells between 0.8×10^6 and 1.2×10^6 cells/ml) with a cell viability of ≥95 %.

7. The anti-foaming agent Pluronic is required to maintain the viability of the HEK293 suspension cells during growth in 24-well blocks.

8. The time required before harvesting depends on the protein being expressed. Intracellular and nuclear located proteins may require only 2–3 days for optimal expression, whereas secreted protein such as receptor ectodomains or antibodies typically require 4–5 days. The time required should be determined empirically for the target class of proteins being investigated.

9. It was found that the addition of urea enhanced the binding of glycoproteins to the nitrocellulose membrane [11].

References

1. Dyson MR, Shadbolt SP, Vincent K et al (2004) Production of soluble mammalian proteins in Escherichia coli: identification of protein features that correlate with successful expression. BMC Biotechnol 4:32
2. Dyson MR (2010) Selection of soluble protein expression constructs: the experimental determination of protein domain boundaries. Biochem Soc Trans 38:908–913
3. Dyson MR, Perera RL, Shadbolt SP et al (2008) Identification of soluble protein fragments by gene fragmentation and genetic selection. Nucleic Acids Res 36:e51
4. Brown MH, Barclay AN (1994) Expression of immunoglobulin and scavenger receptor superfamily domains as chimeric proteins with domains 3 and 4 of CD4 for ligand analysis. Protein Eng 7:515–521

5. Trowitzsch S, Bieniossek C, Nie Y et al (2010) New baculovirus expression tools for recombinant protein complex production. J Struct Biol 172:45–54
6. Bushell KM, Söllner C, Schuster-Boeckler B et al (2008) Large-scale screening for novel low-affinity extracellular protein interactions. Genome Res 18:622–630
7. Gonzalez R, Jennings LL, Knuth M et al (2010) Screening the mammalian extracellular proteome for regulators of embryonic human stem cell pluripotency. Proc Natl Acad Sci USA 107:3552–3557
8. Colwill K, Graslund S (2011) A roadmap to generate renewable protein binders to the human proteome. Nat Methods 8:551–558
9. Dyson MR, Zheng Y, Zhang C et al (2011) Mapping protein interactions by combining antibody affinity maturation and mass spectrometry. Anal Biochem 417:25–35
10. Bradbury ARM, Sidhu S, Dubel S et al (2011) Beyond natural antibodies: the power of in vitro display technologies. Nat Biotechnol 29:245–254
11. Chapple S, Crofts A, Shadbolt SP et al (2006) Multiplexed expression and screening for recombinant protein production in mammalian cells. BMC Biotechnol 6:49
12. Tom R, Bisson L, Durocher Y (2008) Transfection of HEK293-EBNA1 cells in suspension with linear PEI for production of recombinant proteins. CSH Protoc 2008:pdb.prot4977
13. Link AJ, LaBaer J (2008) Construction of Nucleic Acid Programmable Protein Arrays (NAPPA) 3: isolating DNA plasmids in a 96-well plate format. CSH Protoc 2008:pdb.prot5058

Chapter 10

Cell-Free Expression of Protein Complexes for Structural Biology

Takaho Terada, Takeshi Murata, Mikako Shirouzu, and Shigeyuki Yokoyama

Abstract

Cell-free protein synthesis is advantageous for the expression of protein complexes, since it is suitable for the co-expression of two or more components of the target protein complexes. The quantity and the quality of cell-free expressed complexes are generally better than those of protein complexes expressed in conventional cell-based systems, because various parameters, such as the stoichiometry of the component proteins, can be more precisely controlled. In this chapter, we describe techniques for the expression of protein complexes by an *Escherichia coli* cell-free protein synthesis system, which has been successfully utilized in crystallographic structural studies.

Key words Protein complex, Cell-free protein synthesis, *Escherichia coli*, X-ray crystallography, Dialysis

1 Introduction

Cell-free protein synthesis is a convenient method for protein expression. Lysates prepared from the cells of various organisms, such as *Escherichia coli* [1], wheat germ [2], insect [3], and human [4], have been developed and commercialized. Usually, coupled transcription–translation of the DNA template encoding the target protein is performed in the lysate. Plasmids and PCR-amplified DNA fragments can be used as the templates for transcription, e.g., by T7 phage RNA polymerase. The transcribed messenger RNAs are translated into proteins by ribosomes, translation factors, transfer RNAs (tRNAs), and aminoacyl-tRNA synthetases. If the mRNA contains many minor codons, then the lysate is supplemented with minor tRNA species that translate the minor codons, for better protein expression [5]. The lysate may also be supplemented with molecular chaperones, e.g., the bacterial DnaK/DnaJ/GrpE and GroEL/GroES systems, if required for proper protein folding. The substrates for coupled transcription–translation, such as amino acids

Yu Wai Chen (ed.), *Structural Genomics: General Applications*, Methods in Molecular Biology, vol. 1091, DOI 10.1007/978-1-62703-691-7_10, © Springer Science+Business Media, LLC 2014

and nucleoside triphosphates, are continually provided to the reaction mixture, for example, by dialysis against a feeding solution, in order to achieve milligram-level productivity.

Cell-free protein synthesis is becoming one of the standard protein expression methods for structural biology and genomics, because it provides a number of advantages over the conventional recombinant expression of proteins in host cells. First, cell-free protein synthesis is advantageous for high-throughput expression screening and/or large-scale production of various target proteins in structural genomics [6, 7]. Cell-free synthesis can produce sufficient amounts (milligram quantities) of proteins with either selenomethionine substitutions for X-ray crystallography [8] or stable-isotope labels for NMR spectroscopy [9]. Our group has determined ~250 crystal structures and ~1,300 NMR structures, by using cell-free produced protein samples [10]. Cell-free protein synthesis is particularly useful for the production of difficult target proteins, such as physiologically toxic proteins and integral membrane proteins. Membrane proteins can be synthesized in the presence of appropriate detergents and in lipid bilayer environments [11–13].

Furthermore, cell-free protein synthesis is suitable for the expression of protein complexes consisting of two or more different component proteins (or subunits). Such heteromultimers can be synthesized by the co-expression of the DNA templates, each encoding a component of the complex. Based on the results of preliminary small-scale coupled transcription–translation trials, the amounts of the DNA templates in the large-scale expression may be adjusted, to achieve the appropriate expression levels of the component proteins corresponding to their correct stoichiometry in the heteromultimer. By contrast, it is difficult to precisely adjust the expression levels of the component proteins in cell-based expression systems. In cell-free synthesis, the component proteins can be expressed in a particular order by sequentially adding the templates to the reaction mixture. Otherwise, some of the component proteins can be expressed in the presence of the other component proteins that are prepared beforehand and added to the cell-free reaction solution. Moreover, appropriate molecular chaperones can be added to the reaction mixture, to facilitate the proper integration of the component proteins into the complex. In addition, some protein complexes can be reconstituted by co-refolding of the cell-free expressed component proteins.

By using the cell-free protein expression method, we have determined the crystal structures of heterodimeric complexes, including those between Slac2-a/melanophilin and Rab27B GTPase [14], the armadillo repeat domain of *adenomatous polyposis coli* (APC) and the tyrosine-rich domain of Sam68 [15], the Rac guanine nucleotide exchange factor DOCK2 and its partner ELMO1 [16], the extracellular domains of the calcitonin receptor-like receptor (CRLR) and receptor activity-modifying protein 2

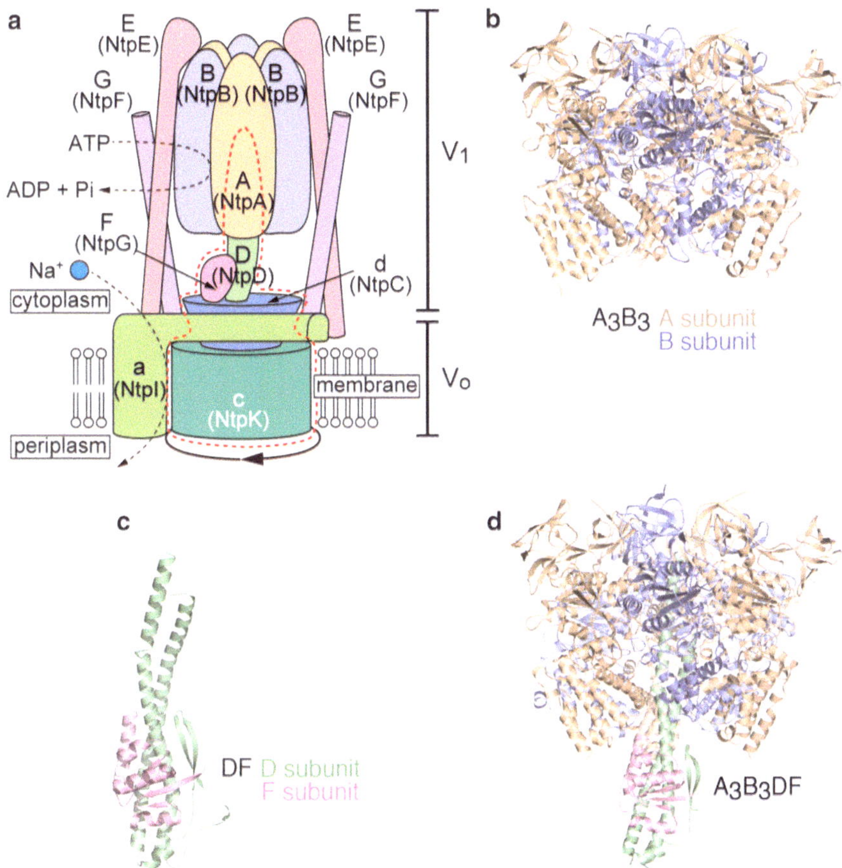

Fig. 1 Crystal structures of *E. hirae* V-ATPase sub-complexes. (**a**) A schematic model of *E. hirae* V-ATPase. The enzyme is composed of nine subunits (Eh-A, -B, -d, -D, -E, -F, -G, -a, -c; previously designated as Ntp-A, -B, -C, -D, -E, -G, -F, -I, -K). Crystal structures of (**b**) Eh-A_3B_3 [19, 21], (**c**) Eh-DF [20] and (**d**) Eh-A_3B_3DF [19, 21]. The Eh-A_3B_3 and Eh-DF complexes were expressed separately, using an *E. coli* cell-free expression system. The Eh-A_3B_3DF complexs were reconstituted from Eh-A_3B_3 and Eh-DF [19–21]

(RAMP2) [the adrenomedullin (AM) receptor] [17], and the extracellular domains of interleukin-5 (IL-5) and the IL-5 receptor α subunit (IL-5RA) [18].

Recently, we reported the crystal structures of the *Enterococcus hirae* V1-ATPase A_3B_3, DF, and A_3B_3DF complexes (Fig. 1), by using cell-free synthesized protein samples [19, 20]. The A_3B_3DF complex was then reconstituted from the A_3B_3 and DF subcomplexes [21]. These component proteins could only be expressed in the soluble forms by co-expression, and they formed the stoichiometric complexes in the cell-free protein synthesis system. The same approach was applied to the expression of human V-ATPase subunits and subcomplexes [22] (Fig. 2). Notably, the qualities of the crystals of the cell-free expressed complexes were much better than those of the recombinant protein complexes expressed in vivo in *E. coli* host cells.

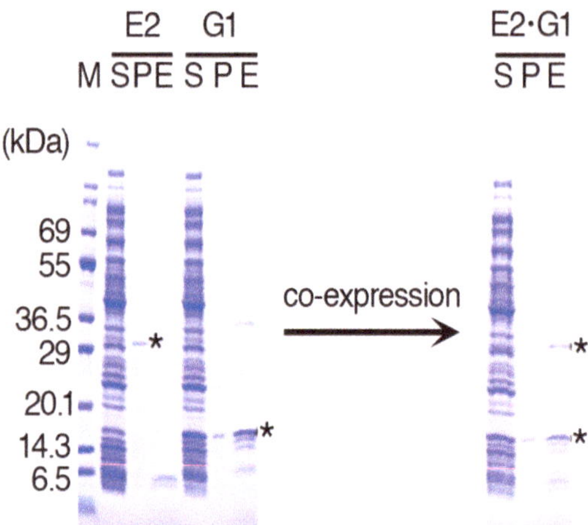

Fig. 2 Cell-free co-expression of the E2 and G1 subunits of human V-ATPase, to form the E2·G1 subcomplex. SDS-PAGE analyses of the individually expressed E2 subunit (*E2*) and G1 subunit (*G1*), and the co-expressed subunits, E2 and G1 (*E2·G1*). *Lane M* molecular weight markers, *lane S* supernatant, *lane P* pellet, *lane E* eluate from the affinity purification column. An *asterisk* indicates the band of each expressed protein, corresponding to its molecular mass. The E2 subunit (26.6 kDa) was observed only in the pellet (*lane P*). In contrast, the G1 subunit (14 kDa) appeared mostly in the supernatant (*lane S*) and only slightly in the pellet (*lane P*), and was eluted well from the column (*lane E*). However, the cell-free co-expressed E2 and G1 subunits were observed mostly in the supernatant (*lane S*), and co-eluted from the column (*lane E*). Therefore, cell-free co-expression of the E2 and G1 subunits successfully resulted in the soluble expression of the E2·G1 subcomplex of human V-ATPase. This figure was prepared by modification of Fig. 2 of Rahman et al. [22]

In this chapter, we describe our protocols for the large-scale expression of protein complexes by coupled transcription–translation, using T7 RNA polymerase and the *E. coli* cell-free protein synthesis system, which are particularly useful for crystallization and X-ray crystallographic analyses.

2 Materials

Prepare all solutions using sterilized ultrapure water and analytical grade reagents.

2.1 Components for Cell-free Synthesis Solutions

1. LMCPY mixture: 160 mM HEPES-KOH buffer (pH 7.5), 4.13 mM L-tyrosine, 534 mM potassium L-glutamate, 5 mM DTT, 3.47 mM ATP, 2.40 mM GTP, 2.40 mM CTP, 2.40 mM UTP, 0.217 mM folic acid, 1.78 mM cAMP, 74 mM ammonium acetate, and 214 mM creatine phosphate. Store below −20 °C (*see* **Notes 1** and **2**).

2. Amino acid mixture: 20 mM each of 19 amino acids, without L-tyrosine, in water. Store at −20 °C (*see* **Notes 2** and **3**).
3. Magnesium acetate solution: 1.6 M magnesium acetate in water. Store at −20 °C.
4. S30 buffer: 10 mM Tris-acetate buffer (pH 8.2), 60 mM potassium acetate, 16 mM magnesium acetate, 1 mM DTT. Store at −20 °C.
5. Sodium azide solution: 5 % (w/v) in water. Store at −20 °C.
6. Creatine kinase solution: Dissolve 500 mg lyophilized creatine kinase powder (Roche Applied Sciences, 127556) in 133.3 mL water. Store at −20 °C (*see* **Note 4**).
7. tRNA solution: Dissolve 500 mg lyophilized tRNA powder (*E. coli* MRE600-derived, Roche Applied Sciences, 109550) in 28.6 mL water. Store at −20 °C (*see* **Note 4**).
8. *E. coli* S30 extract: in the S30 buffer (*see* **Note 5**). Store at −80 °C or in liquid nitrogen (*see* **Note 4**).
9. T7 RNA polymerase: 10 mg/mL in 20 mM Tris–HCl (pH 8.0) buffer containing 100 mM NaCl, 1 mM DTT, 1 mM EDTA, and 50 % glycerol (*see* **Note 6**). Store at −20 °C (*see* **Note 4**).
10. Two or more plasmid DNA templates, each encoding a protein component of the target complex (or subcomplex).

2.2 Apparatus for Large-Scale Cell-Free Synthesis Reactions

1. Constant temperature incubator shaker (Taitec BR-300LF).
2. Dialysis tubing: Spectra/Por 7 (MWCO, 15,000; Sealing Width, 45 mm; Spectrum Laboratories).
3. Dialysis tubing closures (Spectrum Laboratories).
4. 100-mL centrifuge tubes.
5. 50-mL centrifuge tubes.
6. 400-mL square-shaped polystyrene cases.
7. 35-mL conical Oak Ridge tubes (Nalgene).

3 Methods

3.1 Large-Scale Cell-Free Protein Synthesis Reaction

All of the component proteins of the target complex (or subcomplex) can be co-expressed in the *E. coli* cell-free protein synthesis system, by coupled transcription–translation of the plasmid DNA templates encoding the component proteins. The plasmids for cell-free protein expression are designed and prepared as reported [6, 13]. Typically, the protein product consists of the N-terminal tag sequence (e.g., a modified HAT tag), the TEV cleavage site, the linker sequence GSSGSSG, and the target protein [6]. It should be noted that the N-terminal tag sequence is useful not only for affinity purification but also for higher yield. One of the components of the target complex (or subcomplex) should be tagged differently from the others, to facilitate the affinity purification of the complete complex.

The reaction solution contains 2 μg/mL template plasmid(s), 66.7 μg/mL T7 RNA polymerase

30 % (v/v) S30 extract, 0.175 mg/mL tRNA, 1.5 mM each of 19 amino acids without L-tyrosine, 37.3 % (v/v) LMCPY mixture, 0.25 mg/mL creatine kinase, 10 mM magnesium acetate, and 0.05 % (w/v) sodium azide (*see* **Notes 1** and **3**). The feeding solution contains 30 % (v/v) S30 buffer, 1.5 mM each of 19 amino acids without L-tyrosine, 37.3 % (v/v) LMCPY mixture, and 10 mM magnesium acetate (*see* **Notes 1** and **3**).

The typical protocol uses 9 mL of the reaction solution and 90 mL of the feeding solution. The reaction scales can be smaller and larger for expression screening and large-scale purification, respectively. The volume of the feeding solution should be at least ten times larger than that of the reaction solution. Perform all procedures on ice, unless otherwise specified.

1. Set the following parameters for the incubator shaker: 25 °C (*see* **Note 7**), 50 rpm, and 50-mm amplitude.
2. Thaw each component for the cell-free synthesis solution on ice. After thawing, gently shake each reagent tube, to ensure that the solution is homogeneous. Place and keep all reagents on ice during handling.
3. Prepare 90 mL of the feeding solution. First of all, gently shake the tube containing the LMCPY mixture, to make it homogeneous (*see* **Note 2**). Combine the LMCPY mixture (33.6 mL), the amino acid mixture (6.75 mL), the magnesium acetate solution (0.522 mL), the S30 buffer (27 mL), and the sodium azide solution (0.9 mL) (*see* **Notes 1** and **3**). Bring the volume of the mixture solution to 90 mL with water. Place the mixture in a 100-mL centrifuge tube, and mix it thoroughly by turning the tube upside down gently several times.
4. Prepare 9 mL of the reaction solution. Gently shake the tube containing the LMCPY mixture to make it homogeneous (*see* **Note 2**). Combine the LMCPY mixture (3.36 mL), the amino acid mixture (0.675 mL), the magnesium acetate solution (0.052 mL), the sodium azide solution (0.9 mL), and the tRNA solution (0.09 mL) (*see* **Notes 1** and **3**). To this solution, sequentially add the creatine kinase solution (0.6 mL), the T7 RNA polymerase solution (0.06 mL), and the plasmid DNA templates (*see* **Note 8**). Bring the volume of the mixture to 9 mL with water. Place the reaction solution in a 50-mL centrifuge tube and mix it thoroughly by turning the tube upside down gently several times.

5. Place the feeding solution (90 mL from **step 3**) in a 400-mL square-shaped polystyrene case.
6. Seal one end of the dialysis tube (Spectra/Por 7, MWCO = 15,000) with a closure (Spectrum). Place the reaction solution (9 mL) in the tube, remove as much air from the tube as possible, and seal the open end of the tube with a closure.
7. Submerge the dialysis tube in the feeding solution in the polystyrene case. Wrap the case with plastic wrap.
8. Shake the case reciprocally with a incubator shaker (50-mm amplitude, 50 rpm) at 25 °C for 4 h (*see* **Notes 7** and **9**)
9. Transfer the reaction solution from the dialysis tube into a 35-mL conical Oak Ridge tube, and transfer 0.002- and 0.008-mL aliquots into Eppendorf tubes for SDS-PAGE analysis. Centrifuge the 35-mL conical Oak Ridge tube at 20,130 × g for 20 min. Transfer the supernatant to a fresh 50-mL centrifuge tube, and store it on ice.
10. Centrifuge the 0.008-mL aliquot of the reaction mixture in the Eppendorf tube at 20,380 × g for 10 min. Transfer the supernatant to a fresh Eppendorf tube. Suspend the precipitate in an appropriate buffer, using the same volume as that of the supernatant. Analyze 0.001 mL each of the total reaction mixture, the supernatant, and the suspended precipitate by SDS-PAGE.
11. Purify the protein complex in the supernatant, if it is expressed as expected.

4 Notes

1. This protocol is for cellular protein complexes. For cell-free expression and assembly of extracellular protein complexes and extracellular regions of membrane protein complexes with disulfide bonds, the protocol must be modified to lower the DTT concentration, and/or to include the reduced and oxidized forms of L-glutathione (GSH and GSSG, respectively) at an appropriate ratio and a disulfide isomerase, such as *E. coli* DsbC.
2. L-Tyrosine is included in the LMCPY mixture, but not in the "amino acid mixture", as the solubility of L-tyrosine is much lower than those of the other 19 amino acids. The tube containing the LMCPY mixture must be gently shaken just before use.
3. The "amino acid mixture" contains L-methionine. However, when L-selenomethionine, instead of L-methionine, is to be incorporated into the protein complex, an amino acid mixture lacking L-methionine should be used. The L-selenomethionine should be added separately to the reaction solution and the feeding solution.

4. Divide the solution into appropriate volumes, in order to avoid repeated freezing and thawing.
5. The *E. coli* S30 extract is prepared from *E. coli* BL21(DE3) bearing the pMINOR plasmid [5] or BL21-CodonPlus(DE3)-RIL, as reported [1]. A cell-free protein synthesis kit utilizing the S30 extract prepared by our protocol, the Remarkable Yield Translation System Kit, is available from ProteinExpress, Chiba, Japan (http://www.proteinexpress.co.jp/e/index.html). If chaperones are required to facilitate complex (or subcomplex) formation by the component proteins, then the S30 extract may be prepared from *E. coli* BL21 expressing a set of *E. coli* chaperones (usually DnaK/DnaJ/GrpE and/or GroEL/GroES) in addition to the minor tRNAs (e.g., from pMINOR).
6. T7 RNA polymerase is prepared as reported [23].
7. The optimal incubation temperature should be selected from 20, 25, and 30 °C, by a preliminary small-scale expression experiment. If necessary, an incubation at 15 °C may be performed for particularly unstable complexes.
8. The concentrations of the plasmid DNA templates should be determined by a preliminary small-scale expression trial. There will be optimal concentrations of the templates to maximize the coupled transcription–translation reaction, while the ratio of the templates should be adjusted to achieve the correct stoichiometry of the component proteins in the target complex (or subcomplex).
9. An incubation longer than 4 h may be performed in the cases of poor expression, such as at low temperature (20 or 15 °C), when expressing one or more protein components with a large number of residues (>10,000), or with hard to express stretch(s) of amino acid residues.

Acknowledgements

We thank T. Mishima and T. Nakayama for their assistance in manuscript preparation. This work was supported by the RIKEN Structural Genomics/Proteomics Initiative (RSGI), the National Project on Protein Structural and Functional Analyses, and the Targeted Proteins Research Program (TPRP), of the Ministry of Education, Culture, Sports, Science and Technology of Japan.

References

1. Kigawa T, Yabuki T, Matsuda N et al (2004) Preparation of *Escherichia coli* cell extract for highly productive cell-free protein expression. J Struct Funct Genomics 5:63–68
2. Madin K, Sawasaki T, Ogasawara T et al (2000) A highly efficient and robust cell-free protein synthesis system prepared from wheat embryos: plants apparently contain a suicide system directed at ribosomes. Proc Natl Acad Sci USA 97:559–564
3. Wakiyama M, Kaitsu Y, Matsumoto T et al (2010) Coupled transcription and translation from polymerase chain reaction-amplified DNA in *Drosophila* Schneider 2 cell-free system. Anal Biochem 400:142–144
4. Mikami S, Kobayashi T, Masutani M et al (2008) A human cell-derived *in vitro* coupled transcription/translation system optimized for production of recombinant proteins. Protein Expr Purif 62:190–198
5. Chumpolkulwong N, Sakamoto K, Hayashi A et al (2006) Translation of 'rare' codons in a cell-free protein synthesis system from *Escherichia coli*. J Struct Funct Genomics 7:31–36
6. Yabuki T, Motoda Y, Hanada K et al (2007) A robust two-step PCR method of template DNA production for high-throughput cell-free protein synthesis. J Struct Funct Genomics 8:173–191
7. Aoki M, Matsuda T, Tomo Y et al (2009) Automated system for high-throughput protein production using the dialysis cell-free method. Protein Expr Purif 68:128–136
8. Kigawa T, Yamaguchi-Nunokawa E, Kodama K et al (2002) Selenomethionine incorporation into a protein by cell-free synthesis. J Struct Funct Genomics 2:29–35
9. Kigawa T, Yabuki T, Yoshida Y et al (1999) Cell-free production and stable-isotope labeling of milligram quantities of proteins. FEBS Lett 442:15–19
10. Yokoyama S, Terwilliger TC, Kuramitsu S et al (2007) RIKEN aids international structural genomics efforts. Nature 445:21
11. Shimono K, Goto M, Kikukawa T et al (2009) Production of functional bacteriorhodopsin by an *Escherichia coli* cell-free protein synthesis system supplemented with steroid detergent and lipid. Protein Sci 18:2160–2171
12. Wada T, Shimono K, Kikukawa T et al (2011) Crystal structure of the eukaryotic light-driven proton-pumping rhodopsin, Acetabularia rhodopsin II, from marine alga. J Mol Biol 411:986–998
13. Kimura-Soyema T, Shirouzu M, Yokoyama S (2013) Cell-free membrane protein expression. In: Methods in molecular biology: cell-free protein expression and engineering. Submitted
14. Kukimoto-Niino M, Sakamoto A, Kanno E et al (2008) Structural basis for the exclusive specificity of Slac2-a/melanophilin for the Rab27 GTPases. Structure 16:1478–1490
15. Morishita EC, Murayama K, Kato-Murayama M et al (2011) Crystal structures of the armadillo repeat domain of adenomatous polyposis coli and its complex with the tyrosine-rich domain of Sam68. Structure 19:1496–1508
16. Hanawa-Suetsugu K, Kukimoto-Niino M, Mishima-Tsumagari C et al (2012) Structural basis for mutual relief of the Rac guanine nucleotide exchange factor DOCK2 and its partner ELMO1 from their autoinhibited forms. Proc Natl Acad Sci USA 109:3305–3310
17. Kusano S, Kukimoto-Niino M, Hino N et al (2012) Structural basis for extracellular interactions between calcitonin receptor-like receptor and receptor activity-modifying protein 2 for adrenomedullin-specific binding. Protein Sci 21:199–210
18. Kusano S, Kukimoto-Niino M, Hino N et al (2012) Structural basis of interleukin-5 dimer recognition by its α receptor. Protein Sci 21:850–864
19. Arai S, Saijo S, Suzuki K et al (2013) Rotation mechanism of *Enterococcus hirae* V1-ATPase based on asymmetric crystal structures. Nature 493:703–707
20. Saijo S, Arai S, Hossain KM et al (2011) Crystal structure of the central axis DF complex of the prokaryotic V-ATPase. Proc Natl Acad Sci USA 108:19955–19960
21. Arai S, Yamato I, Shiokawa A et al (2009) Reconstitution in vitro of the catalytic portion (NtpA$_3$-B$_3$-D-G complex) of *Enterococcus hirae* V-type Na+-ATPase. Biochem Biophys Res Commun 390:698–702
22. Rahman S, Ishizuka-Katsura Y, Arai S et al (2011) Expression, purification and characterization of isoforms of peripheral stalk subunits of human V-ATPase. Protein Expr Purif 78:181–188
23. Davanloo P, Rosenberg AH, Dunn JJ et al (1984) Cloning and expression of the gene for bacteriophage T7 RNA polymerase. Proc Natl Acad Sci USA 81:2035–2039

Chapter 11

Cell-Free Protein Synthesis for Functional and Structural Studies

Shin-ichi Makino, Emily T. Beebe, John L. Markley, and Brian G. Fox

Abstract

Recent advances in cell-free protein expression systems have made them reliable and practical for functional and structural studies of a wide variety of proteins. In particular, wheat germ cell-free translation can consistently produce target proteins in microgram quantities from relatively inexpensive, small-scale reactions. Here we describe our small-scale protein expression method for rapidly producing proteins for functional assay and techniques for determining if the target is suitable for scale-up to amounts potentially needed for structure determination. The cell-free system is versatile and can be easily customized with the inclusion of additives. We describe simple modifications used for producing membrane proteins.

Key words Cell-free translation, Wheat germ extract, Functional assay, Membrane protein, Liposomes, Transcription

1 Introduction

Rapid and efficient production of high-quality protein for functional and structural studies is not trivial. Although methods for recombinant protein production in cost-effective systems such as bacterial or yeast cells have evolved significantly over the decades [1, 2], they are certainly not suitable for all targets. Similarly, protein production from cell-free systems has improved over the same time period, so that high yields of many types of proteins including toxic and membrane proteins are now possible without the need for the extensive growth optimizations often required by cell-based approaches. The cell-free approach is particularly well-suited for the efficient incorporation of labeled amino acids. In addition, stabilizing compounds or posttranslational reagents can be added to the cell-free protein synthesis reaction mixture. Cell-free expression should be considered as a viable salvage pathway for targets difficult to produce from other systems. Among commercially available cell-free protein expression systems, a wheat germ cell-free extract made

Yu Wai Chen (ed.), *Structural Genomics: General Applications*, Methods in Molecular Biology, vol. 1091, DOI 10.1007/978-1-62703-691-7_11,

from carefully selected embryos has outstanding quality for the durability of the translation reaction [3–5]. The wheat germ system has proven capable of expressing the majority of proteins tested [6] and is less prone to the proteolytic degradation often seen in other cell-free systems [7]. Additionally, because wheat germ extract is of eukaryotic derivation, it is expected to be more suitable than prokaryotic systems for the production of eukaryotic proteins.

The techniques we describe here were developed to identify suitable targets for wheat germ cell-free protein production for structure determination by both NMR spectroscopy and X-ray crystallography. Our screening consists of expression, solubility, and purification testing, requiring only one small-scale reaction. Small-scale purification is a good way to assess the probability of success for downstream applications. Importantly, the translation reaction is reproducible and scalable, which allows efficient planning for the next production. According to our experience, greater than 74 % of the proteins we have tested can be synthesized in μg quantities from a single small-scale reaction, a yield that is often sufficient for functional assays. In many cases, the translation reactions may be used directly for some functional assays by comparison with negative controls in lieu of purification [8].

In this chapter, we describe wheat germ protein synthesis protocols using an expression plasmid containing a gene of interest cloned into the cell-free expression vector, pEU [9]. Unlike many commercial cell-free expression systems that include machineries for both RNA and protein synthesis, the wheat-germ system described here uses uncoupled transcription/translation to increase yield. Because the optimal magnesium requirement for SP6 RNA polymerase is different from that for wheat germ translation, overall yields can be increased by separating the two reactions [4]. Although this may at first glance seem cumbersome, it has the advantage of allowing controlled addition of RNA to the translation. We describe a gel electrophoresis method for checking the quality of the mRNA. Because the cell-free system has no membrane- or cell wall-delineated compartments, there are no barriers for introducing components that can improve protein folding or solubilization. Along with standard methods, we suggest custom reaction modifications for different applications including membrane proteins.

2 Materials

2.1 Reagents and Equipment for In Vitro Transcription and Translation

All reagents must be RNase-free. Ultrapure water (Milli-Q water, Millipore, Billerica, MA) is used for preparation of reagents (*see* **Note 1**). All glassware, stainless steel spatulas, and Teflon-coated magnetic stir bars must be baked at 180 °C for at least 3 h to ensure that they are RNase-free prior to use. Disposable plastic tips and tubes that are certified RNase-free may be used without further treatment. Unless exempted in the method, all buffers should be passed through a 0.2 μm membrane filter to avoid microbial contamination. Reagents should be stored at −20 °C unless otherwise stated.

1. Plasmid clone containing a gene in a pEU-series cell-free expression vector [9] (*see* **Note 2**).
2. Plasmid DNA prepared with a commercially available plasmid isolation kit (*see* **Note 3**).
3. Proteinase K buffer (10×): 100 mM Tris–HCl (pH 8.0), 50 mM EDTA (pH 8.0), and 1 % (w/v) SDS.
4. Proteinase K enzyme solution (Sigma-Aldrich, St. Louis, MO). Store at 4 °C.
5. QIAprep spin miniprep kit (QIAGEN, Valencia, CA), which includes plasmid purification column, Buffer PB, and Buffer PE. Store at ambient temperature.
6. Transcription buffer (TB + Mg, 5×): 400 mM HEPES–KOH diluted from 1 M HEPES–KOH (pH 7.8), 100 mM magnesium acetate, 10 mM spermidine trihydrochloride, and 50 mM DTT (*see* **Note 4**).
7. NTPs mixture: 25 mM each of ATP, GTP, CTP, and UTP, pH adjusted to 7 with 2 N KOH using a pH test strip. Store at −80 °C (*see* **Note 4**).
8. SP6 RNA polymerase and RNase inhibitor (RNasin; Promega Corporation, Madison, WI). For long term storage, keep frozen at −80 °C, but once thawed, store at −20 °C.
9. TAE buffer (50×): 242 g Tris base, 57.1 mL acetic acid, 100 mL of 0.5 M EDTA (pH 8.0), and water adjusted to 1 L. Store at ambient temperature.
10. 1 % agarose gel in 1× TAE buffer for electrophoresis, RNase-free grade. Prepare on the day of use.
11. Electrophoresis chamber and gel casting apparatus (*see* **Note 5**).
12. RNA denaturing buffer (2×): 97 % (v/v) formamide (deionized grade), 10 mM EDTA (pH 8.0), and 0.015 % (w/v) bromophenol blue.
13. Ethidium bromide solution, 10 mg/mL. Store at ambient temperature.
14. UV transilluminator.
15. Dialysis buffer (DB, 5×): 120 mM HEPES–KOH, 500 mM potassium acetate, 12.5 mM magnesium acetate, 2 mM spermidine trihydrochloride, 20 mM DTT, 6 mM ATP, 1.25 mM GTP, 80 mM creatine phosphate, 0.025 % (w/v) sodium azide, and pH adjusted to 7.8 with 2 N KOH using a pH electrode. Store at −80 °C.

16. Amino acids (Advanced ChemTech, Louisville, KY).
17. Mixture of 20 amino acids: each 6 mM in water, and pH adjusted to 7 with 2 N KOH and measured using a pH test strip. Do not filter the preparation because some amino acids are not fully dissolved at this concentration (*see* **Note 6**).
18. Creatine kinase (Roche Applied Sciences, Indianapolis IN). Dissolve in water to make 50 mg/mL and store at −80 °C. Dilute to 1 mg/mL on the day of use.
19. Wheat germ cell-free extract (WEPRO2240H, CellFree Sciences, Co., Ltd, Matsuyama, Japan). The extract concentration is adjusted to 240 absorbance units per mL at 260 nm, and most of the free amino acids have been removed by the manufacturer. The suffix H denotes pretreatment with His-tag affinity resin to subtract wheat germ proteins that co-purify during immobilized metal affinity chromatography (IMAC) purification (*see* **Note 7**). Leftover extract should be flash-frozen using liquid nitrogen and stored at −80 °C (*see* **Note 4**).
20. Polypropylene U-bottom 96-well plates (Greiner Bio-One, Monroe NC).
21. Dialysis cups with a molecular weight cut off value of 12,000 Da (Pin-Hwan Trading Co., Ltd., Taipei, Taiwan) and buffer receptacles which fit the dialysis cups (CellFree Sciences).

2.2 Reagents and Equipment for Analysis and Purification

1. Microcentrifuge.
2. Centrifuge with swinging bucket rotor adapted for microplates.
3. 4–20 % Criterion TGX Stain-free precast 26-well gels (Bio-Rad, Hercules CA) and a stain-free imaging system (Bio-Rad), or any other SDS-PAGE gel and staining systems.
4. 3× SDS sample buffer: 150 mM Tris–HCl (pH 6.8), 37.5 mM EDTA (pH 8.0), 6 % (w/v) SDS, 0.01 % (w/v) bromophenol blue, 6 % (v/v) 2-mercaptoethanol, and 30 % (v/v) glycerol.
5. Precision Plus Protein Unstained Standards (Bio-Rad) or any other protein marker for SDS-PAGE.
6. Ni Sepharose high-performance chromatography resin (GE Healthcare, Piscataway NJ).
7. IMAC binding/washing buffer: 50 mM sodium dihydrogen phosphate, 300 mM NaCl, 25 mM imidazole, and pH adjusted to 8.0 with 2 N NaOH. Store at ambient temperature (*see* **Note 8**).
8. IMAC elution buffer: 50 mM sodium dihydrogen phosphate, 300 mM NaCl, 500 mM imidazole, and pH adjusted to 8.0 by HCl. Store at room temperature.
9. 96-Well filter plate (MultiScreen HTS-HV, 0.45-μm pore, Millipore).

3 Methods

3.1 Proteinase K Treatment for Trace RNase Removal, Small-Scale Method

Plasmid DNA preparations occasionally contain RNase activity, which probably comes from RNase A used in most of the commercially available plasmid purification kits. To ensure no carryover of contaminating RNase in the transcription reaction, Proteinase K treatment of the purified plasmid is recommended (*see* **Note 9**).

1. Small-scale Proteinase K reactions are typically performed in 100 μL volumes. Determine the volume of DNA solution to be treated (*see* **Notes 10** and **11**).
2. Prepare a Proteinase K master mix consisting of 10× Proteinase K buffer, Proteinase K, and Milli-Q water, so that final concentrations of the components in the reaction are 1× Proteinase K buffer and 50 μg/mL Proteinase K.
3. Dispense the master mix and add the DNA solution (*see* **Note 12**).
4. Incubate the reaction for 1–2 h at 37 °C.
5. Add 5 volumes (500 μL) QIAGEN buffer PB to the reaction.
6. Add all 600 μL to a QIAGEN spin column. Centrifuge for 1 min at 14,000 ×*g*. Discard flow-through.
7. Wash the column with 750 μL QIAGEN PE buffer, containing ethanol. Centrifuge for 1 min. Discard flow-through.
8. Centrifuge the column again for 1 min to remove residual ethanol. Transfer the column to a new microcentrifuge tube and allow any remaining ethanol to evaporate for 10 min at room temperature (*see* **Note 13**).
9. Add 20 μL of Milli-Q water to the center of the column filter (*see* **Note 14**). Let the column sit for 1 min.
10. Centrifuge the column assembly at 14,000 ×*g* for 2 min to elute the DNA.
11. Quantify the DNA concentration by reading the absorbance at 260 nm (*see* **Note 15**).

3.2 Transcription

1. For 8 μL transcriptions, calculate the required volume of transcription master mix (2×) according to the following formula (*see* **Notes 16** and **17**):

$$Required\ volume = \{Number\ of\ reactions\} \times 4\ \mu L \times 1.2(20\%\ extra)$$

2. Prepare transcription master mix as directed in Table 1 in the order written from top to bottom.
3. Dispense 4 μL of the transcription master mix into microcentrifuge tubes or a 96-well microplate (*see* **Note 18**).
4. Add 4 μL of DNA solution into the dispensed aliquots (*see* **Notes 19** and **20**).

Table 1
Transcription reaction composition

Components	Volume (μL)	Subcomponents for master mix	Volume (μL)
Transcription Master mix	4	Water	0.88
		5× TB + Mg	1.6
		25 mM NTP (pH 7)	1.28
		SP6 RNA polymerase (80 U/μL)	0.16
		RNase inhibitor (80 U/μL)	0.08
DNA	4		
Total	8		

5. Close the cap of the microcentrifuge tube or seal the plate tightly (*see* **Note 21**).
6. Incubate at 37 °C for 4 h (*see* **Note 22**).
7. The RNA can be used for subsequent translation reaction without purification (*see* **Note 23**).

3.3 Agarose Gel Analysis to Check RNA Integrity

This is an optional technique for checking RNA quality. It is especially recommended for researchers new to RNase-free procedures and is also good for trouble-shooting poor translation results (*see* **Note 24**).

1. After transcription is complete, gently mix the reaction with a micropipette, and transfer 1 μL to a microcentrifuge tube containing 9 μL Milli-Q water and 10 μL 2× RNA denaturing buffer.
2. Heat the sample to at least 65 °C for 2 min to denature the RNA (*see* **Note 25**).
3. Immediately quench the sample on ice to prevent formation of RNA secondary structure that might be caused by slow cooling.
4. Load 2 μL of each RNA sample onto the gel. Load a double-stranded DNA marker with non-denaturing loading dye for a staining control (*see* **Note 26**).
5. Run the electrophoresis in 1× TAE buffer (*see* **Note 27**).
6. Stain the gel in a 0.5 μg/mL ethidium bromide solution in Milli-Q water for 15 min on an orbital shaker (*see* **Note 28**).
7. Destain the gel with Milli-Q water for 10–15 min on an orbital shaker (*see* **Note 29**).
8. Image the gel on a transilluminator. Figure 1 shows an example of transcription analysis by gel.

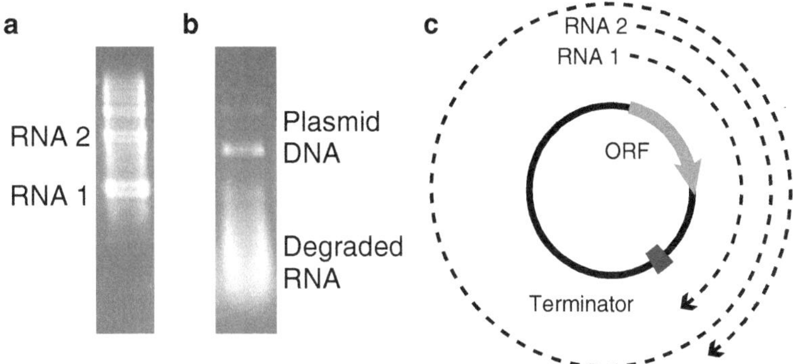

Fig. 1 Agarose gel analysis for transcription. Successful transcription (**a**) appears as a ladder of higher molecular weight bands without low molecular weight species. Degraded RNA (**b**) typically appears as a smear or haze in the low molecular weight range. Fully degraded material or a failed transcription may appear as a single band corresponding to the DNA template (*see* **Note 30**). (**c**) The RNA ladder generated from transcription comes from a length distribution of the transcripts mainly terminating at a specific region on the vector. RNA 1 contains a region between the SP6 promoter and the terminator. RNA 2 and longer transcripts result from one or more rounds of continuous transcription past the terminator. The RNA 1 and 2 correspond to the bands labeled in panel **a**

Table 2
Translation reaction composition

Components	Volume (μL)	Subcomponents for master mix	Volume (μL)
Translation Master mix	20 (22.4)	Water	8.75 (9.8)
		5× DB	2.75 (3.08)
		6 mM Amino acids	1.25 (1.4)
		1 mg/mL Creatine kinase	1 (1.12)
		WEPRO2240H	6.25 (7)
RNA	5 (5.6)		
Total	25 (28)		

Use these volumes if an unpurified transcription reaction will be used as the RNA source (*see* **Note 32**)
The volumes for bilayer reaction setup are shown in parenthesis. In the bilayer method, 25 μL is taken out to inject beneath the feeding buffer

3.4 Translation

Either of two translation methods, bilayer [10] or dialysis, may be selected according to the application. For either method, the reaction has a volume of 25 μL and identical composition (Table 2) (*see* **Note 31**). The bilayer method is suitable for high-throughput experiments, because the reaction can be performed in a 96-well microplate format [10]. The dialysis reaction often results in higher protein yields and is more sensitive to buffer conditions such as pH (Fig. 2), but requires special dialysis cups and receptacles and involves more handling.

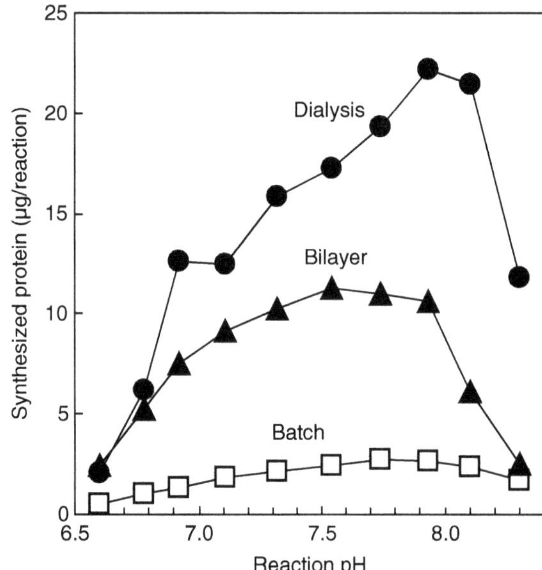

Fig. 2 Comparison of synthesis yields among batch, bilayer, and dialysis methods with changing buffer pH. An N-terminally His-tagged green fluorescent protein (GFP) [11] was synthesized in 25-µL scale translation reactions containing 0.7 mg/mL purified RNA. Synthesized protein was quantified directly by fluorescence (excitation at 488 nm; emission at 509 nm) and calibrated with a purified protein standard. The series of the reaction pH values were obtained by titration of 5× DB preparation with different amounts of KOH

Table 3
Feeding buffer composition

Components	Volume (µL)
Water	600 (93.75)
5× DB (pH 7.8)	160 (25)
6 mM Amino acids, pH 7	40 (6.25)
Total	800 (125)

The volumes for a bilayer reaction are shown in parenthesis

3.4.1 Bilayer Translation

1. Calculate the required volume of feeding buffer according to the following formula:

$$\textit{Required volume} = \{\textit{Number of reactions}\} \times 125\ \mu L \times 1.1 (10\%\ \textit{extra})$$

2. Prepare the feeding buffer as directed in Table 3.
3. Dispense 125 µL of feeding buffer into each well of a U-bottom 96-well microplate.

4. Calculate the required volume of translation master mix according to the following formula:

$$Required\ volume = \{Number\ of\ reactions\} \times 22.4\ \mu L \times 1.2 (20\%\ extra)$$

5. Prepare the translation master mix as described in Table 2, adding components in the order listed.
6. Dispense 22.4 μL of translation master mix into microcentrifuge tubes.
7. Add 5.6 μL of RNA into the dispensed master mix aliquots (*see* **Note 33**).
8. Remove 25 μL of this reaction from the microcentrifuge tube, and slowly inject beneath the feeding buffer dispensed in the 96-well microplate. Take care not to mix the feeding buffer and reaction layers (*see* **Note 34**).
9. Carefully seal the top of the wells using an adhesive film to avoid condensation and/or evaporation.
10. Incubate overnight at ambient temperature (*see* **Note 35**).

3.4.2 Dialysis Translation

1. Insert each dialysis cup into a receptacle, visually inspecting the dialysis membranes for tears or gaps.
2. Equilibrate the membrane for 30 min by adding 500 μL Milli-Q water into the dialysis cups, and 800 μL Milli-Q water into the receptacles (*see* **Notes 36** and **37**).
3. Calculate the required volume of feeding buffer according to the following formula:

$$Required\ volume = \{Number\ of\ reactions\} \times 800\ \mu L \times 1.1 (10\%\ extra)$$

4. Prepare the feeding buffer as directed in Table 3. 15–50 mL conical tubes are useful for this preparation.
5. Remove the water from within the dialysis cups (*see* **Note 38**).
6. Remove water from the receptacles housing the dialysis cups.
7. Immediately dispense 800 μL feeding buffer in each receptacle (*see* **Note 39**).
8. Calculate the required volume of translation master mix according to the following formula:

$$Required\ volume = \{Number\ of\ reactions\} \times 20\ \mu L \times 1.2 (20\%\ extra)$$

9. Prepare the translation master mix as directed in Table 2, adding each component in the order listed.
10. Dispense 20 μL of translation master mix into dialysis cups. Carefully dispense with the pipette tip touching the side wall close to the membrane (*see* **Note 37**).

11. Add 5 µL RNA to dialysis cups, and tightly close the lid (*see* **Note 40**).
12. Seal the top part of the dialysis cup and the receptacles together using Parafilm to avoid evaporation from either compartment.
13. Incubate at room temperature 16–22 h, or overnight (*see* **Notes 35** and **41**).

3.5 Analysis of Expression and Solubility

1. Remove the reactions from the microplate or dialysis cup and transfer to a microcentrifuge tube. Measure final volumes of the reactions using the micropipette (*see* **Notes 42** and **43**).
2. Fractionate supernatant and pellet by centrifuging at 18,000 × *g* for 3 min at ambient temperature.
3. Resuspend the pellet in 150 µL (bilayer) or 25 µL (dialysis) of resuspension buffer by pipetting (*see* **Note 44**).
4. Prepare 18 µL SDS-PAGE samples using 1/25 volume of each fraction, 6 µL 3× SDS sample buffer, and water.
5. Perform SDS-PAGE using 3 µL load volumes. Image the gel after electrophoresis with a stain-free imaging system (Bio-Rad) (*see* **Notes 45** and **46**).

3.6 Affinity Purification

His-tagged proteins can be easily screened for purification efficiency using a portion of a single translation (*see* **Note 47**).

1. Dispense 20 µL of a 50 % suspension of Ni Sepharose resin and 100 µL of IMAC binding/washing buffer into the wells of a 96-well filter plate assembled with a 96-well microplate reservoir.
2. Add 2/3 volume of the supernatant fraction, and shake for 10 min on a microplate shaker.
3. Centrifuge at 2,500 × *g* for 1 min at ambient temperature to pass unbound components through the filter. Discard the flow-through fraction (*see* **Note 48**).
4. Add 150 µL IMAC binding/washing buffer into each well to wash the resin, shake briefly, and centrifuge again. Discard the collected wash.
5. Repeat **step 4** two more times.
6. Replace the 96-well microplate reservoir with a fresh collection plate, and add 50 µL IMAC elution buffer to each well of the filter plate. Shake for 5 min.
7. Centrifuge at 2,500 × *g* for 1 min at ambient temperature to collect the eluate.
8. Prepare SDS-PAGE samples with 12 µL eluate and 6 µL 3× SDS sample buffer, and load 3 µL. Perform SDS-PAGE (*see* **Notes 45** and **46**). Figure 3 shows an optimization of imidazole concentration and an example SDS-PAGE gel image of the purifications.

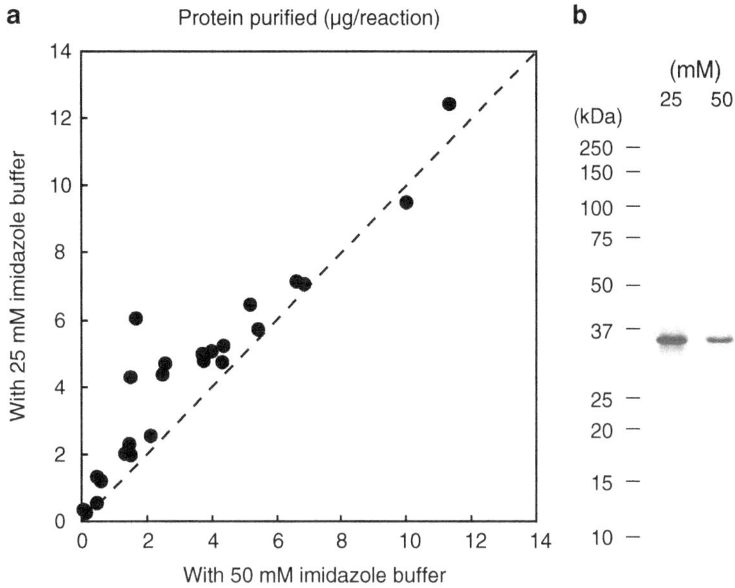

Fig. 3 Purification results using His-tag pretreated extract, WEPRO2240H. (**a**) Purification yield comparison between 25 mM and 50 mM imidazole IMAC binding/washing buffers. 24 proteins were synthesized by the bilayer method and IMAC-purified using 25 mM or 50 mM imidazole-containing buffers. Purified proteins were quantified by band intensity on an SDS-PAGE stain-free gel image accounting for molecular weight and number of tryptophan residues of each protein in comparison with marker bands of known concentrations and number of tryptophan residues. Some of the tested proteins were purified in significantly greater amounts with 25 mM imidazole buffer, which is probably due to a lower affinity for the resin. This result indicates lowering imidazole concentration can expand the target selection for downstream applications. The diagonal dashed line correlates with proteins that purified equally in either buffer. (**b**) A representative SDS-PAGE gel image of purified samples. This protein purified as 6.0 and 1.7 μg per small-scale translation reaction using 25 and 50 mM imidazole IMAC binding/washing buffers, respectively

3.7 Expression of Membrane Proteins

Cell-free translation can be easily modified in several ways to facilitate membrane protein synthesis. Although membrane proteins are partially or fully insoluble after cell-free translation and thus may not be purified directly from the soluble fraction, this method is still very useful in the production of active membrane proteins for functional analysis. The cell-free reaction does not contain intact cell membranes or organelles, but can be supplemented with liposomes to provide an artificial lipid bilayer micro environment, which can encourage protein folding [9]. The resulting proteoliposomes can be isolated by flotation on a discontinuous density gradient after ultracentrifugation [9], or simply pelleted briefly in a microfuge and washed with assay buffer. Pellet isolation also serves to partially purify the membrane protein, because almost all of the proteins from the wheat germ extract have been found to be soluble [4]. If a membrane protein needs to be solubilized and purified, translation reactions are also compatible with some classes of detergents [12], which can solubilize membrane proteins during translation. Once proteins are solubilized, they can usually be purified on affinity media in the same manner as soluble proteins.

3.7.1 Translation with Liposomes or Detergent

Liposomes or detergent can be added to the translation reaction by reducing the corresponding volume of water. A standard 25 μL liposome reaction contains 2 μL of 15 mg/mL liposomes (*see* **Notes 49** and **50**). For detergent-mediated translation, some optimization may be required, as translation may not be compatible with certain types or concentrations of detergent. Because detergents can compromise membrane protein activity, performing a functional assay on protein translated under different detergent conditions is very useful, if practical.

3.7.2 Proteoliposome Purification by Pelleting

1. Centrifuge a membrane protein translation at 18,000 × *g* for 3 min at ambient temperature to produce a visible pellet, containing insoluble protein and lipids (*see* **Note 51**).
2. Wash the pellet and resuspend in assay buffer at the desired concentration (*see* **Note 52**).

3.7.3 Detergent Solubilization Screening

1. After membrane protein translation with or without liposomes, make small aliquots of the translation reaction and centrifuge to obtain the pellets.
2. Suspend the pellet in various buffers with or without detergents (*see* **Note 53**).
3. Centrifuge to separate supernatant and pellet.
4. Check solubility by SDS-PAGE, and functionality if possible.

4 Notes

1. Diethylpyrocarbonate (DEPC)-treated water should not be used, because residual DEPC might negatively affect the reaction.
2. Our standard expression vector, pEU-His-FV, which contains the SP6 RNA polymerase promoter and the tobacco mosaic virus omega translational enhancer is engineered to contain a 6× His N-terminal tag and is compatible with the Flexi-cloning system (Promega). Our expression vector and various plasmid clones are publicly available through the PSI Materials Repository (http://psimr.asu.edu/).
3. PCR-generated fragments can also be used as a transcription template [13]. However, we have seen about 20 % lower protein synthesis yields from PCR DNA templates compared to

plasmid templates in a dialysis mode reaction, and recommend using plasmids for best results.

4. It is advisable to make small aliquots of these reagents to avoid a reduction in synthesis efficiency caused by repeated freezing and thawing.
5. Maintain dedicated electrophoresis equipment for RNA work to reduce the risk of RNase contamination from DNA handling methods.
6. Stable isotope labeled amino acids (^{15}N, ^{13}C, and/or ^{2}H) are available through Cambridge Isotope Laboratories, Andover, MA. If these mixtures are used, prepare a 1 % (w/v) solution and use as 0.1 % (w/v) in the final reaction and buffer.
7. If IMAC purification is not performed, WEPRO2240 should be used. If amino acid labeling is not performed, the WEPRO1240 series extracts, which contain amino acids, can be used. Another type of extract with suffix G is also available, which is designed for expression and purification of proteins with glutathione-*S*-transferase (GST) tags.
8. This imidazole concentration (25 mM) works well only in combination with the His-tag pretreated extract (H-series). If untreated extract is used, increasing the imidazole concentration to 50 mM is required to minimize co-purification of intrinsic wheat germ proteins; however, this condition still permits two wheat germ proteins to co-purify with the target protein. The wheat germ proteins appear as a doublet of around 50 kDa on SDS-PAGE [14]. These proteins are invisible if the stain-free imaging system is used, probably due to a lack of tryptophan residues in the proteins.
9. This step can be skipped if the DNA preparation is consistently pure enough not to show any RNase activity. We employ this step in our standard protocol, because it ensures reliability of the expression screening outcome and minimizes the need to repeat experiments.
10. At least 8 μg DNA should be included for each reaction, because the targeted final concentration is 0.4 mg/mL in 20 μL elution per column. Typical miniprep columns are capable of binding 20 μg DNA. Because there could be substantial loss during purification in a small elution volume, treat the entire DNA miniprep.
11. Proteinase K treatment of maxiprep DNA is usually done by using the same concentration of enzyme and buffer but employs phenol/chloroform extraction and ethanol precipitation as the purification method [14].
12. Transferring the DNA to a new reaction tube is a good way to avoid possible RNase contamination carried over from the wall of the tube used for the preceding DNA purification.

13. This step reduces contamination of the DNA by residual ethanol, which can inhibit transcription.
14. Delivering the water to the filter is important for keeping elution volumes consistent. Avoid trapping water on the plastic ridge surrounding the filter.
15. Due to the small volume (typically around 12 μL), use an instrument that can determine the absorbance from 1 or 2 μL of the eluate, such as a NanoDrop spectrophotometer (Thermo Fisher Scientific, Waltham, MA) or a UV microplate reader.
16. Inclusion of negative and positive control samples is recommended. Typically we use water instead of DNA for a negative control and an expression plasmid encoding green fluorescent protein (GFP) for a positive control. GFP will develop a visible green color during translation, providing an immediate indicator for the success of the translation reaction.
17. The reaction volume, 8 μL, includes 1 μL for electrophoresis and 5 μL (for dialysis) or 5.6 μL (for bilayer) for the translation reaction. If no quality check will be performed, the reaction volume can be minimized down to 5 μL for a dialysis translation, because only 5 μL of the transcription reaction will go into the translation reaction. The slightly larger transcription volume for the bilayer translation ensures sufficient volume to prevent pipetting errors that can prematurely mix the two layers, *see* Subheading 3.4.1 and **Notes 33** and **34**.
18. Either a PCR-plate or U-bottom plate can be used, but make sure it is tightly sealable with a lid or adhesive film.
19. Transcriptions may also be performed by using equal concentrations of DNA. In this case, the DNA and water volumes are adjusted so that the final DNA concentration is 0.2 μg/μL.
20. Mix well by pipetting owing to the small volume.
21. It is critical that transcription plates are well-sealed, or the reaction may evaporate during the course of the incubation. Microcentrifuge tubes are recommended unless multichannel pipettes are used.
22. The reaction solution should start to appear cloudy due to the formation of insoluble magnesium pyrophosphate, a by-product of transcription. For small-volume transcriptions this precipitate may be hard to see, but reference to the clear negative control can be useful for determining whether the transcriptions are progressing properly. It is sometimes helpful to look upwards through the plate or tube at a light source to see the precipitate. Although it is a good indicator for the progress of transcription, the appearance of precipitate does not indicate the absence of RNase contamination.

23. Optionally for certain applications, the RNA can be purified and stored at −80 °C. We use the RNeasy kit (QIAGEN) or PureLink RNA kit (Ambion) for purification (*see* **Note 32** for a modified composition of the translation reaction for use with purified RNA).
24. This protein expression system is highly dependent on the quality of the transcript. Checking RNA by electrophoresis will provide important information regarding RNA quality.
25. Temperatures ranging from 65 to 95 °C are acceptable.
26. The DNA size marker can be used for approximate RNA size evaluation. On a 1 % agarose gel, double-stranded DNA at the 1.5 and 3 kbp positions roughly corresponds to denatured RNA at 2.7 and 6.0 kb, respectively.
27. Voltage will differ depending on the apparatus, but for reference we typically use 100 V at constant voltage for about 25 min.
28. A freshly diluted ethidium bromide solution will provide the most consistent staining.
29. This step reduces background staining of the gel. The time can be extended, if the background is still high.
30. One common reason for a failed transcription is improper removal of ethanol from the Proteinase K-treated plasmid.
31. If amino acid labeling is needed, such as for structural studies, use 0.1 % (w/v) [^{15}N] or [^{13}C/^{15}N] uniformly labeled 20 amino acids mixture for NMR spectroscopy, or 0.6 mM selenomethionine and 0.3 mM of the other 19 amino acids for X-ray crystallography samples.
32. The buffer concentration shown in Table 2 is adjusted to compensate for the effects of carryover components from the transcription reaction. If purified RNA is used, use 7.75 μL water and 3.75 μL 5× DB per 25-μL reaction. The extract can be assumed to contain 1× DB.
33. Mix well by gentle pipetting. Avoid bubbling, which can introduce air into the pipette tip and compromise the bilayer.
34. Establishing distinct feeding and reaction layers at the start of translation is critical to improving translation yield over that seen in a simple batch reaction (Fig. 2). Once the reaction is underlaid beneath the feeding layer, avoid moving or jarring the plate. Bubbles introduced at this step can also cause premature mixing of the layers, so be careful to keep the pipette tip free of air. Do not press the tip to the bottom of the well during injection as it can cause the reaction to dispense under pressure; rather hold the pipette tip near the bottom of the well at a comfortable angle when injecting.

35. Incubation temperatures between 15 and 26 °C produce similar yields. If the room temperature is fairly constant and falls within this range, there is no need to put the microplate in an incubator.
36. This step serves to hydrate the membrane and to remove preservatives. When adding the water to the cup, avoid getting drops on the sides of the cup, as they can be hard to remove and could later dilute the translation reaction. Instead, direct the water close to the membrane. Similarly, avoid trapping air under the membrane when inserting the cup back into the water-filled receptacle. Gently tilting the assembly will remove any air bubbles.
37. Leave the lid loosely closed between manipulations to prevent splashing during lid opening and closing.
38. Be careful not to puncture the membrane.
39. Once equilibrated, do not let the dialysis membrane dry out or it may crack. If a large number of dialysis reactions will be set up, they can be handled in smaller batches to avoid membrane drying.
40. This reaction volume will just cover the surface of the dialysis membrane. This low volume to surface area ratio facilitates good exchange of the reaction with the feeding layer and leads to good translation yields. Increasing the volume of the reaction will not result in proportional increases in protein production; rather, too high a reaction volume can lead to insufficient diffusion of inhibitory translation by-products and result in lower yields. If more protein is needed, we recommend setting up additional 25 μL dialysis reactions.
41. Optionally, the dialysis reaction time can be extended by supplementing RNA and replenishing dialysis buffer [15–17]. This can prolong the activity of the extract and result in higher protein yields.
42. Typical volumes after translation reaction are around 150 μL for a bilayer and 21–28 μL for dialysis reactions. Note that the volume of the bilayer reaction includes both reaction and feeding buffer due to gradual mixing of the two layers.
43. Some applications, such as general expression screening, do not require a precise volume measurement. It can be important for some purposes, such as for the determination of solubility as a function of protein concentration.
44. The volumes and types of buffer for resuspending are modifiable according to the application.
45. If Coomassie staining detection will be used, load 9 μL.
46. Heat-denaturing the SDS sample prior to loading may be necessary for some proteins. For example, tightly folded proteins, such as GFP must be heated to allow molecular weight-

proportional mobility and also to eliminate interference from GFP fluorescence during stain-free imaging. On the other hand, heating can cause some proteins such as integral membrane proteins to aggregate and either not enter the gel, or migrate as a smeared or fuzzy band. A good rule of thumb is to avoid heating unless the protein is known to require heat treatment for full denaturation. Empirical testing of each new target is always a good idea.

47. Proteins containing other affinity tags, such as streptavidin binding peptide [18], Strep(II)-tag [19], or the GST tag [20], can be screened by simply substituting different resins and buffers. Purification efficiency can be monitored using SDS-PAGE.
48. The flow-through fraction can be used to check binding efficiency, if needed.
49. Liposome preparation is described in [9].
50. Liposomes are only added to the translation reaction, not the feeding buffer.
51. The pellet also contains insoluble by-products and co-precipitated proteins.
52. SDS-PAGE can be used to quantify the amount of protein in the pellet. This can help in determining resuspension volume to achieve a desired target concentration [17].
53. Depending on the application, whole translations can be used in solubilization screening instead of pellet fractions.

Acknowledgments

This work was supported by NIGMS Protein Structure Initiative (PSI) grants U54 GM074901 to J.L.M, which funds the Center for Eukaryotic Structural Genomics (CESG), U54 GM094584 to B.G.F., which funds the Transmembrane Protein Center (TMPC), and U01 GM094622 to J.L.M., which funds the Mitochondrial Protein Partnership. The authors thank the many other CESG staff members for their contributions.

References

1. Chen R (2012) Bacterial expression systems for recombinant protein production: *E. coli* and beyond. Biotechnol Adv 30:1102–1107
2. Mattanovich D, Branduardi P, Dao L et al (2012) Recombinant protein production in yeasts. Methods Mol Biol 824:329–358
3. Madin K, Sawasaki T, Ogasawara T et al (2000) A highly efficient and robust cell-free protein synthesis system prepared from wheat embryos: plants apparently contain a suicide system directed at ribosomes. Proc Natl Acad Sci USA 97:559–564

4. Takai K, Sawasaki T, Endo Y (2010) Practical cell-free protein synthesis system using purified wheat embryos. Nat Protoc 5:227–238
5. Carlson ED, Gan R, Hodgman CE, Jewett MC (2012) Cell-free protein synthesis: applications come of age. Biotechnol Adv 30:1185–1194
6. Goshima N, Kawamura Y, Fukumoto A, Miura A, Honma R, Satoh R et al (2008) Human protein factory for converting the transcriptome into an *in vitro*-expressed proteome. Nat Methods 5:1011–1017
7. Rui E, Fernandez-Becerra C, Takeo S, Sanz S, Lacerda MV, Tsuboi T, del Portillo HA (2011) *Plasmodium vivax*: comparison of immunogenicity among proteins expressed in the cell-free systems of *Escherichia coli* and wheat germ by suspension array assays. Malar J 10:192
8. Takasuka TE, Walker JA, Bergeman LF, Vander Meulen KA, Makino S, Elsen NL, Fox BG (2013) Cell-free translation of biofuels enzymes. Methods Mol Biol (in press)
9. Goren MA, Nozawa A, Makino S, Wrobel RL, Fox BG (2009) Cell-free translation of integral membrane proteins into unilamelar liposomes. Methods Enzymol 463:647–673
10. Sawasaki T, Hasegawa Y, Tsuchimochi M, Kamura N, Ogasawara T, Kuroita T, Endo Y (2002) A bilayer cell-free protein synthesis system for high-throughput screening of gene products. FEBS Lett 514:102–105
11. Frederick RO, Bergeman L, Blommel PG, Bailey LJ, McCoy JG, Song J et al (2007) Small-scale, semi-automated purification of eukaryotic proteins for structure determination. J Struct Funct Genomics 8:153–166
12. Chae PS, Rasmussen SGF, Rana RR, Gotfryd K, Chandra R, Goren MA et al (2010) Maltose-neopentyl glycol (MNG) amphiphiles for solubilization, stabilization and crystallization of membrane proteins. Nat Methods 7:1003–1011
13. Sawasaki T, Ogasawara T, Morishita R, Endo Y (2002) A cell-free protein synthesis system for high-throughput proteomics. Proc Natl Acad Sci USA 99:14652–14657
14. Makino S, Goren MA, Fox BG, Markley JL (2010) Cell-free protein synthesis technology in NMR high-throughput structure determination. Methods Mol Biol 607:127–147
15. Makino S, Sawasaki T, Tozawa Y, Endo Y, Takai K (2006) Covalent circularization of exogenous RNA during incubation with a wheat embryo cell extract. Biochem Biophys Res Commun 347:1080–1087
16. Makino S, Sawasaki T, Endo Y, Takai K (2010) *In vitro* dissection revealed that the kinase domain of wheat RNA ligase is physically isolatable from the flanking domains as a non-overlapping domain enzyme. Biochem Biophys Res Commun 397:762–766
17. Jarecki BW, Makino S, Beebe ET, Fox BG, Chanda B (2013) Function of Shaker potassium channels produced by cell-free translation upon injection into *Xenopus* oocytes. Sci Rep 3:1040
18. Keefe AD, Wilson DS, Seelig B, Szostak JW (2001) One-step purification of recombinant proteins using a nanomolar-affinity streptavidin-binding peptide, the SBP-Tag. Protein Expr Purif 23:440–446
19. Voss S, Skerra A (1997) Mutagenesis of a flexible loop in streptavidin leads to higher affinity for the *Strep*-tag II peptide and improved performance in recombinant protein purification. Protein Eng 10:975–982
20. Smith DB, Johnson KS (1988) Single-step purification of polypeptides expressed in *Escherichia coli* as fusions with glutathione *S*-transferase. Gene 67:31–40

Chapter 12

Insoluble Protein Purification with Sarkosyl: Facts and Precautions

Ben Chisnall, Courtney Johnson, Yavuz Kulaberoglu, and Yu Wai Chen

Abstract

When eukaryotic proteins are overexpressed in *Escherichia coli* hosts, they often form inclusion bodies. Natively folded proteins can be extracted from inclusion bodies using mild detergents such as sarkosyl. One common problem is the sequestration of nucleic acid contaminants with the protein of interest. Here we describe methods for monitoring the presence of co-precipitated nucleic acids, and their removal. These procedures are simple to implement and can be easily adapted to a high-throughput format. While sarkosyl is a common chemical, some information such as its UV absorption spectrum and micellar size are absent in the literature or poorly referenced. We review and summarize the properties that are the most relevant to structural biology.

Key words Sarkosyl, Inclusion body, Nucleic acid contamination, Detergent, Insoluble protein, Critical micelle concentration, CMC, Aggregation number, UV absorption spectra, Gel electrophoresis, NanoDrop

1 Introduction

In structural biology projects, one early hurdle is often the protein of interest being insoluble. This problem is common when eukaryotic recombinant proteins are overproduced in bacterial hosts. In many cases, the insolubility problem cannot be overcome by modulating the growth and induction condition of the culture. It has been reported that some proteins, despite depositing as inclusion bodies, retain their native structures and functions. One promising approach is therefore to recover these protein samples from the solid phase employing a mild detergent. However, protein samples extracted from inclusion bodies usually contain nucleic-acid contaminants [1]. Here we describe an easy procedure for monitoring these contaminations and we discuss how these can be removed or avoided.

Yu Wai Chen (ed.), *Structural Genomics: General Applications*, Methods in Molecular Biology, vol. 1091, DOI 10.1007/978-1-62703-691-7_12,

The detergent of choice, sarkosyl (*N*-lauroylsarcosine), has many desirable properties which contribute to its wide usage in protein purification [2, 3]: it is non-denaturing, it forms micelles that are small in size, it does not interfere with spectroscopic concentration measurement, and it is of low cost. In the second half of this chapter, we examine its ultraviolet (UV) absorption spectroscopic properties, the reported values of its critical micelle concentration (CMC) and its average micellar size as these are the most important facts concerning its use in protein purification.

2 Materials

2.1 Agarose Gel Electrophoresis for Proteins

This procedure is executed in exactly the same way as DNA gel electrophoresis except that here protein samples are loaded.

1. Agarose (electrophoresis grade, e.g., from Invitrogen).
2. 1× TAE buffer (40 mM Tris–HCl pH 8.0, 20 mM acetic acid, 1 mM EDTA), 500 ml.
3. Ethidium bromide solution (1 mg/ml).
4. Sample loading buffer (6×) for DNA gel electrophoresis (e.g., TrackIt™ Cyan/Yellow Loading Buffer, Life Technologies).
5. 1 kb DNA ladder.
6. Horizontal gel electrophoresis apparatus (e.g., Bio-Rad Sub-Cell GT System), with gel casting set and combs.
7. UV transilluminator or imaging system (e.g., Bio-Rad Gel Doc XR+).
8. UV spectrophotometer (NanoDrop 1000, Thermo Scientific).

2.2 DNase I and RNase A Digestions

1. 1 M $MgCl_2$ solution.
2. 0.1 M $CaCl_2$ solution.
3. DNase I (AppliChem, 2 mg/ml).
4. RNase A (AppliChem, 2 mg/ml).
5. 50 mM EDTA solution.

2.3 Protein Concentration Measurement in the Presence of Sarkosyl

1. Millipore-grade water.
2. Reference buffer with sarkosyl matching the protein sample (e.g., PBS with 0.1 % sarkosyl, or dialysis buffer, or centrifugation filtrate).
3. Reference buffer without sarkosyl (e.g., PBS).

3 Methods

3.1 Agarose Gel Electrophoresis for Protein

1. Make up a 40 ml (for one gel) solution of 1 % (w/v) agarose, using 1× TAE. Dissolve the powder completely by heating in a microwave oven.
2. Pour into a Bio-Rad Sub-Cell GT gel casting setup, insert an 8-well comb and leave for 1 h to set.
3. Cover the gel with enough 1× TAE buffer to completely submerge the gel (approximately 400 ml).
4. Mix the protein samples with 6× sample loading buffer (25 μl + 5 μl for each sample) and load the mixture into the wells of the gel. Do the same for the DNA marker.
5. Run the electrophoresis at room temperature for 45 min at constant voltage of 70 V.
6. Remove the gel and put it in a tray filled with deionized water. Add 2.5 μl of ethidium bromide solution and place on a shaker to stain for 20 min.
7. Destain in water for 30 min. Examine the gel under a UV transilluminator or imaging system.

3.2 DNase I and RNase A double Digestion

The protein samples with nucleic acid contaminants are subjected to enzyme digestion as follows (in a 50 μl solution):

1. Prepare the protein sample, add $MgCl_2$ to 20 mM and $CaCl_2$ to 1 mM.
2. Add 1 μl of DNase I (2 mg/ml) and 1 μl of RNase A (2 mg/ml).
3. Incubate at room temperature for 1 h.
4. Run a sample on a 1 % agarose gel to check the results (Subheading 3.1).

3.3 Protein Concentration Measurement in the Presence of Sarkosyl

1. Turn on 340 nm normalization in the NanoDrop "Protein A280" measurement mode.
2. Measure spectrum of protein sample using reference buffer with sarkosyl as blank.
3. Measure spectrum of protein sample using reference buffer without sarkosyl as blank. This spectrum is reported.
4. Measure spectrum of reference buffer with sarkosyl using buffer without sarkosyl as blank to check the background absorption of sarkosyl. This can be used to estimate the concentration of sarkosyl in the buffer.

4 Discussion

4.1 Monitoring Nucleic Acid Contaminants in Protein Samples

A single band in sodium dodecyl sulfate–polyacrylamide gel electrophoresis (SDS-PAGE), stained with Coomassie blue (Fig. 2, right panel "S" lane), is no guarantee of a protein purified to homogeneity when using sarkosyl solubilization. When such a sample was examined using a UV spectrophotometer, such as the NanoDrop 1000 which scans from 220–350 nm, the spectrum typically lacks a peak at the characteristic 280 nm for proteins. Instead, the peak shifts to somewhat below 280 nm due to the contribution of the nucleic acids which absorb maximally at 260 nm (Fig. 1, dark and light grey spectra) or even worse, there may be no peak in this region but just a shoulder. One needs to perform an agarose gel electrophoresis, which shows that the sample is contaminated with both DNA and RNA (Fig. 2, left panel "S" lane).

4.2 Reducing/ Removal of Nucleic Acid Contaminants

Typical results of DNase I and RNase A single and double digestions are shown in Fig. 2 left panel, confirming the identities of the major contaminants. The nucleic acids are digested into smaller pieces but they still need to be removed by a further step of, for example, size exclusion chromatography, which is outside the scope of this chapter. However, it is rather straightforward to incorporate this double digestion step during cell lysis and before

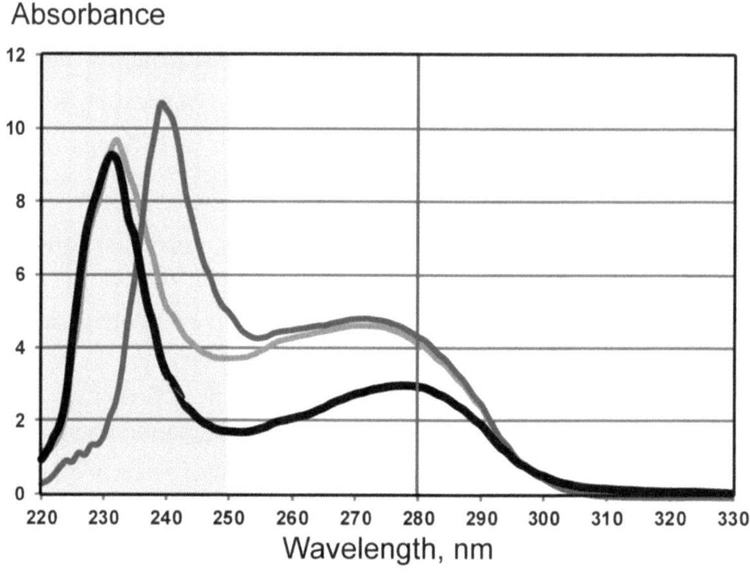

Fig. 1 UV spectra of protein samples purified from inclusion bodies using sarkosyl. *Dark and light grey*: samples solubilized in 10 % and 1 % sarkosyl buffers, respectively. *Black*: sample digested with DNase I and RNase A before solubilization in 1 % sarkosyl. These spectra were recorded against a reference buffer without sarkosyl (*see* Subheading 3.3). The shaded area is where the absorption due to sarkosyl dominates

centrifugation. Following this digestion, the sample extracted from the inclusion bodies has much reduced contamination and shows a good protein peak at 280 nm (Fig. 1, black spectrum).

5 Properties of Sarkosyl

In this section, we reviewed and summarized some basic knowledge of sarkosyl which are important factors to consider in protein purification. We found that while its data are reported widely, some references may not be reliable.

5.1 Chemical Properties

The chemical structure of sarkosyl (synonyms: *N*-methyl-*N*-(1-oxododecyl)-glycine, sodium salt; sodium *N*-lauroyl sarcosinate) is shown in Fig. 3. The free acid (CAS number 97-78-9) has a chemical formula of $C_{15}H_{29}NO_3$ and a molecular mass of 271.4 g; whereas its sodium salt (CAS number 137-16-6) has a chemical formula of $C_{15}H_{28}NO_3Na$ and a molecular mass of 293.4 g. The solubility of sarkosyl in water is reported by several vendors to be ≥100 mg/ml (≥10 % w/v; 0.34 M). In our hands and other protein purification works, it has been used at up to 10 % [2].

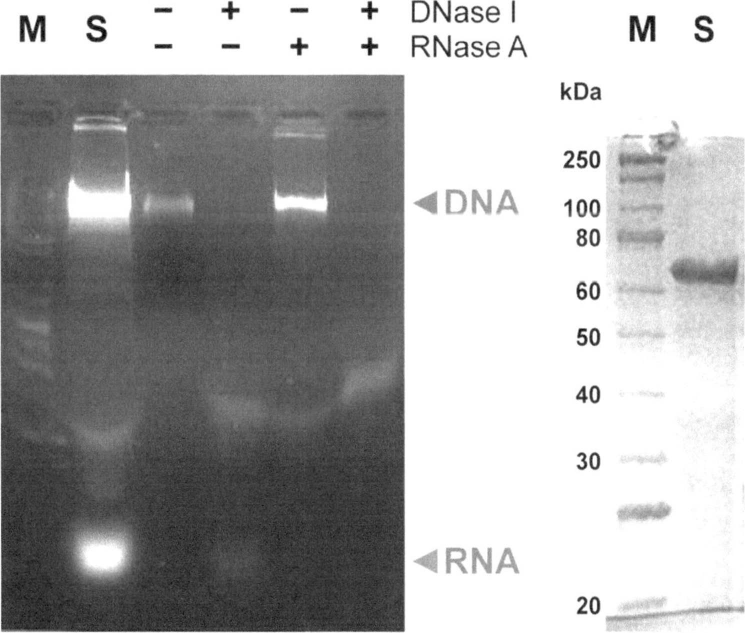

Fig. 2 Agarose gel electrophoresis and SDS-PAGE of protein samples. A protein sample solubilized in sarkosyl was run in 1 % agarose gel and stained with ethidium bromide (*left*). M: 1kB DNA ladder; S: sample extracted with 1 % sarkosyl; the rest are samples digested with the respective enzymes. The DNA and RNA contaminants are indicated. The same sample (S) was analyzed in 12 % SDS-PAGE (*right*) and stained with Coomassie blue. M: protein ladder (10–250 kDa, New England Biolabs)

5.2 Spectroscopic Properties

Sarkosyl has no chiral center so it does not show circular dichroism.

It would be most useful to know the absorption characteristics of sarkosyl in the UV range which are used for monitoring protein concentration. We made a concentration series of sarkosyl solutions with water and measured their spectra (Fig. 4a) on a NanoDrop 1000 spectrophotometer which is commonly used in many protein labs. The most important is that sarkosyl absorbs minimally at >260 nm (*see* **Note 1**). Even at 10 %, its presence will not interfere significantly with the measurement of protein and nucleic acid concentrations. However, it should be noted that with >1 % sarkosyl, the region 220–260 nm are not interpretable (*see* **Note 2**). The best practice would be to measure the sample against a reference buffer without sarkosyl (*see* Subheading 3.3; Fig. 4b; **Note 3**).

5.3 Micellar Properties

The knowledge concerning sarkosyl micelle formation becomes very important when it is employed in preparing samples for biophysical studies that are sensitive to the overall shape and size of the solution species, e.g., in solution scattering. There are several independent determinations of the CMC of sarkosyl in the literature [4–6] (*see* **Note 4**) and from vendors (Anatrace, AppliChem) using different methods, resulting in values ranging from 9.5 to 15 mM (0.28–0.44 %) at 20/25 °C. Sigma-Aldrich product information reported the micellar average molecular weight to be 600 (aggregation number is 2) but without a reference to the original literature. According to this information, sarkosyl could be easily removed by simple dialysis or filtration using devices with a pore size (MWCO) of 3,000, even when its concentration is >CMC. However, one should be aware that sarkosyl may bind the protein tightly and form mixed micelles which cannot be removed by dialysis (*see* **Note 5**).

6 Notes

1. Sigma-Aldrich product information lists that sarkosyl has absorption maxima (λ_{max}) at 220 and 265 nm, and a molar extinction coefficient (ε) at 280 nm of 3 M^{-1} cm^{-1} (catalogue number L5777). The reference cited was "Data for biochemical research" by Dawson et al. [7], a popular data book from which the origin of these data could not be traced. We

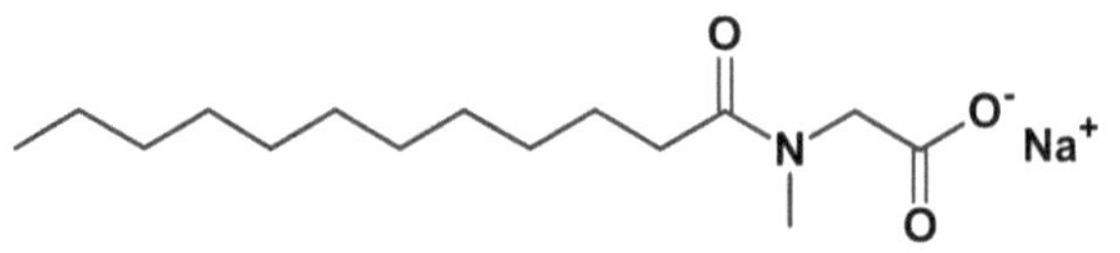

Fig. 3 Chemical structure of sarkosyl

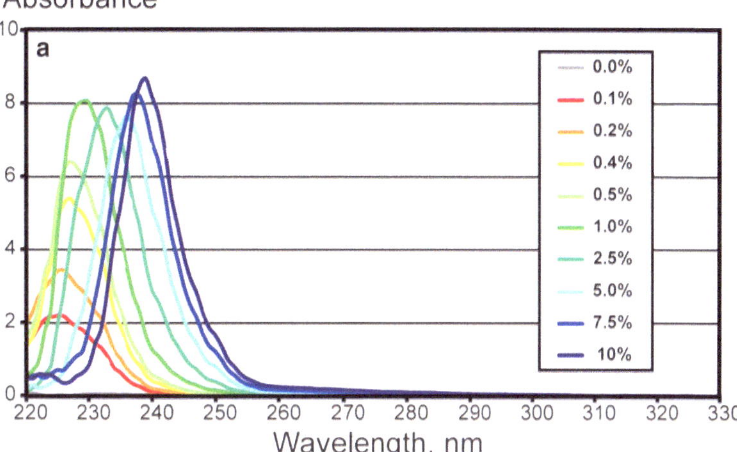

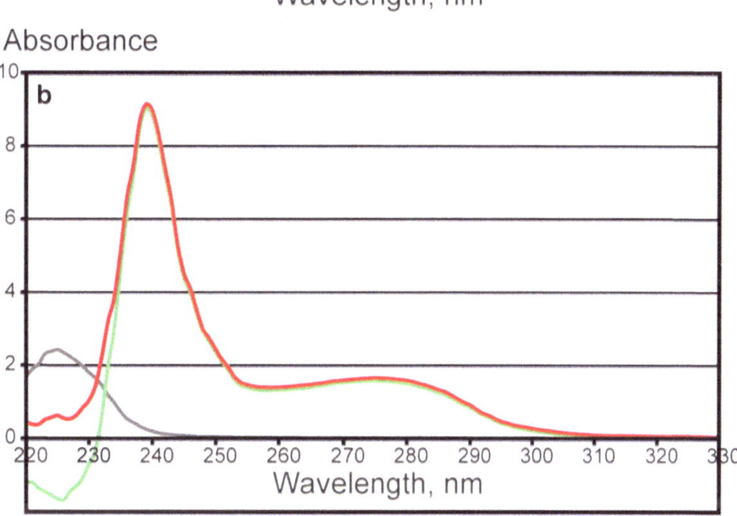

Fig. 4 The UV spectra of sarkosyl as a function of concentration and that of a typical protein sample with sarkosyl. (**a**) Spectra of sarkosyl solutions in water. (**b**) *Red*: protein sample in 0.1 % sarkosyl measured against buffer without sarkosyl; *Green*: sample against buffer with 0.1 % sarkosyl; *Grey*: buffer with 0.1 % sarkosyl against buffer without sarkosyl (Color figure online)

searched in the literature for the UV spectrum of sarkosyl without success. Our results agree with a low value of molar extinction coefficient (ε) at 280 nm but contradict with a reported λ_{max} of 265 nm.

2. The shape of these NanoDrop spectra are misleading because they show a shift of the peak towards longer wavelengths, as the concentration increases beyond 1 %. This is an artifact of the instrument when the optics is overloaded with signal, as a result reducing the effective usable wavelength to around 260 nm.

3. Because of the apparent shift of the absorption peak, a sample measured against a reference buffer, both with sarkosyl, will show a spectrum having negative values at the lower wavelengths in the 220–240 nm region (Fig. 4b, green spectrum).
4. It should be noted that ref. 6, the most cited reference, is inaccessible. The journal article was not available online, nor can we obtain the contact details of the two authors.
5. Shown in Fig. 4b is a sample first solubilized in 10 % sarkosyl, subsequently diluted to 0.1 % sarkosyl, and concentrated. The grey buffer (filtrate during concentration) spectrum is consistent with the 0.1 % sarkosyl spectrum in Fig. 4a. However, the protein sample (red and green spectra) still has a prominent peak at 240 nm. When compared with Fig. 4a, this suggests that the protein sample still contains a large amount of (5–10 %) sarkosyl. This calls for further purification steps to be performed.

Acknowledgements

B.C. was supported by a King's College London (2012) summer vacation studentship. Mahima Kalra and Florence Thomas also took part in implementing sarkosyl solubilization methods.

References

1. McNally E, Sohn R, Frankel S et al (1991) Expression of myosin and actin in *Escherichia coli*. Methods Enzymol 196:368–389
2. Tao H, Liu W, Simmons BN et al (2010) Purifying natively folded proteins from inclusion bodies using sarkosyl, Triton X-100, and CHAPS. Biotechniques 48:61–64
3. Frankel S, Sohn R, Leinwand L (1991) The use of sarkosyl in generating soluble protein after bacterial expression. Proc Natl Acad Sci USA 88:1192–1196
4. Bordes R, Tropsch J, Holmberg K (2010) Role of an amide bond for self-assembly of surfactants. Langmuir 26:3077–3083
5. Gad EAM, ElSukkary MMA, Ismail DA (1997) Surface and thermodynamic parameters of sodium N-acyl sarcosinate surfactant solutions. J Am Oil Chem Soc 74:43–47
6. Venkataraman NI, Subrahmanyam VVR (1985) Effect of structure on surfactance of sodium salts of N-acylamino acids in aqueous solutions. J Ind Chem Soc 62:507–512
7. Dawson RMC (1986) Data for biochemical research, 3rd edn. Clarendon, Oxford, xii, p 580

Part II

Experimental Structure Determination and Characterization

Chapter 13

Estimation of Crystallization Likelihood Through a Fluorimetric Thermal Stability Assay

Vincent Mariaule, Florine Dupeux, and José A. Márquez

Abstract

Construct design and sample formulation are critical in structural biology projects. Large numbers of sample variants are often produced and analyzed for a single target and significant effort is dedicated to sample characterization in order to identify at an early stage the most promising samples to help save manpower and time. Here, we present a method based on a thermal stability assay that can help estimate the likelihood of biological samples to produce crystals. This assay is rapid, inexpensive and consumes very small amounts of sample. The results can be used to prioritize certain constructs at an early stage or as an objective test to help decide when to undertake other type of approaches addressed at improving sample properties.

Key words Thermofluor, Differential scanning fluorimetry, Thermal denaturation, Crystallization likelihood

1 Introduction

Biological samples that are stable, monodisperse and that lack unfolded regions show a higher tendency to crystallize [1]. However, it is not always possible to anticipate which samples will have these properties. A standard approach in structural biology projects consists in producing a number of sample variants for each single target originating from multiple constructs or from different species that are assayed for crystallization either sequentially or in parallel [2, 3]. This strategy has often proven successful. However, up-scaling protein production for a number of different samples requires manpower and time. For this reason, a significant amount of effort is often dedicated to identifying at an early stage those constructs or sample variants that are more likely to produce crystals [4]. A number of experimental techniques like gel filtration,

Yu Wai Chen (ed.), *Structural Genomics: General Applications*, Methods in Molecular Biology, vol. 1091, DOI 10.1007/978-1-62703-691-7_13, © Springer Science+Business Media, LLC 2014

mass spectrometry, light scattering, analytical ultracentrifugation, or NMR, among others, can help identify those samples with optimal properties or at least identify those that are unlikely to produce results [1, 4]. However, some of these experimental approaches consume appreciable quantities of sample or require costly equipment.

Fluorimetric thermal stability assays, also called Differential Scanning Fluorimetry (DSF) or Thermofluor assays, have been extensively used to identify ligands or buffer components that promote sample stability and increase crystallization success rate [5–7]. This assay was originally developed for the high throughput screening of small molecules in drug discovery and relies on the use of an environmentally sensitive fluorescent probe [8]. Such probes are weakly florescent in water, but show increased quantum yields in organic solvents or hydrophobic environments [9]. In a typical thermofluor assay, the fluorescent probe is added to the sample, which is then subjected to a stepwise increase in temperature, while the fluorescence signal is being recorded (with a RT-PCR machine). Initially, no or only weak interactions are expected to occur between sample and probe, as the surface of proteins tends to be dominated by hydrophilic side chains. However, as the protein unfolds it will expose its hydrophobic core, to which the fluorophore can bind. Hence, protein denaturation can be monitored through the sharp increase of fluorescence expected to occur as the protein unfolds. The midpoint or the inflection point in this transition is often used as an approximation to the melting temperature of the sample. The authors have used this method to evaluate the melting temperature of 657 unique samples in a reference buffer. These samples were also subjected to extensive crystallization screening. The results of this analysis show a critical value of T_m below which crystallization success rate decreases very rapidly [10]. This assay requires very small amounts of sample, uses standard equipment, and is very rapid. The results can be used to prioritize certain constructs or help make decisions in biological crystallography projects.

2 Materials

Prepare all solutions using ultrapure water (prepared by purifying deionized water to attain a sensitivity of 18 MΩ cm at 25 °C) and analytical grade reagents.

1. Reference Thermal Stability Assay buffer (RTSA buffer): 100 mM HEPES pH7.5, 150 mM NaCl. Add 10 mL of water in a 50 mL graduated cylinder. Weigh 1.2 g HEPES and then 0.44 g NaCl and transfer to the graduated cylinder. Add 35 mL of water and mix. Adjust the pH to 7.5 with HCl. Make up to 50 mL with water.

2. SYPRO® Orange 9.5× mix solution: Thaw SYPRO® Orange 5,000× stock solution (Invitrogen, catalogue No: S6651, 10× 50 μL) at room temperature. Add 3.9 μL of SYPRO® orange 5,000× stock to 2.0121 mL of RTSA buffer. Mix by pipetting and protect from light.
3. qPCR machine (Stratagene, Mx3005P).
4. PCR plate (Thermo scientific, catalogue No: AB-0600).
5. Clear tape for qPCR (Bio-Rad, catalogue No: 223–9444).

3 Methods

3.1 The Thermal Stability Assay

1. Turn on the qPCR machine and the corresponding software 20 min before the run to pre-warm the lamp.
2. Prepare the sample solution: Calculate the protein solution needed for a final assay concentration of 10 μM with the following formula:

$$V_{protein} = \frac{400 \times MW_{protein} \times 10^{-3}}{C_{protein}}$$

$V_{protein}$ per single assay in μL, $MW_{protein}$ in kDa, $C_{protein}$ in mg/mL. Preferably, use a maximum 2 μL of $V_{protein}$ per assay. Increase the concentration of your protein if needed.
As the SYPRO Orange fluorescence signal depends on the hydrophobicity of the medium, adding lipids and detergents to the sample buffer is not recommended. However, in some cases it is possible to obtain results under these conditions.
3. Calculate the volume of RTSA buffer to add in each well of the PCR plate.

$$V_{RTSA\,buffer} = 20 - V_{protein}$$

$V_{RTSA\,buffer}$ in μL.
4. Add first the calculated volume of RTSA in each well then add the protein volume calculated above. Mix gently by pipetting avoiding air bubbles.
5. Add 21 μL of SYPRO® Orange 9.5× mix solution avoiding the formation of air bubbles.
6. Seal the plate with clear tape for PCR.
7. Set parameters for the thermal shift assay:
Excitation/emission filters: 492 nm/ 568 nm.
1 min plateau.

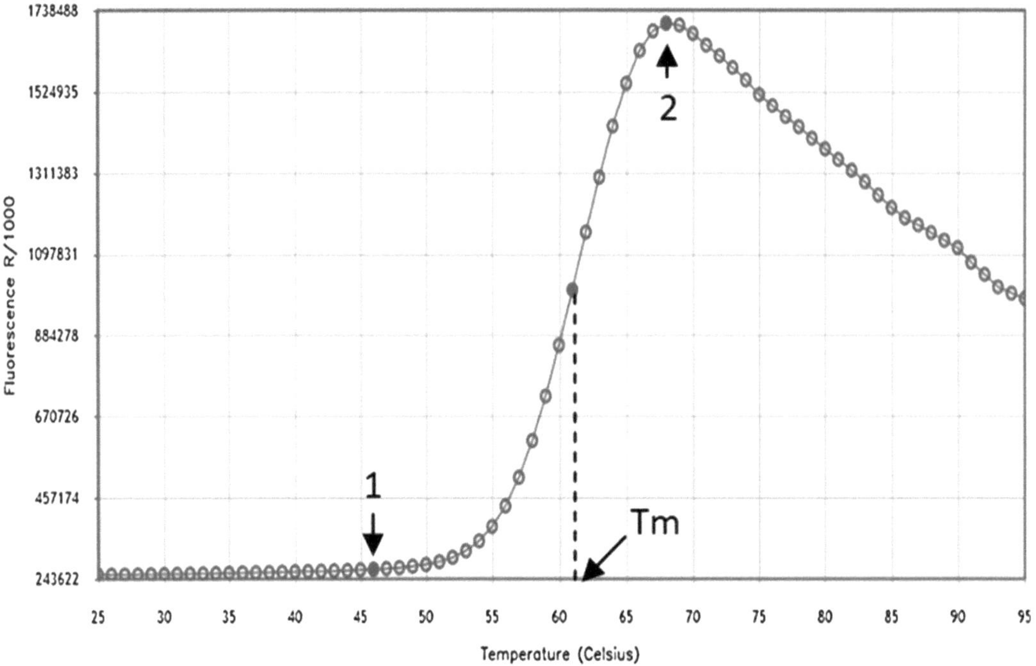

Fig. 1 Thermal stability assay. Optimal denaturation curve

1 °C step increase.

Start temperature: 25 °C.

72 cycles.

8. Export raw data as an Excel file for analysis.

3.2 Data Analysis

1. Import the data file from the qPCR machine into Excel or a similar program. Plot the fluorescence value as a function of the temperature for each well. An optimal denaturation curve (Fig. 1) has an approximately sigmoidal shape with a baseline, and a rapid transition region that represents the denaturation process. However, after this rapid transition the fluorescence signal tends to show a rapid decrease, likely arising from the precipitation of the unfolded protein.

   To calculate the melting temperature, select a point in the base line (Fig. 1, point 1). If the curve does not display a constant baseline use the lowest point before the temperature transition region. Select a second point corresponding to the maximum of fluorescence signal after the transition region (Fig. 1, point 2). The melting temperature (T_m) can be approximated as the temperature corresponding to the midpoint between maximum and baseline signals. To compare results from different assays, it is recommended to use normalized data instead of raw data (*see* **Note 1**).

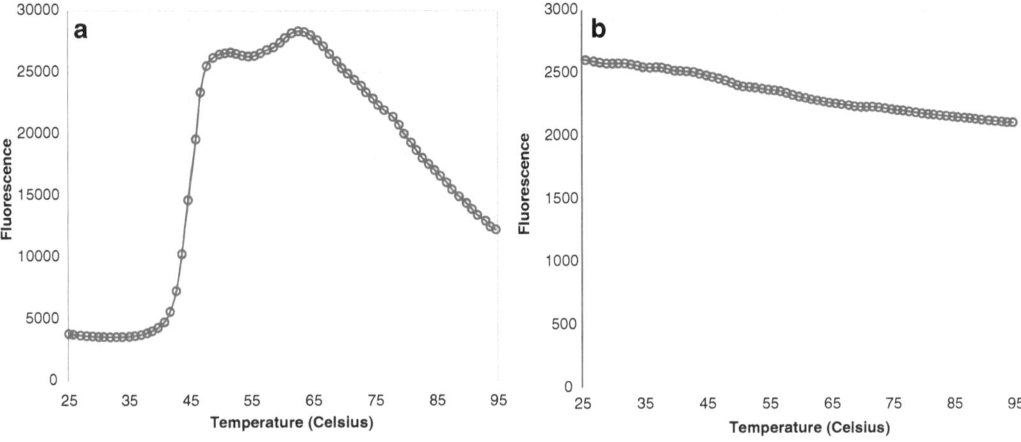

Fig. 2 Multiphasic (**a**) and monotonic (**b**) curves in thermal shift assay

2. Some samples produce a denaturation curve with multiple transitions (Fig. 2a). This may be indicative of sample heterogeneity.
3. Some samples produce a continuous decrease of the fluorescence signal with no clear transition regions (Fig. 2b). No clear T_m can be estimated in this case.

3.3 Interpretation of Results

1. For samples with an optimal denaturation curve a T_m can be easily estimated. Proteins with T_m above 45 °C show a higher tendency to crystallize, while crystallization likelihood decreases rapidly for values of T_m below 45 °C [10]. This criterion can be used to decide which constructs or samples should be prioritized.
2. For samples with a T_m comprised between 29 and 44 °C crystallization can be attempted at 4 °C. In general incubating the crystallization experiments at least 25 °C below the estimated T_m is likely to maximize crystallization likelihood [10].
3. For samples with very low T_m, a screening for buffer conditions or ligands that increase protein stability, as those described in refs. 5–7 can be carried out.
4. Multiphasic denaturation curves may be indicative of sample heterogeneity. The causes can be varied. For example, it may indicate the presence of multiple proteins in the sample that unfold independently. They could also be due to the presence of different domains or both folded and unfolded regions in the same protein. Partial saturation with a ligand may sometimes produce this type of curves reflecting the differences in stability between the bound and unbound forms

of the protein. Although these curves may be indicative of some type of heterogeneity, which could be deleterious for crystallization, they could be very informative and an effort to interpret the results in the context of the particular sample may prove fruitful. In some cases thermofluor assays can be used to study protein–protein or protein–ligand interactions [11, 12].

5. Some samples produce a continuously decreasing fluorescent signal with no clear transition regions (Fig 2b). Often these samples show a high fluorescence signal at low temperature, which may indicate binding of the fluorophore to the folded protein or the presence of hydrophobic components in the sample buffer. However, this type of results is not very informative on the likelihood of these samples to crystallize [10]. Other florescent probes may be tested in these cases (*see* **Note 2**).
6. The presence of detergents in the sample buffer may interfere with the fluorescent signal (*see* **Note 3**).
7. In some cases, natural fluorescent cofactors can be exploited as intrinsic probes for thermal denaturation experiments [13].

4 Notes

1. Data normalization can be obtained by applying the following formula:

$$x_{normalized,T} = \frac{x_T - x_{\min}}{x_{\max} - x_{\min}}$$

$x_{\text{normalized},T}$: normalized fluorescence value at temperature T, x_T: raw fluorescence value at temperature T, $x_{\min}$: minimal raw fluorescence value, $x_{\max}$: maximal raw fluorescence value.

2. Some samples may fail to produce results with SYPRO® Orange. Other environmentally sensitive fluorescent probes can be tested in those cases [5, 9].
3. The presence of detergents in the sample may interfere with the fluorescent signal. Specific thermofluor methods have been developed for membrane proteins [14].

References

1. Zulauf M, D'Arcy A (1992) J Cryst Growth 122:102–106
2. Banci L, Bertini I, Cusack S et al (2006) Acta Crystallogr D 62:1208–1217
3. Graslund S, Nordlund P, Weigelt J et al (2008) Nat Methods 5:135–146
4. Geerlof A, Brown J, Coutard B et al (2006) Acta Crystallogr D 62:1125–1136
5. Niesen FH, Berglund H, Vedadi M (2007) Nat Protoc 2:2212–2221
6. Ericsson UB, Hallberg BM, Detitta GT et al (2006) Anal Biochem 357:289–298
7. Vedadi M, Niesen FH, Allali-Hassani A et al (2006) Proc Natl Acad Sci USA 103: 15835–15840
8. Pantoliano MW, Petrella EC, Kwasnoski JD et al (2001) J Biomol Screen 6:429–440
9. Hawe A, Sutter M, Jiskoot W (2008) Pharm Res 25:1487–1499
10. Dupeux F, Rower M, Seroul G et al (2011) Acta Crystallogr D 67:915–919
11. Matulis D, Kranz JK, Salemme FR et al (2005) Biochemistry 44:5258–5266
12. Heads JT, Adams R, D'Hooghe LE et al (2012) Protein Sci 21:1315–1322
13. Forneris F, Orru R, Bonivento D et al (2009) FEBS J 276:2833–2840
14. Alexandrov AI, Mileni M, Chien EY et al (2008) Structure 16:351–359

Chapter 14

CrystalDirect™: A Novel Approach for Automated Crystal Harvesting Based on Photoablation of Thin Films

José A. Márquez and Florent Cipriani

Abstract

The last years have seen a major development in automation for protein production, crystallization, and X-ray diffraction data collection, which has contributed to accelerate the pace of structure solution and to facilitate the study of ever more challenging targets through macromolecular crystallography. This has led to a considerable increase in the numbers of crystals produced and analyzed. However the process of recovering crystals from crystallization supports and mounting them in X-ray data collection pins remains a manual and delicate operation. Here we present a novel approach enabling full automation of the crystal mounting process and describe the operation of the first-automated CrystalDirect harvesting unit. Implications for crystallography applications and for the future operational integration of automated crystallization and data collection resources are discussed.

Key words Macromolecular crystallography, Crystallization, Automation, High-throughput, Automated crystal harvesting, CrystalDirect

1 Introduction

Over the last decade automated approaches for protein production and crystallization have become common tools in structural biology in general and in macromolecular crystallography in particular [1–5]. This has not only contributed to decrease the time for structure determination, but has allowed to undertake the study of ever more challenging targets, like multi-protein complexes and membrane proteins that often require the production and analysis of large numbers of crystals [2, 3]. In parallel synchrotron beam lines and data processing software have been progressively automated increasing their throughput significantly [2, 6–9]. These developments have also contributed to facilitate the use of crystallography in drug design, by facilitating the production and X-ray analysis of large numbers of crystals with potential ligands [10, 11]. However, automated crystallization and data collection facilities remain separated by the need to harvest crystals out of crystallization supports,

Yu Wai Chen (ed.), *Structural Genomics: General Applications*, Methods in Molecular Biology, vol. 1091, DOI 10.1007/978-1-62703-691-7_14, © Springer Science+Business Media, LLC 2014

typically crystallization plates, and mount them in supports compatible with X-ray data collection at cryogenic temperatures. This introduces a manpower intensive step that contributes to slow down the process of crystal analysis. Moreover, manual crystal recovery and flash cryo-cooling sometimes results in a loss of diffraction power due to mechanical damage, sample loss, or improper cryo-cooling. Automated crystal harvesting is needed to bridge the gap between automated crystallization and data collection units and to improve the reliability and reproducibility of the whole process [12].

Several approaches have been proposed to fill this gap. Rupp and coworkers have developed a semiautomated crystal mounting device capable of recovering crystals through an operator-guided robotic system [12–14]. Another approach involves exposing crystals to X-rays in situ, that is directly in the crystallization support where they grow [15]. Robotic beam line systems able to present either crystallization plates or micro-fluidic devices to X-ray beams at synchrotron beam lines have been developed and used for the rapid discrimination between salt and protein crystals as well as for multi-crystal data collection both at standard and micro focus beam lines [9, 16–18]. However, this approach has two major drawbacks. First, there is significant X-ray scattering background associated to the plastics used in the crystallization plates and micro-fluidic chips, which may complicate the analysis of the data. Second, since the experiments are performed at room temperature, data collection may be rapidly limited by radiation damage.

In this chapter we describe the operation of the first CrystalDirect harvesting unit based on a novel concept designed to enable full automation of the crystal mounting process [19]. In the CrystalDirect approach crystals are grown on the surface of a very thin film in a specially designed vapor diffusion crystallization plate. This film has been selected to be both compatible with crystal growth and with data collection (i.e., produces very low X-ray scattering background). In order to recover the crystals, instead of "fishing" them out of the crystallization drop, the film area containing the crystal is excised and attached to a pin (Fig. 1). Rather than using a mechanical system which could introduce vibrations, the CrystalDirect system uses a laser to excise the film through photoablation. Laser photoablation is commonly used in industry for precise machining of materials. In addition to its high precision it has the advantage of removing matter from the ablated zone without affecting the adjacent areas. The use of this approach in macromolecular crystallography has been recently illustrated [19].

In addition to providing full automation of crystal harvesting the CrystalDirect approach has the advantage of reducing the mechanical stress to the crystals as, contrary to the standard manual process, no tool enters the crystallization experiment during the harvesting operation. Moreover, it works similarly well with either

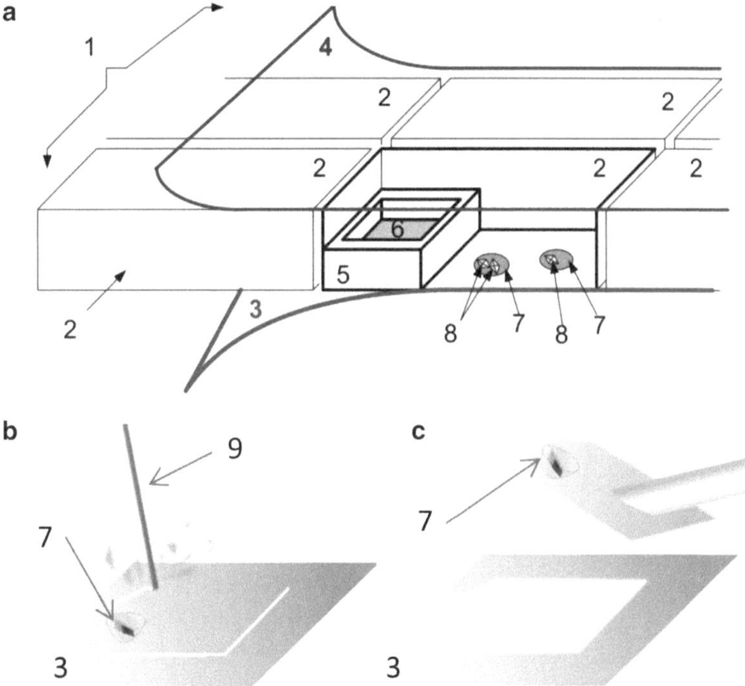

Fig. 1 The CrystalDirect concept. (**a**) Schema of the CrystalDirect 96-well vapor diffusion crystallization plate. Numbers designate the different components: *1*: body of the 96-well plate, *2*: crystallization cells, *3*: CrystalDirect crystallization film, *4*: top sealing film, *5*: solution reservoir, *6*: crystallization solution, *7*: crystallization drops deposited on the CrystalDirect film, *8*: Crystals. (**b**) Film areas containing one or multiple crystals can be excised with the aid of a laser [9]. (**c**) The excised film containing the crystals is attached to a pin compatible with X-ray data collection. Reproduced from Acta Crystallographica [19], a journal of the International Union of Crystallography

large or microcrystals, as the cut area can be adapted to match the size of the crystal. Alternatively, multiple crystals can be recovered on the same support and analyzed by applying serial X-ray data collection strategies. Finally, the laser ablation process is extremely precise and can also be used to ablate parts of crystals without affecting the other parts [19]. This advantage could be used to remove areas of the crystals containing defects before data collection.

The CrystalDirect approach opens the prospect for the full automation of the crystal harvesting process and could contribute to the operational integration of automated crystallization and X-ray data collection facilities. This could contribute significantly to increase their productivity and decrease significantly the delay between crystal identification and data analysis, bringing a new level of efficiency in macromolecular crystallography. Moreover the CrystalDirect approach could contribute significantly to the advancement of very challenging projects in structural biology, like those involving the study of membrane proteins, multi-protein complexes or large ligand screening efforts. To realize its full potential though, the CrystalDirect system will need to be combined with automated protocols for the delivery of cryo-protectants and other chemicals to crystals. Some cryo-protection methods potentially compatible with the CrystalDirect system have been already proposed [14, 20, 21]. These and other methods are currently under investigation.

2 Materials

1. CrystalDirect 96-well vapor diffusion crystallization plates (Fig. 1) conform to the Microplate standard of the Society for Biological Screening (SBS) and are compatible with all of the popular crystallization robots. Each of the cells in a CrystalDirect plate consist of two parts, one formed by a plastic reservoir that holds the crystallization solution and a second part consisting of an "open window" through the plastic frame. This window is closed by applying a very thin film at the bottom of the plate that represents the surface that will hold the crystallization drops (Fig. 1). Once crystallization experiments have been set up, either manually or using standard robotic equipment, a standard film is applied to the top of the plate to seal the crystallization cells (*see* **Note 1**).
2. The CrystalDirect harvesting unit [19] (Fig. 2) consists of a motorized table that holds the crystallization plate. A laser source, a scanner that directs and focuses the beam of the laser

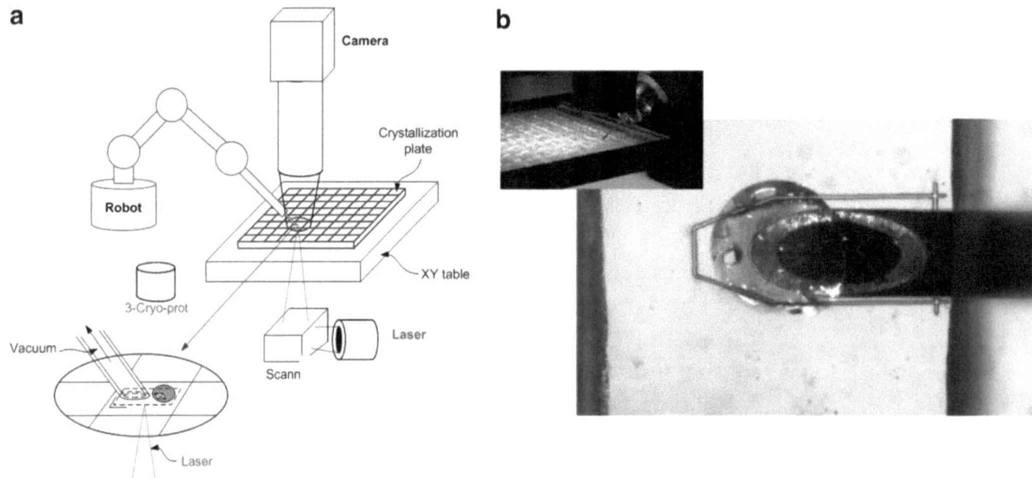

Fig. 2 The CrystalDirect Harvesting unit. (**a**) Schematic view of the components of the harvesting unit. (**b**) Detail of the film excision and pin attachment process as seen through the camera of the harvesting unit. The *inlet* shows a CrystalDirect plate on the plate support of the harvesting unit

to specific areas, a pin handling robot and a cryo-cooling unit. All the process is monitored through a camera and controlled through a computer.

3. The CrystalDirect software allows selecting any position on the crystallization plate and drawing cutting areas around the crystals.
4. CrystalDirect pins have been designed to be compatible with standard synchrotron beam lines and with the CrystalDirect process. They are similar to current crystal SPINE mounting pins [22], except that they are cut in bias at the tip, to provide a better surface of attachment between the pin and the film (Fig. 2). In addition to this, the shaft of the pin is hollow allowing the application of vacuum through it, which ensures proper attachment of the film to the pin during the harvesting process.

3 Methods (*See* Note 7)

3.1 Setting Up Crystallization Experiments in CrystalDirect Plates

1. The CrystalDirect plates are similar to current vapor diffusion plates and can be used to set up crystallization experiments either manually or with the help of robotic equipment and standard protocols. The plates are compatible with all of the popular robotic systems for crystallization, liquid handling, and plate imaging. They have been designed for high-quality imaging with both visible and ultraviolet light (*see* **Note 2**).

3.2 Definition of the Harvesting Area and Automated Crystal Harvesting

1. Once crystals have grown, the plates are transferred to the plate holder of the CrystalDirect harvesting unit. Using the specific software accessible from the control computer the operator can select any position in the crystallization plate and visualize the crystallization drop through the camera of the system (Fig. 2). With the help of the software the operator locates the crystals and selects the cutting area and location of the pin (*see* **Note 3**). Once the harvesting information is validated the process follows in a fully automated way (**steps 2–5**).
2. Once the harvesting information is validated The CrystalDirect system prepares a pin by automatically applying a small amount of glue to its tip. The Pin is then moved close to the harvesting area.
3. The tip of the pin is set in contact with the film while applying vacuum through the shaft of the pin. This is to facilitate attachment between the pin and the film. Vacuum can be released once contact has occurred.
4. The laser is activated and the scanner drives the focused laser beam through the predefined cutting shape to excise the film.
5. The pin handling unit transfers the pin, with the film attached to its tip into a cryo-stream for flash-freezing (*see* **Note 4**). Alternatively the pin can be maintained at room temperature.

3.3 Crystal Recovery

1. Either fresh or frozen samples ready for X-ray analysis can be recovered from the harvesting system (*see* **Note 5**). An automated crystal recovery and storage unit allowing continuous operation with multiple samples is currently under development (*see* **Note 6**).

4 Notes

1. Although initially designed to provide a support for the CrystalDirect harvesting process, the CrystalDirect plates are also very well suited for in situ X-ray diffraction experiments. This is because crystals grow on a very thin film that generates negligible X-ray scattering as compared to standard crystallization plates in which crystals grown on a thick plastic support producing significant X-ray background. The low background of the CrystalDirect plates could be an advantage for the analysis in situ of microcrystals, or weakly diffracting samples.
2. CrystalDirect plate definitions for the most common robotic equipment can be obtained from the HTX lab at the EMBL Grenoble Outstation (htx@embl.fr).
3. Remote operation of the CrystalDirect system is also possible. A web-based system allowing specifying and drawing cutting areas directly on images produced by standard crystallization imaging robots is currently under development.
4. A number of cry-cooling methods compatible with the crystal direct system have been proposed [14, 20, 21]. Methods for fully automated delivery of cryo-protectants and other chemicals to crystals are currently under development.
5. After the harvesting process crystals are mounted in standard pins compatible with most X-ray data collection facilities. The CrystalDirect harvesting unit can be operated off-line, i.e., harvesting and storing multiple frozen crystals in standard pucks that will be later transferred to a beam line and loaded in a standard automated sample changer. Alternatively, the system could be integrated as a beam line component supplying fresh or frozen crystals for on-line analysis with X-rays.
6. Future versions of the CrystalDirect harvesting unit will include a plate hotel to store multiple crystallization plates and a crystal storage system to allow continuous operation harvesting multiple crystals from multiple crystallization plates.
7. A video illustrating the automated CrystalDirect harvesting process is available at https://embl.fr/htxlab/.

Acknowledgments

We are thankful to Dr. Ulrich Zander, Martin Röwer, Gael Seroul, Christoph Landret, Julien Huet, Alexandre Gobbo, Gergely Papp, and Frank Felisaz for their contributions and to the European Molecular Biology Laboratory for support to the CrystalDirect project. We want to thank INSTRUCT and the E.C.-funded P-CUBE project (FP7/2007-2013; grant No 227764) for financial support.

References

1. Edwards A (2009) Annu Rev Biochem 78:541–568
2. Abola E, Kuhn P, Earnest T et al (2000) Nat Struct Biol 7(Suppl):973–977
3. Banci L, Bertini I, Cusack S et al (2006) Acta Crystallogr D 62:1208–1217
4. Graslund S, Nordlund P, Weigelt J et al (2008) Nat Methods 5:135–146
5. Rupp B, Segelke BW, Krupka HI et al (2002) Acta Crystallogr D 58:1514–1518
6. Lamzin VS, Perrakis A (2000) Nat Struct Biol 7(Suppl):978–981
7. Wasserman SR, Koss JW, Sojitra ST et al (2012) Trends Pharmacol Sci 33:261–267
8. Arzt S, Beteva A, Cipriani F et al (2005) Prog Biophys Mol Biol 89:124–152
9. Axford D, Owen RL, Aishima J et al (2012) Acta Crystallogr D 68:592–600
10. Murray CW, Blundell TL (2010) Curr Opin Struct Biol 20:497–507
11. Blundell TL, Jhoti H, Abell C (2002) Nat Rev Drug Discov 1:45–54
12. Viola R, Carman P, Walsh J et al (2007) J Struct Funct Genomics 8:145–152
13. Viola R, Carman P, Walsh J et al (2007) J Appl Crystallogr 40:539–545
14. Viola R, Walsh J, Melka A et al (2011) J Struct Funct Genomics 12:77–82
15. McPherson A (2000) J Appl Crystallogr 33:397–400
16. Jacquamet L, Ohana J, Joly J et al (2004) Structure 12:1219–1225
17. Ng JD, Clark PJ, Stevens RC (2008) Acta Crystallogr D 64:189–197
18. Yadav MK, Gerdts CJ, Sanishvili R et al (2005) J Appl Crystallogr 38:900–905
19. Cipriani F, Rower M, Landret C et al (2012) Acta Crystallogr D 68:1393–1399
20. Pellegrini E, Piano D, Bowler MW (2011) Acta Crystallogr D 67:902–906
21. Kim CU, Kapfer R, Gruner SM (2005) Acta Crystallogr D 61:881–890
22. Cipriani F, Felisaz F, Launer L et al (2006) Acta Crystallogr D 62:1251–1259

Chapter 15

Methods to Refine Macromolecular Structures in Cases of Severe Diffraction Anisotropy

Michael R. Sawaya

Abstract

Diffraction anisotropy is characterized by variation in diffraction quality with reciprocal lattice direction. In the example presented here, diffraction extended to 2.1 Å resolution along a* and c* directions but only to 3.0 Å along the b* direction. Severe anisotropy such as this is often associated with lack of detail in electron density maps, stalled model improvement, and poor refinement statistics. Published methods for overcoming these difficulties have been combined and implemented in the diffraction anisotropy server. Specifically, the server offers information to diagnose the degree of anisotropy, and then applies ellipsoidal resolution boundaries, anisotropic scaling, and B-factor sharpening to the data set to compensate for the deleterious effects of diffraction anisotropy. Here, I offer advice on implementing these methods to facilitate refinement of macromolecular structures in cases of severely anisotropic data.

Key words Diffraction anisotropy, Crystallographic refinement

1 Introduction

Difficulties in refining structures with anisotropic data can be reduced by adopting two strategies. The first strategy is to refrain from discarding valuable reflection measurements in the outer resolution shells. For example, during data collection, the user may notice strong reflection intensities recorded at 2.1 Å along one axis of the detector but relatively weak along the orthogonal axis. The quality of the measurements near this axis in the 2.1 Å resolution shell will appear diminished by their grouping with the remaining reflections enclosed by the same *spherical* shell. It would be self-defeating to eliminate these strong reflections based on these misleading statistics since they are actually well measured and offer a valuable contribution to the detail of the electron density map. The anisotropy server offers tools to examine the appropriate resolution limits along each of the principal axes and then redefine the boundaries of a data set as an *ellipsoid*, congruous with

Yu Wai Chen (ed.), *Structural Genomics: General Applications*, Methods in Molecular Biology, vol. 1091, DOI 10.1007/978-1-62703-691-7_15, © Springer Science+Business Media, LLC 2014

the anisotropic limits of the diffraction signal. In this way, strong reflections are no longer systematically grouped with weak in the highest resolution shells, and statistics at high resolution more accurately reflect data quality.

The second strategy is to sharpen electron density maps for model building. Sharpening compensates for side effects caused by anisotropic scaling, involuntarily employed by refinement programs to eliminate directional dependence of $<|F|>$. Anisotropic scaling diminishes the magnitudes of the high resolution reflections in the strongest diffracting direction, and in so doing, diminishes the level of detail in electron density maps. The server offers options for sharpening the structure factors, restoring the level of detail in the map, and facilitating model-building efforts.

2 Methods

The procedures outlined in Fig. 1 and below are illustrated from my experiences in refining the structure of a PE-PPE complex from *Mycobacterium tuberculosis* [1]. Visually, diffraction extended to 2.1 Å resolution along a* and c* directions but only to 3.0 Å along the b* direction.

2.1 Data Preparation

1. Process the data using a higher resolution limit than might seem warranted by the R_{merge} or I/σ statistics. For example, include shells of data to $I/\sigma > 1$ or even lower. The recently validated $CC_{1/2}$ statistic [2] could instead be used as a criterion. Include data to $CC_{1/2}$ of 0.20 or 0.10. Use the server as a tool to carve the final ellipsoidal resolution boundary for the data set.
2. Keep the data unmerged. Unmerged data permits calculation of the R_{merge} and $CC_{1/2}$ statistics after rejecting the poor measurements that fall outside the chosen ellipsoidal resolution boundary. The use of merged data still permits diagnosis of anisotropy, but does not permit the user to observe or report updated R_{merge} or $CC_{1/2}$ statistics.
3. If using Denzo/Scalepack [3], adjust the error model so that $0.9 < \chi^2 < 1.1$ in each resolution bin. If neglected, the $|F|/\sigma$ values calculated by the server will give a false indication of the resolution limit.

2.2 Server Versions

There are currently three versions of the anisotropy server; each accepts a different data format.

1. Use the XDS version (http://services.mbi.ucla.edu/anisoscale/anisoscale_xds) if the data is in XDS ahkl format [4]. It is the most comprehensive and convenient to use of the three versions. If provided with unmerged data, the server will report R_{merge} and I/σ statistics after imposing the ellipsoidal boundary, offering a more accurate view of data quality.

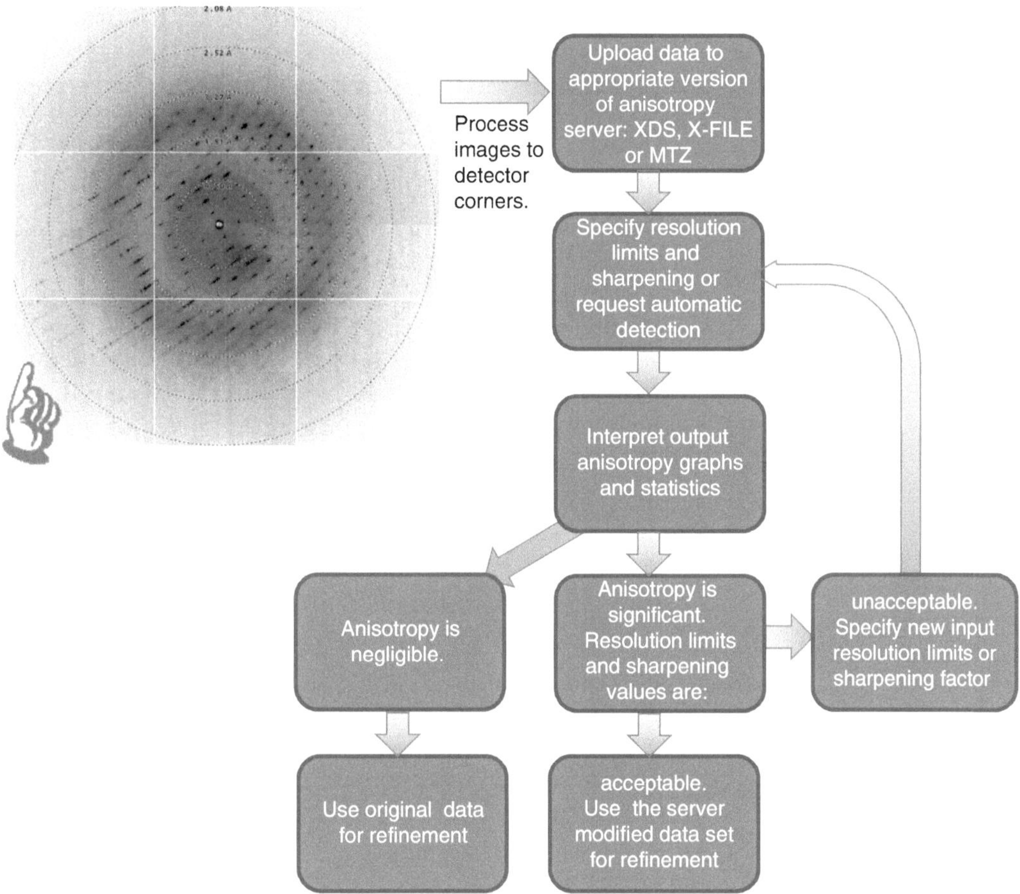

Fig. 1 An overview of the steps involved in using the anisotropy server. (1) Process the data with care not to omit strong reflections at high resolution even though they may not fill the entire shell due to anisotropy, such as those indicated in the diffraction image. If in doubt about the choice of high resolution cutoff, integrate over the entire detector surface. (2) Choose the version of anisotropy server which is compatible with data format. (3) Submit data to the server with appropriate input parameters selected. The server reports anisotropy analyses, suggests anisotropic resolution limits, applies these boundaries, applies optional B-factor sharpening, and reports statistics. (4) From these reports, conclude one of the following: (a) anisotropy is negligible and the data is best used in its original form, (b) that the suggested modifications are appropriate and should be implemented in the following refinement steps, or (c) some modifications are justified but with different specifications. In this last case, resubmit the data with specified instructions to change the resolution limits or sharpening factor

2. Use the Denzo/HKL version (obtainable from a link at http://services.mbi.ucla.edu/anisoscale/) if the data is in x-file format [3]. It offers the advantages of working with unmerged data, but requires the user to download the scripts and install them on the user's computer.
3. Use the MTZ version (http://services.mbi.ucla.edu/anisoscale/) if just a diagnosis of anisotropy is needed. It accepts CCP4's .mtz files [5] and therefore imposes no restrictions on

the choice of data processing software since it is usually simple to convert any format to mtz. This version offers no update of statistics after drawing ellipsoidal resolution boundaries. Supply data as structure factor amplitudes not intensities.

2.3 Input Parameters

1. Specify the location of the data set for analysis using the "Browse" button.
2. Specify resolution limits for the three principal axes or request automatic limits detection. I recommend automated detection for an initial analysis. Request it by submitting default values of "0.0" for limit 1, limit 2, and limit 3 fields. If the user concludes from the output analysis that a change in limits is desired, he can adjust them by running the server a second time with specified values. Enter them (Å) in any order (*see* **Note 1**). The server will apply the limits to the appropriate axes.
3. Specify B-factor sharpening (Å^2) or request automatic sharpening. The automatic sharpening factor is derived from the most negative principal component of the anisotropic part of B affecting the observed amplitudes (*see* **Note 2**).
4. Specify data labels for $|F|$ and σ values if the user is using the MTZ version of the server. Be sure the labels correctly match those in the uploaded .mtz file. If not, the server will produce no output. σ values are important for the accuracy of the anisotropy analysis and must be included.
5. Push the "Submit Data" button.

3 Interpreting Anisotropy Analysis

3.1 Asses the Degree of Anisotropy

The server reports "almost no", "mild", "strong", or "severe" anisotropy based on the spread in components of the anisotropic B, calculated by the program PHASER [6]. If the spread is less than 25 Å^2 the server will indicate no anisotropy or mild anisotropy. If the spread is greater than 25 Å^2 the server will indicate strong or severe anisotropy. The spread is printed in the first graphic and depicted as a reading on a thermometer (Fig. 2). This diagnosis is only an estimate and the user should verify the result by interpreting the $<|F|>/\sigma$ falloff graph, a more reliable diagnostic, discussed below.

3.2 Asses the Need for an Ellipsoidal Resolution Boundary

The server offers a suggestion for resolution limits along each of the three principal axes, if automatic detection is requested. These are printed at the bottom of the first graphical output (Fig. 2). These limits correspond to the resolutions at which the $<|F|>/\sigma$ values dip below a value of 3.0, along each of the three principal axes, calculated by the CCP4 program TRUNCATE [5].

1. If the spread in the three suggested resolution limits is less than 0.25 Å, an ellipsoidal resolution boundary is unlikely to be

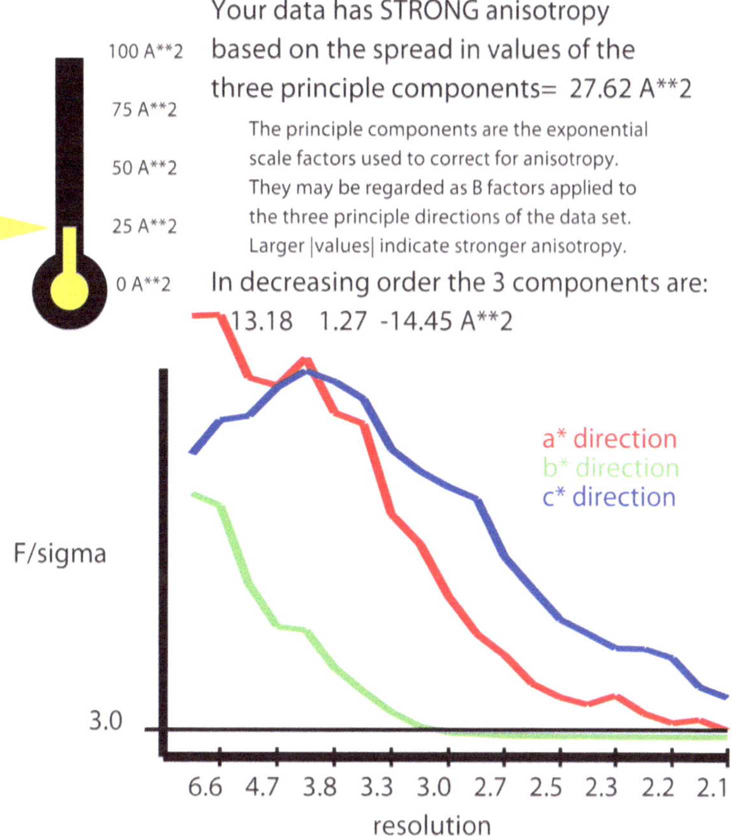

Fig. 2 "Strong anisotropy" is indicated by the spread in components of the anisotropic B-factor, 27.6 Å^2. The suggested resolution limits are 2.1 Å, 3.0 Å, and 2.1 Å corresponding to the resolution values at which $<|F|>/\sigma$ dips below a value of 3.0 near the a*, b*, and c* axes, respectively. The spread in resolution limits is large, 0.9 Å, suggesting the anisotropy is severe and that defining an ellipsoidal boundary appears justified for this case

beneficial. In this case, use a conventional spherical resolution boundary drawn by the data processing program of choice. There is no need to regard the remaining output from the anisotropy server.

2. If the spread is more than 0.25 Å, an ellipsoidal resolution boundary may be beneficial. The larger the spread, the more justified is the use of an ellipsoidal resolution boundary. In this case, continue with anisotropy analysis.

4 Improving the Structure Model

1. From the server, use the link provided to download the structure factors (mtz format) trimmed to the requested ellipsoidal resolution limits, anisotropically scaled, and sharpened by the requested factor.
2. Continue atomic refinement as customary, but use the downloaded structure factors. The drop to expect in R-factors is proportional to the severity of anisotropy. In the PE-PPE project, the drop was 6 %.
3. View refined maps and models in a graphics display program.
4. Adjust map sharpening. The program COOT [7] offers an interactive means of sharpening the map by adjusting a slider, currently available under the "calculate" menu. If the electron density map lacks features, specify a more negative value. If the map appears overly detailed with noisy features, specify a more positive value. It will not affect atomic B-factors.
5. Build in newly visible features. In the PE-PPE project, sharpening permitted the observation of additional details not otherwise observable, such as water molecules, carbonyl bumps, and side chain rotamers. Additional building improved the model, leading to an additional 5 % drop in R-factors. Similar experiences were reported by others [8–10].

5 Reporting Data Statistics After Drawing Ellipsoidal Resolution Limits

5.1 Resolution

1. State in the abstract that the diffraction was "anisotropic" with a maximal resolution that the user specified. Do not specify only the maximal resolution without the qualifying word "anisotropic"; it would be misleading. Alternatively, specify the spread in resolution limits among the three principal directions. For example, "the diffraction was anisotropic with resolution limits between 2.1 Å and 3.0 Å."
2. Explain in the methods section that an ellipsoidal resolution boundary was used. State the three values for the limits and to which axes they correspond to. For example, "an ellipsoidal resolution boundary was drawn with limits of 2.1 Å along a* and c* and 3.0 Å along b*."

5.2 Data Quality Statistics

1. Report the data statistics before and after drawing the ellipsoidal resolution boundary. Do not report only the initial statistics; it could be considered misleading. The XDS version of the server offers a log file from XSCALE, showing the data statistics after the ellipsoidal resolution boundary is drawn (Table 1).

Table 1
A table of data quality statistics provided by the XDS version of the server reveals improvement in R_{merge} and I/σ statistics after carving the ellipsoidal boundary

	Observed		Redundancy		Completeness		R_{merge}		I/σ	
Resolution	Before	After	Before	After	Before (%)	After (%)	Before (%)	After (%)	Before	after
9.35	3,440	3,387	7.5	7.5	87.1	87.0	3.4	3.4	41.5	41.7
6.61	7,167	7,085	9.5	9.5	100.0	100.0	4.0	4.0	40.7	40.8
5.40	9,105	9,050	9.6	9.6	100.0	100.0	5.5	5.5	32.9	33.0
4.68	10,474	10,373	9.8	9.8	99.9	100.0	5.4	5.4	33.8	33.7
4.18	12,192	12,047	9.9	9.9	100.0	99.9	5.2	5.2	33.5	33.7
3.82	13,539	13,446	10.1	10.1	99.9	99.9	5.9	5.9	30.2	30.4
3.53	14,571	14,336	10.0	10.0	100.0	100.0	6.8	6.7	26.4	26.4
3.31	15,261	15,202	9.9	9.9	100.0	100.0	9.2	9.1	20.3	20.4
3.12	15,632	15,522	9.8	9.8	99.7	99.7	11.5	11.3	16.2	16.4
2.96	16,495	16,554	9.6	9.6	99.5	99.5	14.1	14.2	13.3	13.2
2.82	17,196	16,649	9.4	9.2	99.3	94.8	16.1	15.8	11.2	11.7
2.70	16,784	15,737	9.0	8.6	99.1	85.7	24.5	22.6	8.0	9.4
2.59	17,203	16,175	8.9	8.3	99.0	79.0	33.1	30.3	6.2	7.8
2.50	17,597	15,131	8.5	7.4	98.9	69.5	40.7	36.9	5.0	6.7
2.41	16,933	13,907	8.1	6.8	98.5	61.9	49.6	40.3	4.0	6.4
2.34	17,693	13,585	8.2	6.3	99.1	56.0	55.4	42.8	3.6	6.2
2.27	17,080	11,657	7.7	5.3	98.4	46.6	62.1	44.9	3.1	6.1
2.20	17,854	10,592	7.7	4.7	98.2	40.4	75.1	49.4	2.7	5.8
2.15	16,724	7,929	7.3	3.4	98.1	30.2	100.1	59.4	2.0	4.9
2.09	13,913	5,289	5.6	2.2	82.7	19.9	135.2	70.4	1.4	4.0
Total	286,853	243,653	8.6	7.4	97.8	71.9	10.6	9.2	12.5	16.9

These updated values reflect more accurately the quality of the data in the high resolution bins. I/σ improved from 1.4 to 4.0 in the highest resolution shell. R_{merge} improved from 135 % to 70 %. A decrease in completeness (83–20 %) is an unavoidable consequence of imposing the ellipsoidal resolution boundary; it does not infer the need to discard the shell

Showing both sets of statistics indicates to reviewers that incompleteness in the high resolution shell is due to anisotropy rather than poor data collection skills.

2. Do not exclude a high resolution shell due to poor completeness. Completeness necessarily decreases as a consequence of using an ellipsoidal resolution boundary. If the I/σ or $CC_{1/2}$

statistic is acceptable, the resolution shell should be retained, even if the completeness is only 20 %. A well-measured reflection doesn't lose its worth simply because its neighboring reflections are poorly measured.

6 Notes

1. Adjust the Ellipsoidal Resolution Limits: There are some instances when the automatically detected resolution limits are inaccurate or do not meet the user's personal criteria. The user can specify new resolution limits for the three principal axes by rerunning the server and filling in the appropriate fields.
   (a) If any of the three curves cross the threshold more than once, then the suggested resolution limits will be too low. The server selects (incorrectly here) the lowest resolution crossing point for each $<|F|>/\sigma$ curve. Rerun the server using the highest resolution crossing for each $<|F|>/\sigma$ curve.
   (b) If any of the three curves do not cross the threshold value of 3.0, reprocess the data to higher resolution and resubmit it to the anisotropy server. Otherwise, the anisotropy analysis may be inaccurate and useful, well-measured high resolution reflections may be discarded.
   (c) If the $<|F|>/\sigma$ cutoff of 3 is considered to be too conservative, as may be justified by a recent report from Karplus and Diederichs [2], rerun the server with higher resolution limits.
   (d) If large improvements in R_{merge} and I/σ statistics are observed after imposing the ellipsoidal boundary (Table 1), it may be necessary to rerun the server with higher resolution limits.
2. Adjust the Sharpening Factor: I do not recommend using a value more negative (sharpening) than the suggested value, printed in the second graphical output of the server (Fig. 3). Excessive sharpening will lead to arbitrarily low atomic B-factors upon further atomic refinement, thus falsely indicating higher model quality. If the user wants the atomic B-factors to be unaffected, specify zero for the sharpening factor, then submit the data to the server again. Use COOT [7] to sharpen the map interactively as described below.

Acknowledgment

I am grateful to Duilio Cascio for helpful comments regarding this manuscript and the diffraction anisotropy server.

286853 reflections were in the initial data set. 43200 were discarded because they fell outside the specified ellipsoid with dimensions 1/ 2.1, 1/ 3.0, 1/ 2.1 $Å^{-1}$ along a*,b*,c*, respectively. These discarded reflections had an average F/sigma of 0.24.

243653 reflections remain after ellipsoidal truncation. Anisotropic scale factors were then applied to remove anisotropy from the data set. Lastly, an isotropic B of -29.34 $Å^2$ was applied to restore the magnitude of the high resolution reflections diminished by anisotropic scaling. The following pseudo precession images illustrate the individual steps.

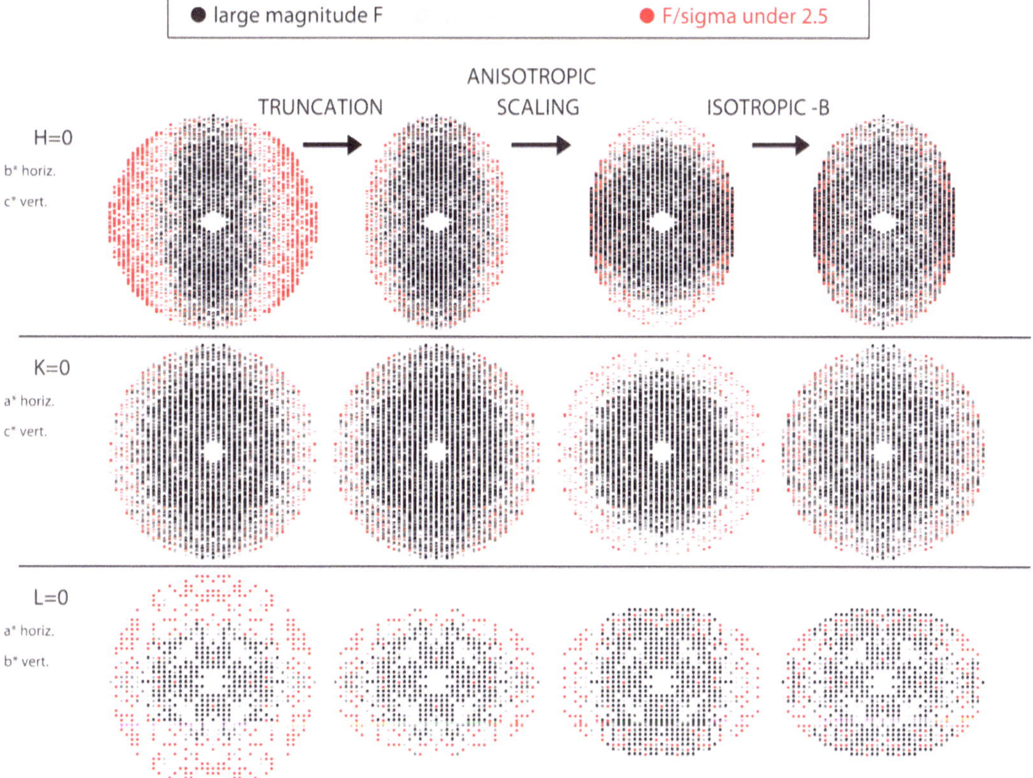

Fig. 3 The server carves out an ellipsoidal resolution boundary by rejecting reflections falling outside the server suggested or user requested anisotropic resolution limits. The carving is illustrated (Strong et al. [1]) with views of the $h=0$, $k=0$, and $l=0$ planes. Compare column 1 with column 2. Structure factor magnitudes are plotted in *grayscale*, with the largest $|F|$s represented in *black*. Structure factors with $|F|/\sigma$ less than 2.5 are colored red. The server reports the number of reflections discarded (43,200) and the average $|F|/\sigma$ of these reflections (0.24). If the limits are chosen correctly, the $|F|/\sigma$ will be less than 3.0 and the majority of the *red colored* reflections will have been removed, as is the case here. These discarded reflections had an average $|F|/\sigma$ of 0.24. The effect of the anisotropic scaling and B-factor sharpening can also be seen. Note in the $h=0$ precession image that the high resolution reflections near the vertical axis were diminished by the anisotropic scaling procedure (i.e., they appear *lighter gray* compared to the previous panels). After sharpening, these reflections are restored to their former magnitude as evidenced by comparing the similarity in darkness of these reflections in the *left* and *right columns*. The server suggests a sharpening factor (–29.3 $Å^2$) to compensate for the side effects of anisotropic scaling

References

1. Strong M, Sawaya MR, Wang S et al (2006) Toward the structural genomics of complexes: crystal structure of a PE/PPE protein complex from Mycobacterium tuberculosis. Proc Natl Acad Sci USA 103:8060–8065
2. Karplus PA, Diederichs K (2012) Linking crystallographic model and data quality. Science 336:1030–1033
3. Otwinowski Z, Minor W (1997) Processing of X-ray diffraction data collected in oscillation mode. In: Carter CW Jr, Sweet RM (eds) Methods in enzymology: macromolecular crystallography, part A, vol 276. Academic, New York, pp 307–326
4. Kabsch W (2010) XDS. Acta Crystallogr D 66:125–132
5. Winn MD, Ballard CC, Cowtan KD et al (2011) Overview of the CCP4 suite and current developments. Acta Crystallogr D 67:235–242
6. McCoy AJ, Grosse-Kunstleve RW, Adams PD et al (2007) Phaser crystallographic software. J Appl Crystallogr 40:658–674
7. Emsley P, Lohkamp B, Scott WG et al (2010) Features and development of Coot. Acta Crystallogr D 66:486–501
8. Lee M, Maher MJ, Guss JM (2007) Structure of the T109S mutant of Escherichia coli dihydroorotase complexed with the inhibitor 5-fluoroorotate: catalytic activity is reflected by the crystal form. Acta Crystallogr F 63:154–161
9. Suits MD, Sperandeo P, Dehò G et al (2008) Novel structure of the conserved Gram-negative lipopolysaccharide transport protein LptA and mutagenesis analysis. J Mol Biol 380:476–488
10. de Chiara C, Rees M, Menon RP et al (2013) Self-assembly and conformational heterogeneity of the AXH domain of ataxin-1: an unusual example of a chameleon fold. Biophys J 104:1304–1313

Chapter 16

Applications of NMR-Based PRE and EPR-Based DEER Spectroscopy to Homodimer Chain Exchange Characterization and Structure Determination

Yunhuang Yang, Theresa A. Ramelot, Shuisong Ni, Robert M. McCarrick, and Michael A. Kennedy

Abstract

The success of homodimer structure determination by conventional solution NMR spectroscopy relies greatly on interchain distance restraints (less than 6 Å) derived from nuclear Overhauser effects (NOEs) obtained from ^{13}C-edited, ^{12}C-filtered NOESY experiments. However, these experiments may fail when the mixed ^{13}C-/^{12}C-homodimer is never significantly populated due to slow homodimer chain exchange. Thus, knowledge of the homodimer chain exchange kinetics can be put to practical use in preparing samples using the traditional NMR method. Here, we described detailed procedures for using paramagnetic resonance enhancements (PREs) and EPR spectroscopy to measure homodimer chain exchange kinetics. In addition, PRE and EPR methods can be combined to provide mid-range (<30 Å) and long-range (17–80 Å) interchain distance restraints for homodimer structure determination as a supplement to short-range intrachain and interchain distance restraints (less than 6 Å) typically obtained from ^{1}H-^{1}H NOESY experiments. We present a summary of how to measure these distances using NMR-based PREs and EPR-based double electron electron resonance (DEER) measurements and how to include them in homodimer structure calculations.

Key words NMR, EPR, Spectroscopy, Homodimer, Chain exchange, Structure determination

1 Introduction

Structural genomics provides three dimensional (3D) structures of proteins on a genome-wide scale by emphasizing high-throughput methods of structure determination. Owing to the numerous contributions from scientists working on the Protein Structure Initiative (PSI) project, the number of protein structures solved and deposited in the Protein Data Bank (PDB) increased dramatically in the past decade. Complete genome sequence information allows almost every open reading frame (ORF) to be cloned and expressed as a protein. The 3D structure of many of these proteins can then be determined at the atomic level using two main

Yu Wai Chen (ed.), *Structural Genomics: General Applications*, Methods in Molecular Biology, vol. 1091, DOI 10.1007/978-1-62703-691-7_16, © Springer Science+Business Media, LLC 2014

techniques: X-ray crystallography and Nuclear Magnetic Resonance (NMR) spectroscopy. In particular, the NMR method plays an important role for structure determination of proteins for which single crystals could not be obtained, making it a complementary method to X-ray crystallography [1]. Protein NMR has become a routine method for structure determination for solution proteins with molecular mass under 25 kDa. However, structure determination of homodimer, oligomeric, or protein–protein/DNA/RNA complexes is still challenging. Notwithstanding, the Northeast Structural Genomics (NESG) consortium has solved the structure of 26 oligomeric proteins out of 299 targets (306 pdb submissions) during the second phase of the PSI project (http://spine.nesg.org/nmr_progress.cgi). Given that about 80 % of proteins from *Escherichia coli* have been predicted to exist as homo-oligomers based on a survey of their protein annotations in the Swiss-Prot database [2], and more than 86 % of all proteins in the Swiss-Prot database were predicted to exist naturally as oligomeric complexes based on a computational prediction [3], more powerful and robust methods are needed for structural characterization of dimers or oligomers.

Prior to preparing the exchanged ^{13}C-/^{12}C-labeled sample to measure interchain nuclear Overhauser effects (NOEs) using the conventional NMR method [4–6] for homodimer structure determination, it is useful to have knowledge of the chain exchange kinetics. These can be obtained from techniques such as electronic paramagnetic resonance (EPR) spectroscopy and paramagnetic relaxation enhancement (PRE). In this chapter, we present protocols for making NMR-based PREs and EPR-based double electron electron resonance (DEER) measurements of homodimer chain exchange rate kinetics. The kinetics and thermodynamics of subunit exchange in multimeric proteins is of great potential significant to understanding functions in the cell that require exchanging binding partners in order to carry out specific biochemical or cellular functions [7, 8].

In addition to traditional solution-state NMR spectroscopy using NOE-based methods, protein structure determination can integrate information obtained from other techniques such as EPR spectroscopy [9], PRE [9–11], residue dipolar coupling (RDC) [12], and small-angle X-ray scattering (SAXS) [13]. In this chapter, we present a method for combining traditional NMR NOE restraints for homodimer structure determination with mid-range and long-range distance restraints using NMR-based PREs and EPR-based DEER measurements, respectively. The methods are demonstrated using the homodimer protein Dsy0195 from strictly anaerobic bacterium *Desulfitobacterium hafniense*. The details on the protein Dsy0195 constructs, expression, purification, and (*1-oxy-2*, 2, 5, 5-tetramethyl-D-pyrroline-3-methyl)-methanethiosulfonate (MTSL) spin labeling were previously published [9, 14, 15].

2 Materials

2.1 Equipment

In our laboratory, three NMR spectrometers were used for the structure determination studies and for measurement of chain exchange for homodimer Dsy0195: BrukerAvance III 850 and 600 MHz spectrometers equipped with conventional 5 mm probes, and a Varian Inova 600 MHz with a 5 mm HCN cold probe (*see* **Note 1**). All DEER experiment data were collected on the spectrometer (Bruker ELEXSYS E580) with a SuperQft Q-band bridge and Q-band (EN 5107D2) probe (34 GHz), and analyzed by the software package DeerAnalysis2011 (requires Matlab) [16]

2.2 Protein Samples

The locations of MTSL introduced into the homodimer protein of interest by site-direct spin labeling are critical to both PRE and DEER experiments. In this work, single cysteine mutants S36C and S52C of Dsy0195 were produced for DEER and PRE experiments (*see* **Note 2**). Protein samples of Dsy0195 prepared in this work included: wild type (*wt*) Dsy0195, ^{15}N-Dsy0195, MTSL-Dsy0195-S52C, Dsy0195-S36C, and MTSL-Dsy0195-S36C. The protein concentrations were 1.0 mM for PRE experiments and 0.2 mM for DEER experiments, respectively (*see* **Note 3**).

2.3 Reagents and Buffers

Prepare all solutions using 18 MΩ deionized water (LABCONCOPRO PS, Kansas city, MO, USA) and analytical grade reagents unless otherwise stated. Store all buffers and protein samples prepared at 4 °C before any EPR or NMR experiment.

1. 35 mM MTSL (Toronto Research Chemicals, Inc.) in methanol.
2. 150 mM ascorbic acid (Acros Organics).
3. PRE sample buffer: 20 mM ammonium acetate, 100 mM sodium chloride, 5 mM calcium chloride, 0.02 % (*w*/*v*) sodium azide, 50 μM DSS (for internal chemical shift referencing), 5 % (*v*/*v*) D_2O, pH 4.5.
4. DEER sample buffer: 20 mM ammonium acetate, 100 mM sodium chloride, 5 mM calcium chloride, 0.02 % (*w*/*v*) sodium azide, 30 % (*w*/*v*) glycerol, pH 4.5 (*see* **Note 4**).

3 Methods

3.1 Chain Exchange Rate of Dsy0195 Measured by DEER Experiments

1. Dilute 0.2 mM MTSL-Dsy0195-S36C to 0.1 mM with the DEER buffer to generate the control sample. Transfer the control sample into an EPR capillary tube (*see* **Note 5**), and freeze carefully by dipping into liquid nitrogen. All DEER data were

collected on a Bruker ELEXSYS E580 spectrometer with a SuperQft pulse bridge, 10 W amplifier and Q-band (EN 5107D2) probe at 80 K.

2. Find the resonator dip (*see* **Note 6**). In order to ensure that consistent levels of microwave power were achieved, a screen capture of the dip in the tuning window within the Bruker Xepr software was taken once the tuning was optimized for the first sample. For subsequent samples, the sample was manipulated within the resonator and the dip was compared to the screen capture until an identical frequency and coupling was obtained.
3. DEER experimental setup. The relaxation times T_1 and T_2 were measured to optimize the shot repetition time and DEER end point, respectively. The pulse lengths were optimized to be as short as possible to yield a maximum modulation depth (*see* **Note 7**). In this case, 10 ns ($\pi/2$) and 20 ns (π) probe pulses and a 24 ns (π) pump pulse were used; however, this is spectrometer dependent and will have to be determined in each case. The data were collected with an 80 MHz frequency separation between the pump and probe frequencies. Collect the DEER data.
4. Repeat the DEER experiment of the control sample four more times to obtain the measurement reproducibility. Figure 1a shows the background subtracted and scaled time domain data for five repeated DEER measurements of the control sample (*see* **Note 8**). The modulation depth is measured as the percent difference between the maximum intensity at time zero and the first minimum, which in this case was 14 %. The repeated results shown in Fig. 1a demonstrate the small experimental deviation (Fig. 1a).
5. Mix 10 μl of 0.2 mM Dsy0195-S36C and 10 μl of 0.2 mM MTSL-Dsy0195-S36C, transfer into EPR tube, and freeze in liquid nitrogen at a dimer exchange time of approximately 1 h to attempt to estimate the exchange rate. Collect the DEER data as in **step 3**. Based on the obtained modulation depth from this experiment, determine a range of appropriate time points for the remaining experiments.
6. Repeat **step 5** with a series of exchange time times as determined in the previous step, in this case, 2.5, 5, 10, 15, 25, 35, 52.5, 112.5, and 155 min (*see* **Note 9**).
7. The scaled modulation depth for each measurement was derived as in **step 4**. All modulation depths were scaled to the value of the control sample, in which the depth of 0.14 was scaled to 1.

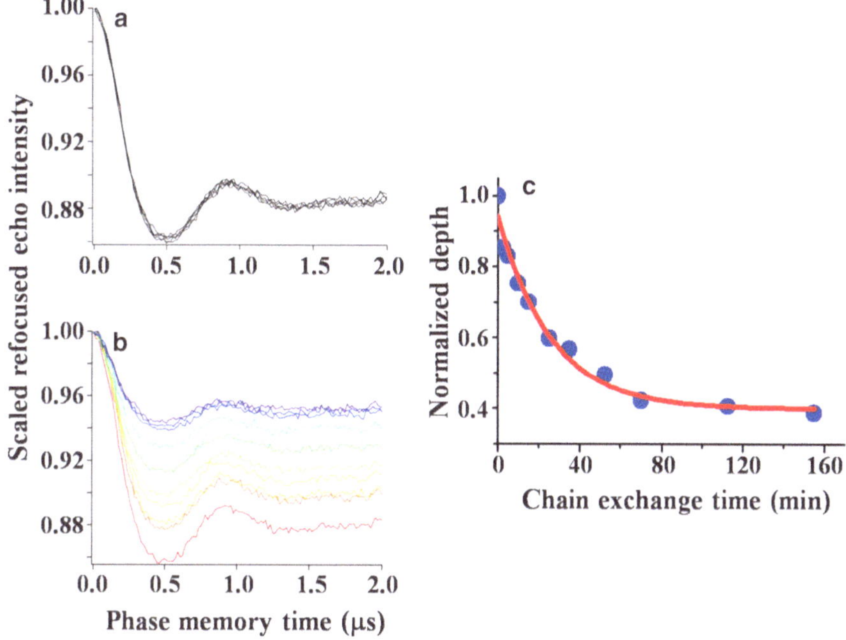

Fig. 1 Homodimer chain exchange rate measurement using DEER experiment. (**a**) Overlay of scaled refocused echo intensity versus phase memory time for the control sample (repeated five times). (**b**) Time-domain DEER signals of 1:1 mixture of Dsy0195-S36C-MTSL/^{15}N-Dsy0195-S36C at a series of exchange time points: 0, 2.5, 5, 10, 15, 25, 35, 52.5, 70, 112.5, 155 min. (**c**) Plot of the normalized modulation depth versus chain exchange time following the first-order exponential decay. The *blue filled circle* stands for the experimental depth and *red solid line* for the fitting curve following the first-order exponential decay

8. Plot the curve of the normalized modulation depth versus the exchange time. Curve fitting obeys a first-order exponential decay (Fig. 1c), and the homodimer chain exchange rate of 0.039 min^{-1} is derived from the fit.

3.2 Chain Exchange Rate of Dsy0195 Measured by PRE Experiments (See Note 10)

1. Prepare the control sample without chain exchange. Dilute 150 μl of 1.0 mM ^{15}N-Dsy0195 with the PRE buffer to a final concentration of 0.5 mM as the control sample. Transfer the control sample into the NMR tube and put into the NMR spectrometer.
2. Setup the NMR acquisition parameters. Load the standard 2D SOFAST-HMQC pulse sequence sfhmqcf3gpph, and set pulse sequence parameters by default values including the flip angle for the first proton pulse of 120°, the band-selective proton pulse centered at 8.0 ppm covering a bandwidth of 4.0 ppm. The acquisition parameters were setup as follows: spectrum width, $SW_1 \times SW_2 = 29 \times 15$ ppm for ^{15}N and ^{1}H (optimized from the chemical shift dispersions in both dimension); time

domain data set, $TD_1 \times TD_2 = 80 \times 1{,}024$; number of scan, ns = 4; acquisition time for the ^1H dimension of 47 ms, and the maximum acquisition time for the ^{15}N dimension of 23 ms; recycle time, d1 = 100 ms. The total experimental time is 58 s. Collect the 2D SOFAST-HMQC spectrum.

3. Process the NMR data with NMRPipe [17]. The time domain data were Fourier transformed in both dimensions after zero fill in the t1 × t2 dimensions to 128 × 2,048 points, applying a window function of a 90° shifted sine-bell and mirror image linear prediction (LP) in the ^{15}N dimension. The processed spectra were visualized using SPARKY [18] and peak intensities were obtained from SPARKY peak heights.
4. Peak intensity normalization. The cross peak intensities were normalized with the average intensity from the seven backbone peaks K44, E41, G55, E56, K57, L58, and G59, for which distances between amide protons of interest in one chain to the Cβ of S52 in the other chain are greater than 30 Å based on the crystal structure of Dsy0195 (*see* **Note 11**).
5. Estimate the peak intensity deviations of control sample. Repeat the data collection four times at an interval time of approximately 10 min. Figure 2a showed the average peak intensity and standard deviation versus individual amino acid (39 backbone peaks without overlap). The plot indicated the errors in peak intensity reproducibility could be up to 15 %.
6. Mixed sample preparation and NMR data collection. Mix 150 μl of 1.0 mM *U*-^{15}N-Dsy0195 and 150 μl of 1.0 mM MTSL-Dsy0195-S52C. Transfer the mixed sample into NMR tube, put into NMR magnet. Lock the signal, tune and match the probe, shim the field, determine the solvent carrier frequency and 90° pulse width as quickly as possible (can be done within 5 min). Collect the spectra at a series of chain exchange times 6, 8, 10, 15, 20, 25, 35, 45, 52.5, 60, 70, 90, 112.5, 140, 155, 180, and 210 min (*see* **Note 12**).
7. Process the data and normalize the peak intensities as in **steps 3** and **4**.
8. Repeat **steps 6** and **7** four more times. The average peak intensity and standard deviation for each peak can be obtained from the five repeated measurements.
9. Derive the homodimer chain exchange rate. Plot the peak intensity versus exchange time for each residue that had a decrease in peak intensity. This curve for residue E84 is given as an example (Fig. 2b). The curve fitting followed the first-order exponential decay and indicated that the chain exchange rate of Dsy0195 was 0.037 ± 0.008 min^{-1} (*see* **Note 13**).

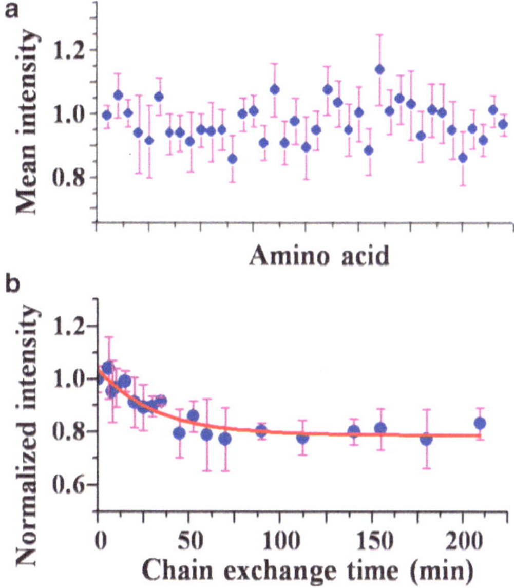

Fig. 2 Homodimer chain exchange rate measurement using PRE experiment. (**a**) Plot of the normalized peak intensity (average from five repeats) of 39 backbone amide peaks without overlap in 2D SOFAST-HMQC spectra of ^{15}N-Dsy0195. The *blue circle* stands for the average intensity, and the *pink line* for the standard deviation. (**b**) Plot of the normalized intensity (average from five repeats) versus chain exchange time for a given residue (E84) as an example. The *blue filled circle* stands for the peak intensity, *pink line* for standard deviation, and the *red solid line* for the fitting curve following the first-order exponential decay

3.3 Structure Determination of Homodimer Dsy0195 by Combining NMR and EPR Derived Interchain Distance Restraints

3.3.1 Overview

The initial Dsy0195 monomer structure was determined following the conventional protocols of our laboratory, referred to in recent structure publications [19–22]. The monomer structure quality was assessed using software package PSVS 1.4 [23]. Although the overall tertiary structure was correct, many distance violations were present in both N-terminus and C-terminus of Dsy0195 that are located at the homodimer interface known from its crystal structure. These violations could result from some interchain NOEs were automatically assigned as interchain NOEs during structure calculations using CYANA [24] with input of peak lists of NOESY spectra. An additional problem in the structure determination of the homodimer Dsy0195 by NMR spectroscopy was that few interchain NOEs were detected from the ^{13}C-edited, ^{12}C-filtered NOESY experiment, which led to either poor quality of the homodimer structure solved by NMR method alone [14]. These problems were overcome by using the mid-range and long-rang interchain distance restraints that were derived from PRE and DEER experiments [9, 14]. After these interchain distance restraints were used as input, more interchain NOEs were assigned in the conventional ^{13}C-edited NOESY spectra by the CYANA program [9]. In this chapter, we focus on deriving these mid-range and long-range distance restraints by PRE and DEER experiments and using them in the structure determination of homodimer Dsy0195 along with conventional NMR data using the CYANA program.

3.3.2 Long-Range Distance Restraints Derived from DEER Experiments

1. Sample preparation. Two samples each of 0.2 mM MTSL-Dsy0195-S36C and MTSL-Dsy0195-S52C were prepared for DEER experiments.
2. Collect the DEER data as in Subheading 3.1 **step 3**. Distances were derived using the DEERAnalysis2011 package from the Jeschke laboratory at ETH using a Tikhonov regularization. As mentioned above, this software is well documented and a detailed description of the analysis is out of the scope of this chapter.
3. Two distances of 34 and 20 Å between two nitroxides in the MTSL-labeled homodimers S36C and S52C were derived, respectively. Considering that the free radical of the nitroxide is typical 8–10 Å from its own backbone amide, the error bar for DEER distance restraints is ±5 Å in the following structure calculation.

3.3.3 Mid-range Distance Restraints Derived from PRE Experiments

1. Mixed sample with PREs (referred to as paramagnetic state or reduced state) preparation. Mix 150 μl of 1.0 mM ^{15}N-labeled Dsy0195 and 150 μl of 1.0 mM MTSL-Dsy0195-S52C. Transfer the mixture into the NMR tube and put into the NMR spectrometer. The experiment was carried out at 293 K.
2. NMR data collection. Collect 2D ^{1}H-^{15}N HSQC in the reduced state (*see* **Note 14**). After the experiment, add about 3–4 μl 150 mM ascorbic acid (correspondingly two- to three-fold molar excess to MTSL) to get rid of MTSL from the protein (referred to as diamagnetic state or oxidized state). Collect the 2D $^{1H\text{-}15N}$ HSQC in the diamagnetic state with the same parameters as the paramagnetic state.
3. Process the NMR data with NMRPipe. The time domain data were Fourier transformed in both dimensions after zero fill at $t1 \times t2$ dimensions of $512 \times 2{,}048$ and application of a 90° shifted sine-bell window function. The processed spectra were visualized and peak picked using SPARKY.
4. Peak intensity normalization. Measure the intensity of each resonance from the spectrum of mixed sample before the addition of ascorbic acid, recorded as I_{para*}, that of the mixed sample after the addition of ascorbic acid, recorded as I_{dia}. Peak intensities were normalized as the same in Subheading 3.2 **step 4**. Then, the peak intensity of solely paramagnetic state I_{para} is equal to $2(I_{para*} - I_{dia}/2)$ (equation 3 in ref. 11), where I_{dia} is the intensity for diamagnetic state (*see* **Note 15**).

5. Mid-range interchain distance restraints. Based on the ratio of peak intensity (I_{para}/I_{dia}) for each resonance, the interchain distances were derived following the equations (4) and (5) in the literature [11]. Based on the calculation, the distances for peak intensity ratios of 0.9 and 0.1 are about 22 and 13 Å, respectively. Three classes of distance restraints were used for structure calculation. Peaks with an intensity ratio >0.9 were restrained only with a lower bound of 22 Å; peaks with an intensity ratio <0.1 were restrained only with an upper bound of 13 Å; peaks with intensity between 0.1 and 0.9 were restrained with ± 4 Å for upper and lower bound, respectively, for structure calculation. For Dsy0195, 67 lower-bound and 35 upper-bound interchain restraints were derived from PRE data.

3.3.4 Interchain NOEs Determination and Homodimer Structure Calculation

As additional input, the interchain restraints from both PRE and DEER experiments were in the structure calculation. We used the CYANA program to calculate the homodimer structure with the chemical shift assignments, NOESY peaklists, dihedral-angle restraints, backbone hydrogen-bond restraints. The homodimer structure calculation resulted in a converged ensemble of structures. We assessed the structure quality using PSVS 1.4 and checked the NOE assignments manually based on the NOE and dihedral-angle violations.

*3.3.5 Homodimer Structure Refinement (See **Note 16**)*

Convert the final NOE-derived distance restraints from CYANA, dihedral-angle restraints, and hydrogen-bond restraints to Xplor/CNS format using PDBStat [23]. These restraints were used to calculate 100 structures using Xplor-NIH followed by refinement of the 20 lowest energy structure using restrained molecular dynamics in explicit water with CNS 1.2 (CNSw) [25]. Use PSVS 1.4 to assess the structure quality and check the restraints violation. The final NMR ensemble of 20 structures with the lowest energy was deposited in the PDB (PDB ID 2KYI).

4 Notes

1. Although we report the NMR spectrometers that were specifically used in our lab, any NMR spectrometer with 500 MHz or greater could be used for these experiments. Since our PRE experimental data were collected on a Bruker Avance III 600 MHz spectrometer, the parameter names given for the 2D SOFAST-HMQC in this chapter are consistent with the Bruker NMR software package TopSpin.
2. Distances between protons of interest and the MTSL nitroxide that are within 30 Å can be derived by PRE experiments, whereas the measureable distances between two nitroxides

range from 17 to 80 Å for DEER experiments. In this work, the distance between the Cβ of S36 in the two chains is about 33.5 Å based on the crystal structure, which is suitable for DEER measurements, while the distance for S52 is about 15 Å, suitable for PRE experiments.

3. In general, higher protein concentrations (~1 mM or greater) will provide the best signal-to-noise ratios and shorter data collection times for NMR and PRE experiments as long as protein does not aggregate. In contrast, protein concentrations of about 0.1–0.2 mM are optimal for the DEER experiment using a Q-band probe for good signal-to-noise with the benefit of avoiding the strong background signal of interchain molecular electron–electron interaction.
4. Homogenous distributions of homodimers are achieved by addition of 30 % glycerol as a cryoprotectant.
5. For these measurements, it is critical that the sample height within the tube is kept consistent so that identical levels of tuning can be obtained from sample to sample.
6. As the primary objective of these measurements is the precise determination of the modulation depth in the DEER experiment, achieving consistent coupling is essential. In a typical Q-band setup, the coupling is less overcoupled than in an X-band experiment. As such, the pump and probe frequency separation is on the order of the width of the resonator bandwidth, making the microwave power achieved at the two frequencies very sensitive to the tuning.
7. In a typical DEER experiment, one collects data to as long an end point as possible given the limitation of the T_2 of the sample. This yields the most precise distance distribution information. In these exchange experiments, since it is only the modulation depth which is being determined, one can shorten the end point of the experiment to enhance the signal-to-noise and sample throughput as a precise distance determination is not the primary goal.
8. Quantitative determination of the modulation depth from sample to sample depends on consistent analysis of the data, particularly the subtraction of the intermolecular background signal. This was accomplished using the DEERAnalysis2011 package from the Jeshcke laboratory [16]. A detailed description of the analysis is out of the scope of this chapter, but the included documentation with the software is extensive.
9. These selected incubation times for chain exchange were appropriate for the system studied, but these can be adjusted based on the exchange rate of each individual system. The first data point is limited to the time it takes to mix the MTSL-labeled and unlabeled protein, load the tube and freeze the

sample in liquid nitrogen. In practice, this could potentially be as low as a 1 min given conventional methods. If shorter times are needed in the case of very rapid dimer exchange rates, rapid freeze quench (RFQ) methods could potentially be adapted.

10. The mixture of ^{15}N-labeled Dsy0195 and MTSL-labeled Dsy0195-S52C in 1:1 ratio were prepared (the ratio can be different) [15], and fast NMR data collection using 2D SOFAST-HMQC [26] experiments were carried out at a series of mixing times for dimer chain exchange. Samples were put into 5 mm Shigemi NMR tubes with magnetic susceptibility matched to solvent D_2O (BMS-003, Shigemi, Inc., Allison Park, PA, USA). All spectra were carried out on a Bruker Avance III 600 MHz spectrometer at 293 K.

11. In the NMR spectra of mixed samples, NMR signals are coming from two dimer forms: ^{15}N-Dsy0195 and ^{15}N-Dsy0195/MTSL-Dsy0195-S52C that was created by the chain exchange during the mixing time. PREs can only be observed for those amide protons in the ^{15}N-labeled chain of dimer ^{15}N-Dsy0195/MTSL-Dsy0195-S52C that are less than 30 Å away from the nitroxide in the other chain. For the other amide protons without PREs, the peak intensities in 2D HMQC will remain unchanged during the chain exchange process. These are candidates for normalization of peaks in the entire spectrum.

12. The optimal exchange times will vary significantly for each dimer depending on the chain exchange rate.

13. Although the plot of peak intensity versus chain exchange time of each individual peak can be plotted, the successful curve fitting following the first exponential decay is not guaranteed. Especially, for the residues with no PREs or weak PREs, the curve fitting will not follow a first-order exponential decay due to no or little change of the peak intensity.

14. The acquisition parameters are setup as follows: spectrum width: $SW_1 \times SW_2 = 29 \times 18$ ppm (optimized from the chemical shift dispersions in both dimension), time domain data set, $TD_1 \times TD_2 = 256 \times 1{,}024$; acquisition time (AQ) at ^1H dimensions of 34 ms and the maximum acquisition at ^{15}N dimension of 51 ms; number of scan, NS = 128; recycle time, d1 = 1 s. The total experimental time is about 10 h.

15. The intensity of each peak ($I_{para*} - I_{dia}/2$) in paramagnetic state comes only from the dimer ^{15}N-Dsy0195/MTSL-Dsy0195-S52C, which accounts for 50 % of the sample in the mixture of 1:1 ^{15}N-Dsy0195 and MTSL-Dsy0195-S52C [11].

16. This step was done following our laboratory protocols at the time when the structure of Dsy0195 was determined. The protocols of protein structure calculation and refinement are different from lab to lab.

Acknowledgments

This work was supported by the National Institute of General Medical Sciences, Grant Number: U54-GM074958; National Science Foundation, Grant Number CHE-0645709; BrukerBiospin, Miami University and Ohio Board of Reagents.

References

1. Wüthrich K (1986) NMR of proteins and nucleic acids. Wiley, New York
2. Goodsell DS, Olson AJ (2000) Structural symmetry and protein function. Annu Rev Biophys Biomol Struct 29:105–153
3. Shen HB, Chou KC (2009) Quatldent: a web server for identifying protein quaternary structural attribute by fusing functional domain and sequential evolution information. J Proteome Res 8:1577–1584
4. Otting G, Wüthrich K (1989) Extended heteronuclear editing of 2D 1H NMR spectra of isotope-labeled proteins, using the X(ω1, ω2) double half filter. J Magn Reson 85: 586–594
5. Lee W, Revington MJ, Arrowsmith C et al (1994) A pulsed field gradient isotope-filtered 3D 13C HMQC-NOESY experiment for extracting intermolecular NOE contacts in molecular complexes. FEBS Lett 350:87–90
6. Folmer RH, Hilbers CW, Konings RN et al (1995) A (13)C double-filtered NOESY with strongly reduced artefacts and improved sensitivity. J Biomol NMR 5:427–432
7. Sobott F, Benesch JL, Vierling E et al (2002) Subunit exchange of multimeric protein complexes. Real-time monitoring of subunit exchange between small heat shock proteins by using electrospray mass spectrometry. J Biol Chem 277:38921–38929
8. Pan J, Rintala-Dempsey AC, Li Y et al (2006) Folding kinetics of the S100A11 protein dimer studied by time-resolved electrospray mass spectrometry and pulsed hydrogen-deuterium exchange. Biochemistry 45:3005–3013
9. Yang Y, Ramelot TA, McCarrick RM et al (2010) Combining NMR and EPR methods for homodimer protein structure determination. J Am Chem Soc 132:11910–11913
10. Battiste JL, Wagner G (2000) Utilization of site-directed spin labeling and high-resolution heteronuclear nuclear magnetic resonance for global fold determination of large proteins with limited nuclear overhauser effect data. Biochemistry 39:5355–5365
11. Rumpel S, Becker S, Zweckstetter M (2008) High-resolution structure determination of the CylR2 homodimer using paramagnetic relaxation enhancement and structure-based prediction of molecular alignment. J Biomol NMR 40:1–13
12. Wang X, Bansal S, Jiang M et al (2008) RDC-assisted modeling of symmetric protein homo-oligomers. Protein Sci 17:899–907
13. Grishaev A, Wu J, Trewhella J et al (2005) Refinement of multidomain protein structures by combination of solution small-angle X-ray scattering and NMR data. J Am Chem Soc 127:16621–16628
14. Yang Y, Ramelot TA, Cort JR et al (2011) Solution NMR structure of Dsy0195 homodimer from Desulfitobacterium hafniense: first structure representative of the YabP domain family of proteins involved in spore coat assembly. J Struct Funct Genomics 12: 175–179
15. Yang Y, Ramelot TA, Ni S et al (2013) Measurement of rate constants for homodimer subunit exchange using double electron–electron resonance and paramagnetic relaxation enhancements. J Biomol NMR 55: 47–58
16. Jeschke G, Chechik V, Ionita P et al (2006) DeerAnalysis2006—a comprehensive software package for analyzing pulsed ELDOR data. Appl Magn Reson 30:473–498
17. Delaglio F, Grzesiek S, Vuister GW et al (1995) NMRPipe: a multidimensional spectral processing system based on UNIX pipes. J Biomol NMR 6:277–293
18. Goddard TD, Kneller DG (2008) SPARKY 3. University of California, San Fransisco
19. Yang Y, Ramelot TA, Cort JR et al (2012) Solution NMR structure of hypothetical protein CV_2116 encoded by a viral prophage element in Chromobacterium violaceum. Int J Mol Sci 13:7354–7364
20. Yang Y, Ramelot TA, Cort JR et al (2011) Solution NMR structure of photosystem II reaction center protein Psb28 from

Synechocystis sp. Strain PCC 6803. Proteins 79:340–344

21. Ramelot TA, Yang Y, Xiao R et al (2012) Solution NMR structure of BT_0084, a conjugative transposon lipoprotein from Bacteroides thetaiotamicron. Proteins 80:667–670
22. Feldmann EA, Ramelot TA, Yang Y et al (2012) Solution NMR structure of Asl3597 from Nostoc sp. PCC7120, the first structure from protein domain family PF12095, reveals a novel fold. Proteins 80:671–675
23. Bhattacharya A, Tejero R, Montelione GT (2007) Evaluating protein structures determined by structural genomics consortia. Proteins 66:778–795
24. Guntert P (2004) Automated NMR structure calculation with CYANA. Methods Mol Biol 278:353–378
25. Linge JP, Williams MA, Spronk CA et al (2003) Refinement of protein structures in explicit solvent. Proteins 50:496–506
26. Schanda P, Brutscher B (2005) Very fast two-dimensional NMR spectroscopy for real-time investigation of dynamic events in proteins on the time scale of seconds. J Am Chem Soc 127:8014–8015

Chapter 17

A Cost-Effective Protocol for the Parallel Production of Libraries of $^{13}CH_3$-Specifically Labeled Mutants for NMR Studies of High Molecular Weight Proteins

Elodie Crublet, Rime Kerfah, Guillaume Mas, Marjolaine Noirclerc-Savoye, Violaine Lantez, Thierry Vernet, and Jerome Boisbouvier

Abstract

There is increasing interest in applying NMR spectroscopy to the study of large protein assemblies. Development of methyl-specific labeling protocols combined with improved NMR spectroscopy enable nowadays studies of proteins complexes up to 1 MDa. For such large complexes, the major interest lies in obtaining structural, dynamic and interaction information in solution, which requires sequence-specific resonance assignment of NMR signals. While such analysis is quite standard for small proteins, it remains one of the major bottlenecks when the size of the protein increases.

Here, we describe implementation and latest improvements of SeSAM, a fast and user-friendly approach for assignment of methyl resonances in large proteins using mutagenesis. We have improved culture medium to boost the production of methyl-specifically labeled proteins, allowing us to perform small-scale parallel production and purification of a library of $^{13}CH_3$-specifically labeled mutants. This optimized protocol is illustrated by assignment of Alanine, Isoleucine, and Valine methyl groups of the homododecameric aminopeptidase PhTET2. We estimated that this improved method allows assignment of ca. 100 methyl cross-peaks in 2 weeks, including 4 days of NMR time and less than 2 k€ of isotopic materials.

Key words Methyl group, Isotopic labeling, High molecular weight proteins, NMR spectroscopy, SeSAM, Assignment, Site-directed mutagenesis

1 Introduction

Supramolecular systems are involved in many of the key processes that occur in cells. Therefore, understanding their local structure and dynamics is critical. For such investigations, NMR spectroscopy is a technique of choice and now allows studies of assemblies up to 1 MDa [1]. This was made possible by the development of protocols for the selective protonation of methyl groups in perdeuterated proteins [2–6]. This strategy is based on some very favorable relaxation properties of methyl groups in proteins that show

Yu Wai Chen (ed.), *Structural Genomics: General Applications*, Methods in Molecular Biology, vol. 1091, DOI 10.1007/978-1-62703-691-7_17, © Springer Science+Business Media, LLC 2014

increased sensitivity compared to backbone amide proton. Moreover, methyl group containing residues are usually common and well dispersed within the polypeptide sequences, covering homogeneously the protein space. Thus, methyl groups are excellent probes of protein structure, dynamics, and interactions, particularly for very large proteins. In addition to studying naturally occurring methyl groups, methyl-containing amino acids can also be used to replace solvent-exposed residues as NMR reporters of protein interaction in order to, for instance, validate an expected binding site [7, 8].

In all cases, analysis of structural and dynamic information yielded by methyl groups requires sequence-specific assignment of methyl resonances. Conventional through bonds assignment approaches [9] to assign backbone and side chain methyl groups resonances work efficiently for small proteins but cease to be applicable to proteins over 100 kDa. In this case, alternative approaches are required. Several examples of different methyl group assignment procedures that have been successfully applied to large proteins are provided below. To date, many of the supramolecular systems studied by NMR spectroscopy are multimeric; an option to assign such assemblies is thus to try to split the quaternary complex into smaller fragments. This “divide-and-conquer” technique [1, 10] relies on disassembling the oligomeric system and transferring the resonances assignment to the full-size complex. This method however requires considerable optimization to find conditions that destabilize oligomeric interfaces without significantly disrupting the structure of the monomer or domain. Another alternative to overcome size limitation is solid-state NMR spectroscopy [11], in which the linewidth is independent of the molecular weight [12]. Yet, this approach requires crystal preparation giving high-quality spectra similar to solution state NMR, and time-consuming analysis of complex ^{13}C–^{13}C correlation spectra acquired using solid-state NMR. Methyl-methyl Nuclear Overhauser Enhancement (NOE) experiments can also be used in combination with chemical shift prediction programs to assign methyl groups in proteins [13]. This method is so far limited to small systems due to the complexity of detecting long range NOE at more than 7 Å in very large protein assemblies. Another approach to assign large proteins is to analyze paramagnetic relaxation enhancements (PRE) induced by nitroxide spin-labels in combination with an available 3D structure [14].

Alternatively, several groups have reported a mutagenesis-based approach for assignment of some methyl resonances in large proteins. Using Leu or Val mutations along with stereospecific $^{13}CH_3$ labeling of Leu/Val residues, A. Seven and J. Rizo [15] were able to assign methyl resonances to a 73 kDa protein domain. Similarly, 15 of 17 methyl resonance frequencies of methionine methyl groups of an RNA polymerase were assigned by site-by-site mutation strategy [16]. A similar approach was used to obtain some Ile, Leu, or Val methyl group assignments in the protease ClpB [17]. In parallel, the assignment-by-mutagenesis strategy has also been applied to a 468 kDa supramolecular protein oligomer, for the first time in a highly systematic way [18]. The method called SeSAM (Sequence-Specific Assignment of methyl groups by Mutagenesis) is based on automated molecular biology techniques, small-scale parallel preparation of residue-specific isotope-labeled samples, and sensitivity-optimized NMR experiments. Each mutant construct is expressed on a small-scale using fully perdeuderated expression media supplemented with isotope-labeled metabolic precursors designed for the specific protonation of a single class of methyl group [2–5, 10, 19–21]. A conservative mutation of one methyl-containing residue to another nonlabeled one causes the disappearance of its NMR correlation from NMR spectrum of a specifically methyl-labeled sample. This systematic strategy led to complete resonance assignment of the 34 isoleucine-δ1 and 30 alanine-β methyl groups in less than 2 months.

Although very effective, this method remains difficult to implement. The production step is achieved from 50 mL culture medium per mutant, i.e., one culture flask for each mutant. It therefore requires a lot of manipulation from the user and may be the source of handling mistakes. Moreover, the purification step is time consuming because each mutant is purified sequentially. Therefore, we attempted to optimize and simplify this strategy (Fig. 1). First we improved the culture medium to enhance cell density, allowing us to decrease the culture volume and perform all the cultures in parallel, in 24 deep-well plates. Then, all the mutants were purified in parallel, on a 96-well plate format, therefore enabling purification of the samples in a few hours instead of weeks. Using this improved approach, we were able to reduce experiment time by a factor of 4 and isotope cost by a factor of 2 compared to previously published implementation [18].

2 Materials

2.1 Expression of Methyl-Specifically Labeled Proteins

1. Freshly transformed *E. coli* cells (BL21(DE3), BL21(DE3) RIL… etc) to overexpress protein of interest.
2. LB broth.
3. 2× M9 medium prepared in H_2O (for 1 L: 20 g of Na_2HPO_4, $7H_2O$; 6 g of KH_2PO_4; 1 g of NaCl; 2 g of NH_4Cl). Autoclave to sterilize.
4. M9 prepared in D_2O (for 1 L: 5.3 g of anhydrous Na_2HPO_4; 3 g of anhydrous KH_2PO_4; 0.5 g of NaCl; 1 g of NH_4Cl). Use sterile D_2O.
5. Oligo-elements (for 1 L of M9 medium: 1 mL of 1 M $MgSO_4$, 1 mL of 0.1 M $CaCl_2$, 1 mL of 0.1 M $MnCl_2$, 1 mL of 50 mM

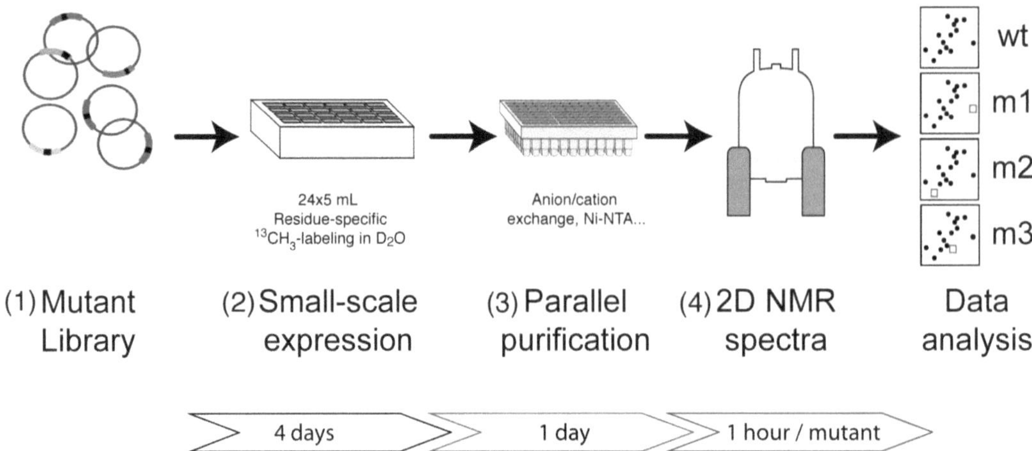

Fig. 1 The Principle of improved SeSAM. Schematic illustration of the parallel mutation-based NMR assignment strategy. (1) Each methyl-containing residue in the target sequence is mutated, on a site-by-site basis, to another similar methyl containing amino acid (e.g., Val-to-Ala) (*see* Table 1). (2) Mutant constructs are expressed on a small-scale in 24 deep-well plates, using M9 medium prepared in D_20 and supplemented with 2 g/L ^{2}H-cell extract. One hour before induction, fully perdeuderated expression medium is complemented with isotope-labeled metabolic precursors designed for the specific incorporation of $^{13}CH_3$ isotopes of a single class of methyl groups. The volume of culture is adjusted to ensure a minimal yield of 0.3–0.5 mg of purified protein. (3) Cell pellets are then lysed in parallel using chemical lysis buffer and proteins are purified in 96-well plates filled with anion exchange (or any other suitable) resin. Each mutant sample is dialyzed against H_20, lyophilized, and dissolved in NMR suitable buffer. (4) NMR spectra can be acquired using the SOFAST-methyl TROSY pulse sequence and an NMR spectrometer operating at high magnetic field. Sequence-specific assignment of each NMR signal is inferred by comparing the 2D ^{1}H-^{13}C correlation spectrum of each member of the mutant library with a spectrum of the native protein. A conservative mutation of one methyl-containing residue to another nonlabeled one causes the disappearance of the methyl group signal from NMR spectra recorded for a specifically methyl-labeled sample

$ZnSO_4$, 0.5 mL of 100 mM $FeCl_3$). Sterilize on 0.22 μm filter. Stocks solutions should be prepared in H_2O when used in M9 100 % H_2O or 50 % H_2O/D_2O (for the precultures) and in D_2O when used in M9 100 % D_2O. In this case, all powders should be dissolved and lyophilized twice in D_2O to remove residual water before preparing stock solutions.

6. Vitamin cocktail (for 50 mL: 25 mg of pyridoxine, 25 mg of biotin, 25 mg of panthothenate hemi-calcium, 25 mg of folic acid, 25 mg of choline chloride, 25 mg of niacineamide, 2.5 mg of riboflavin, 125 mg of thiamine). Solubilize by adjusting the pH around 7, sterilize on 0.22 μm filter and decrease the pH around 5 for long-term storage. Use 2 mL for 1 L of M9 medium. Vitamins should be prepared in H_2O for precultures and in D_2O when used in M9 100 % D_2O (*see* above).

7. Isotopes: D_2O ($^2H \geq 99.8$ %), D-(^{2}H, ^{12}C)-glucose ($^2H \geq$ 98 %), deuterated rich cell extract. Several sources of cell extract are commercially available (Spectra 9 (CIL), Celtone® Complete Medium (CIL), BioExpress® 1000 (CIL), Silantes®

E.Coli-OD2 (Silantes), Isogro® (Isotec), etc.) In this study, we chose Isogro® (noted as ^{2}H-cell extract in the following text), but other cell extracts are likely to give the same results.

8. IPTG (1 M in D_2O).
9. $^{13}CH_3$-methyl-specifically labeled precursors were purchased on a deuterated form ready for direct introduction into the culture medium (NMR-Bio): ^{2}H-$^{13}CH_3$-Alanine ($^{13}C \geq 99$ %; $^{2}H \geq 98$ %), ^{2}H-$^{13}CH_3$-2-ketobutyric acid ($^{13}C \geq 99$ %; $^{2}H \geq 98$ %), ^{2}H-$^{13}CH_3$-2-hydroxy-2-methyl-3-oxo-4-butanoic acid ($^{13}C \geq 99$ %; $^{2}H \geq 95$ %), ^{2}H-L-Isoleucine ($^{2}H \geq 98$ %), ^{2}H-L-Leucine ($^{2}H \geq 98$ %), ^{2}H-α-ketoisovalerate ($^{2}H \geq 98$ %).
10. 10 mL- 24 deep-well plates.
11. Gas permeable adhesive seals.

2.2 Purification

1. 10× BugBuster® buffer (Merck-Millipore).
2. DNAse, RNAse, lysozyme (Euromedex).
3. 96-well filter plates (Macherey-Nagel).
4. 2,2 mL-96 deep-well plates.
5. Aluminum seals.
6. QIAvac 96 Vacuum manifold (Qiagen) or any vacuum manifold for processing 96-well plates.
7. Appropriate resin (for nontagged proteins: any ion exchange resin (Q sepharose, SP sepharose, etc.) or affinity resin (Protein A, Protein G, Heparin sepharose, etc.); for tagged proteins: Ni-NTA, Talon resin, Strep-tactin sepharose. etc.). In this study, we used Q sepharose resin (GE Healthcare Life Sciences).
8. Anion exchange chromatography equilibration/washing buffer: 20 mM Tris–HCl pH 7.5, 160 mM NaCl.
9. Anion exchange chromatography elution buffer: 20 mM Tris–HCl pH 7.5, 350 mM NaCl.
10. Dialysis membrane (Gebaflex, Dialysis system for small-volume samples).

2.3 NMR Spectroscopy

NMR spectra were recorded on a Agilent Direct Drive spectrometer operating at a proton frequency of 800 MHz equipped with a 5 mm cryogenically cooled triple resonance pulsed field gradient probe head. Samples were loaded in a 2.5 mm shigemi tube inserted coaxially into a 5 mm tube.

3 Methods

The goal of the approach is both to improve yields and simplify the procedure previously published [18]. For that purpose, all the small-scale parallel production, lysis, and purification steps should first be setup on uniformly deuterated native protein before applying the

Table 1
Suggested substitutions for methyl containing residues

Mutated CH_3-containing residue	Suggested substitution	Acceptable substitution
Ala	Ser	Val [18], Gly, Cys, Thr
Ile	Val [30, 32]	Leu [17, 18, 32], Met
Leu	Ile [17], Met	Val, Phe
Val	Ile [17]	Met, Leu, Ala [25], Thr
Thr	Ser	Ala, Asn, Val
Met	Leu	Ile, Val

Adapted from BLOSUM 62 substitution matrix [29]
The BLOSUM (BLOcks SUbstitution Matrix) matrix is a substitution matrix used for sequence alignment of proteins [29]. BLOSUM matrices are used to score alignments between evolutionarily divergent protein sequences. They are based on local alignments. In this table, suggested/acceptable substitutions for each of the 6 methyl-containing amino acids, for the purpose of assignment using SeSAM strategy [18], are listed along with an application reference when it exists

protocol to the whole isotopically labeled library of mutants. The hypothesis is made that all mutants behave identically to the native protein and that methyl labeling does not change the purification profile (same expression level, same purification conditions, etc.).

3.1 Generation of Mutant Plasmids Libraries

Constructs carrying single point mutations can be purchased commercially or prepared using an automated molecular biology platform. Here, they were generated by in-house automated molecular biology Platform (RoBioMol—Institut de Biologie Structurale, Jean-Pierre Ebel) using an automated PCR-based protocol adapted from the QuikChange site-directed mutagenesis method. PCR amplification was performed with Phusion Hot Start enzyme (Finnzymes) using the expression plasmid pET41c-PhTET2 as template and the specific mutagenic primers. Products were purified and digested by DpnI. Final mutations were selected by transformation and verified by sequencing. Valine residues were mutated into Alanine residues, Alanine into Valine, Isoleucine into Leucine. Amino acids must be substituted by an isosteric one to avoid considerable changes in the structure and minimize secondary shift effects (*see* **Note 1** and Table 1).

3.2 Protein Labeling for Methyl Detection

Selective methyl protonation provides excellent probes for monitoring interactions and dynamics, and high-quality spectra can be recorded in very large systems. For assignment, a prerequisite is to restrict labeling to only one type of methyl-containing residue at once, in order to minimize secondary chemical shift perturbations due to the mutation. Currently, methods are available to label Isoleucine (δ1 [2] or γ2 [22]), Alanine [3, 23], Threonine [24], or Methionine [10, 20] methyl groups efficiently. For Leucine and Valine, however, the labeling scheme is more difficult because these amino acids share the same biogenesis pathway and each residue has two prochiral methyl groups. Consequently, compared to other residues, for a similar spectral window, the number of methyl group resonances is four times higher, potentially resulting in peak overlap and loss of resolution. Thus, until recently, the assignment-by-mutagenesis strategy was not attractive for Valine residues. Since then, a stereospecific isotopic labeling method was developed for Leucine/Valine labeling [21]. More recently, our group has also developed the stereoselective labeling of Valine without Leucine using coincorporation of pro-*S* Acetolactate and ^{2}H-Leucine [25]. This new isotopic labeling strategy increased the resolution and reduced overlaps and potential secondary chemical shifts by a factor of 4 on average. Thus, it enabled easy assignment-by-mutagenesis of Valine residues in large protein assemblies, compared to standard labeling using 2 keto-acids [26].

In this section we describe the protocol to specifically label large proteins for methyl detection and produced them on a small-scale using enriched cell extract.

1. Bacteria must first be trained to adapt progressively in 100 % ^{2}H medium. All these adaptation steps will be performed in 24-well plates covered with gas permeable adhesive seal. Pick a fresh colony of transformed cells from each mutant of an LB plate (*see* **Note 2**) and start a 2-mL bacterial culture of LB medium prepared in H_2O. Once OD_{600nm} reaches 3–4 (6–8 h) at 37 °C, transfer 50 μL of the cell culture into 2 mL of M9 medium prepared in H_2O. The starting OD_{600nm} of this new culture should be about 0.1. Incubate the culture in a shaking incubator (220–250 rpm) at 37 °C overnight (the OD_{600nm} should be around 2–2.5) and transfer 100–200 μL of the cell culture into 2 mL of M9 medium prepared in 50 % D_2O (*see* **Note 3**) (starting OD_{600nm} = 0.2). Let the culture grow at 37 °C until the OD_{600nm} reaches 2–2.5 and transfer 300 μL of the cell culture into 3 mL of M9 medium in 100 % D_2O (starting OD_{600nm} = 0.25). Incubate the culture at 37 °C overnight until OD_{600nm} is 1.5–2 and use this cell culture as your starting culture.
2. Prepare the volume of M9 needed for the whole culture (2 × 5 mL for each mutant) in a sterile flask (M9 in D_2O + oligo-elements in D_2O + vitamins in D_2O + 2 g/L of ^{12}C, ^{2}H glucose + antibiotics). Add 2 g/L of ^{2}H-cell extract (*see* **Notes 4–6**).
3. Fill a 24 deep-well block (*see* **Note 7**) with 3.5 mL of this medium. Inoculate each well with the overnight culture at a

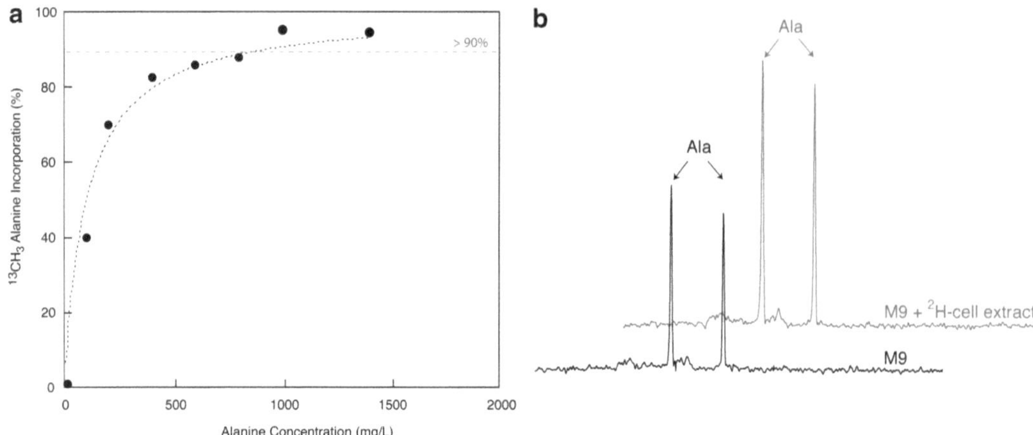

Fig. 2 Level of incorporation of $^{13}CH_3$-alanine in proteins expressed in M9 medium supplemented with ^{2}H-cell extract. Level of incorporation of ^{13}C at the Cβ-alanine position in overexpressed ubiquitin as a function of the amount of the exogenous alanine added in culture medium 1 h prior to induction. Ubiquitin was expressed in *E. coli* in M9/D_2O culture medium supplemented with 2 g/L ^{2}H, ^{12}C glucose, and 2 g/L ^{2}H, ^{12}C Isogro®. Different amounts of $^{13}CH_3$ L-alanine (0–1.4 g/L) were added 1 h before induction. ^{2}H-Isovalerate (400 mg/L) and ^{2}H-Isoleucine (120 mg/L) were added to prevent scrambling from alanine. Fixed amount of $^{13}CH_3$ L-methionine (0.5 g/L) was used as an internal reference. The level of incorporation was determined by analyzing the intensities of one alanine methyl resonance with respect to signal of methyl group of one methionine residue. A level of incorporation in Alanine side chains ≥90 % is obtained by adding 1 g of alanine per liter of culture medium. (**b**) Comparison of 1D ^{13}C-filtered NMR spectra of U-[^{12}C, ^{2}H], Ala-[$^{13}CH_3$]β ubiquitin expressed in equal volume of M9 medium (M9) or M9 medium supplemented with ^{2}H-cell extract (M9+^{2}H-cell extract). Data were recorded at 37 °C using a 2.5 mm Shigemi tube, on an 800 MHz NMR spectrometer equipped with a cryogenic probe head. Comparison of 1D spectra shows an increase of the signal-to-noise ratio by a factor of 1.65. These results are in agreement with an increase of both the cell density (OD_{600nm} × 1.7 with ^{2}H-cell extract) and the culture yields (yields × 1.75 with ^{2}H-cell extract), indicating that ^{13}C-Alanine is fully incorporated in these conditions

starting OD_{600nm} of 0.3. Cover with gas permeable adhesive seal and grow at 37 °C until OD_{600nm} reaches 1.5.

4. Add the precursors (*see* **Note 8**) diluted in 1.5 mL of M9/D_2O, 1 h prior to IPTG induction. The amount of $^{13}CH_3$-precursors to add was optimized to ensure complete labeling of proteins overexpressed in rich medium (Fig. 2a). The yield of the protein was improved by a factor of 1.7 when ^{2}H-cell extract was added to the culture medium, while the frequency of labeling was still almost 100 %, proving there is no isotopic dilution by ^{2}H-cell extract (Fig. 2b).
5. Let the culture grow for 1 h. The OD_{600nm} should reach a value of 1.5.
6. Add IPTG (in D_2O) to 0.5 mM to induce protein expression. Continue incubation at 37 °C for 4 h (depending on your protein).

7. Centrifuge the entire plate in a swing-out rotor for 20 min at 3,250 × *g*. Discard the supernatant and store the pellet at −80 °C or process directly.

3.3 Lysis

Protein purification and lysis strategy may vary according to the protein of interest (presence and type of fusion tag). Here we described the lysis strategy optimized for PhTET2 expressed in BL21(DE3)RIL.

1. Lyse the cells (*see* **Note 9**). Add 250 μL of lysis buffer into each well in the 24-well culture plate (1× BugBuster®, 20 mM Tris–HCl pH 7.5, 100 mM NaCl (*see* **Note 10**), 5 mM $MgCl_2$, 2 μg/mL DNAse, 10 μg/mL RNAse, 0.3 mg/mL Lysozyme) and resuspend the cells by pipetting up and down. Pool the 2 pellets of the same mutant (*see* **Note** 7) and incubate 40 min at room temperature with (occasional) shaking.
2. Heat the crude extract at 85 °C (*see* **Note 11**) for 15 min by making the plate directly float in a water bath.
3. Centrifuge the plate at 3,250 × *g* for 20 min. Both soluble- and whole-cell pellets should be analyzed (*see* Subheading 3.4 **step 6**).

3.4 Small-Scale Parallel Purification

In this section, we describe a typical protein purification procedure using an anion exchange resin, but alternative resins can be used (*see* **Note 12**).

1. Prepare the purification 96-well filter plate. Resuspend the Q sepharose resin thoroughly. Pipet 800 μL of resin suspension (bed volume of 400 μL) into each well of the plate. Wash each well twice with 1 mL of water and 3 times with 1 mL of equilibration buffer using a vacuum manifold (or alternatively a centrifuge with a swing-out rotor).
2. Transfer the clear supernatants (Subheading 3.3 **step 3**) into the 96-well filter plate containing the resin. Seal the block with aluminum seal and caps to avoid leaks and place at room temperature for 1 h with gentle shaking.
3. Place the 96-well filter plate over a 96 deep-well plate in the vacuum manifold. Remove the aluminum seal and caps (only over the used wells) and let the samples flow through the resin first by gravity, then by applying vacuum until the samples have been completely drawn through the plate.
4. Place a drain deep-well plate and wash the resin by adding 400 μL of washing buffer to each well and then apply vacuum as above. Repeat the wash 4 times.
5. Place the filter plate on top of a new deep-well plate and add 400 μL of elution buffer in each well. Incubate 2 min and proceed as above. Repeat the elution 4 times and store the eluate.

6. Analyze by SDS-PAGE the total, soluble and eluted fractions for some of the mutants, according to standard procedures.

3.5 NMR Spectroscopy

To assign a large protein such as PhTET2 (468 kDa, 353 residues per subunit), the SeSAM strategy [18] is followed.

The type of probe and the tube configuration can affect the required amount of material. For small sample amounts, users should always choose the most sensitive probe available (best signal-to-noise per mg of protein). Using a standard 5-mm cryogenically cooled probe, the protein should be concentrated to optimize the amount of spins present in the most sensitive area of the active volume (i.e., near the axial symmetry axis). This can be achieved simply by placing the sample in a small diameter tube (1 to 3 mm) centered inside a standard 5-mm tube. As a result, compared to a 5-mm NMR tube, the sensitivity gain is ca. a factor of 2 per mg of protein. Alternatively, using cryoprobes optimized for small volumes (3 mm coldprobe (Agilent); 1.7 mm microcryoprobe (Bruker)) can further increase sensitivity twofold allowing a reduction of the culture volume (one 5-mL well for each mutant) or a division of the acquisition time by a factor of 4. In this study, the samples were loaded in a 2.5 mm shigemi tube placed coaxially to a regular 5-mm NMR tube as a sample holder and NMR spectra were recorded on a spectrometer equipped with a 5-mm cryoprobe.

1. Prepare methyl-labeled mutants as described above. Transfer the protein in an NMR suitable buffer. In our case, the protein was extensively dialyzed in H_2O, lyophilized and resuspended in 60 μL (± 0.15 mM) of 20 mM Tris–DCl, 20 mM NaCl pH 7.4 in 100 % D_2O. Alternatively (if the protein is not stable in pure H_2O), buffer can be exchanged by a series of dilution/concentration steps in D_2O buffer using a concentration unit with an appropriate cut-off.
2. Record ^{1}H-^{13}C SOFAST-methyl-TROSY spectra (more sensitive experiment by unit of time [27]) for each mutant sample and native protein. Here, taking advantage of the thermostability of PhTET2, NMR data are recorded at 50 °C. The angle of proton excitation pulse is set to 30° and the recycling delay is optimized to 0.4 s. The length of each NMR experiment is adjusted depending on the concentration of the sample (typically 1 h for 9 nmol of sample).
3. All data are processed and analyzed with NMRPipe [28]. Sequence-specific assignments of each NMR signal are inferred by comparing the 2D ^{1}H-^{13}C correlation spectrum recorded for each member of the mutant library with a spectrum recorded for the wild-type protein (Fig. 3). Conceptually, assignment-by-mutagenesis is straightforward. In practice, however, the overlap of resonances and the occurrence of secondary chemical

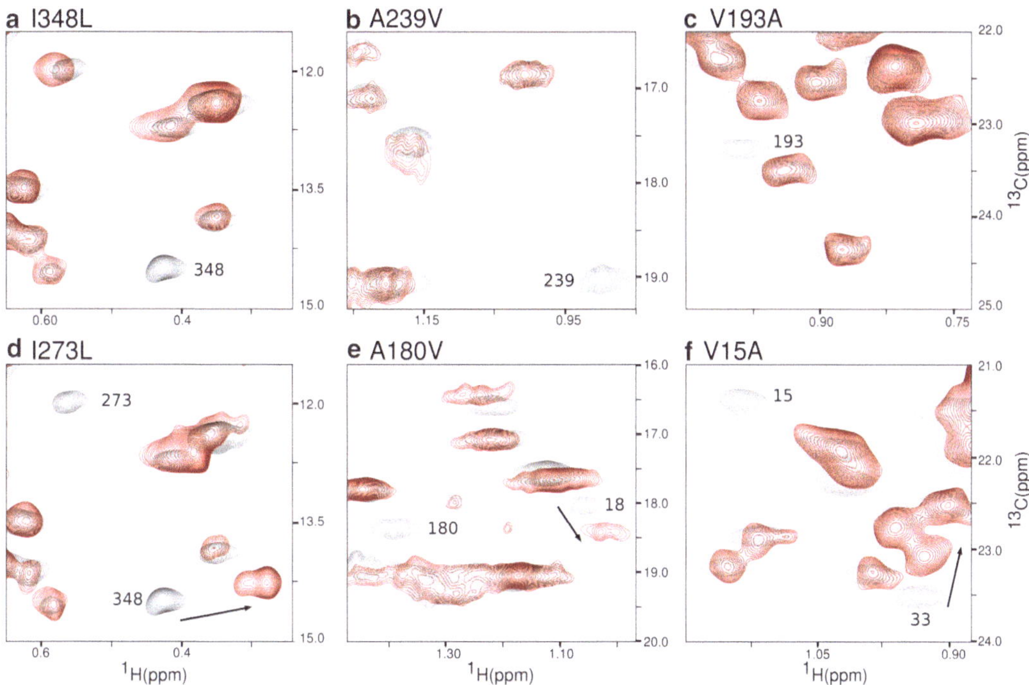

Fig. 3 Examples of spectra of mutants with specific Ile-δ1, Ala-β, or Val pro-*S*-labeled methyl probes. Spectra of mutants displaying modest secondary chemical shift perturbations (**a–c**) were initially chosen for analysis and, when possible, sequence-specific assignment of methyl groups were made. This initial set of unambiguous assignments assisted the analysis of spectra displaying larger chemical shift perturbations (**d–f**). Spectra of this figure are extracted from the work of Amero et al. [18] and Mas et al. [25]. SOFAST Methyl-TROSY spectra were recorded at 50 °C, on an 800 MHz NMR spectrometer equipped with a cryogenic probe head, using samples of U-[^{12}C, ^{2}H], Ile-[$^{13}CH_3$]δ^1 (**a** and **d**) [18], U-[^{12}C, ^{2}H], Ala-[$^{13}CH_3$]β (**b** and **e**) [18], or U-[^{12}C, ^{2}H], Val-[$^{13}CH_3$]$^{pro\text{-}S}$ (**c** and **f**) [25] labeled mutant PhTET2 protein (*red*). Each spectrum was overlaid with the reference spectrum of the native particle (*black*). Experimental acquisition times were adjusted to the sample protein concentration (from 0.26 to 0.42 mg/sample, i.e., 7–11 nmol of PhTET2 monomer) with a maximum experimental time set to 1 h. The assignment inferred for the missing resonance in the mutant spectrum is indicated and secondary chemical shift perturbations are annotated with *arrows*

shifts changes can confuse the analysis. Perturbations are likely to occur but can be minimized (using conservative mutations). In the first round, all spectra with only one missing peak are considered for a straightforward assignment of a first set of resonances. Then, more complex spectra are studied with consideration of structure, first set of unambiguous assignment and the whole set of spectra taken into account, in order to analyze secondary chemical shift perturbations. That is why considering the full library of single-site mutations greatly simplifies the process of resonance assignment by cross-validating the results several times (*see* **Note 13**).

4 Conclusion

Thanks to the development of protocols and molecules that allow residue-specific protonation of methyl groups in highly perdeuterated proteins, it is now feasible to apply solution NMR techniques to protein systems as large as 1 MDa. Structural and dynamic information yielded by methyl groups is most useful when a sequence-specific assignment for the probe is known. Easily obtaining these assignments remains the major hurdle in many studies of large proteins. Here we describe a fast, efficient, and user-friendly protocol for resonance assignment that has allowed us to assign up to 100 methyl cross-peaks in about 2 weeks with 4 days of NMR time and a isotopic cost of less than 2 k€. We demonstrated the feasibility of this protocol on samples labeled on Alanine, Isoleucine, or Valine. This method can also be extended to Methionine- or Threonine-labeled proteins. As for Valine labeling, the methyl-specific labeling of Leucine residues will require the development of a stereoselectively labeled amino acid (or precursor) to label Leucine residues independently from Valines.

5 Notes

1. Amino acids should typically be substituted by an isosteric amino acid to prevent significant changes in the local environment and protein packing, which could introduce significant chemical shift perturbations. We chose to exchange Valine to Alanine according to Amero et al. [18], who mutated Alanine into Valine. However, according to BLOSUM matrices for amino acid substitutions [29], the Valine to Isoleucine mutations (score 3) would have been wiser and most likely to have minimal effect on the structure and NMR spectra. Indeed, Valine differs from Isoleucine by the loss of only one CH_2 group, whereas it differs from Alanine by the loss of $CH–CH_3$. According to these matrices, Chan et al. [30] generated a set of mutants in which each Isoleucine was substituted by a Valine (score 3). The resulting chemical shift perturbations were smaller than those from the corresponding Ile to mutations Ala (score −1) facilitating the assignment (Table 1).
2. Antibiotics are not specified. Put appropriate antibiotics considering the plasmid and the cells used. However, ampicillin is not suitable because it can be inactivated by the β-lactamases produced by the cell, resulting in plasmid loss and drop of protein expression. This effect seems to be worse in D_2O medium because of the successive adaptation steps that promote loss of selective pressure. Here, we routinely used the pET-41 plasmid, which contains the kanamycin resistance gene.

3. M9 50 % D_2O is prepared with 2× M9 H_2O and same volume of D_2O. The oligo-elements, vitamins, antibiotics, and glucose used at this step are still prepared in water.
4. To increase cell density, we supplemented M9 medium with ^{2}H-cell extract. We optimized the protocol with Isogro® but many similar rich bacterial cell growth media are commercially available (Spectra 9, Celtone® Complete Medium, BioExpress® 1000, Silantes® E.Coli-OD2, etc.). We did not test them but they can most likely be used to replace Isogro®, after suitable optimization.
5. Rich medium can be the source of isotopic dilution as it contains $^{12}C^2H_3$-labeled amino acids. To minimize this, we tested different concentrations of ^{2}H-cell extract (2, 3, or 10 g/L). Growth is significantly increased from 0 to 2 g/L; however higher concentrations of ^{2}H-cell extract did not show further significant improvement. Moreover, ^{2}H-cell extract is added at the start of the culture, whereas methyl specifically labeled precursors are added 1 h before induction. This early addition reduces isotopic dilution because most of deuterated amino acids are consumed for the cell growth, while $^{13}CH_3$-precursors are incorporated for the production of the labeled protein. Optimization was done using Isogro®, but similar results are expected for other types of commercial ^{2}H-cell extracts.
6. For ubiquitin, yields with 2 g/L of ^{2}H-cell extract were increased by a factor of 1.7 (Fig. 2b).
7. Considering both yields, parallelization of the method and simplification of the purification steps, we were able to reduce the volume of production to 10 mL for each mutant. One 24 deep-well plate is thus suitable for 12 mutants (2 wells of 5 mL for each mutant). The final concentration in a 2.5 mm Shigemi tube is around 0.15 mM (9 nmol or 0.35 mg of PhTET2 monomer) in 60 μL.
8. Incorporation level of $^{13}CH_3$-alanine was determined using 2 g/L of ^{12}C, ^{2}H-glucose, and 2 g/L of ^{2}H-cell extract in culture medium. Using higher concentrations of glucose may cause isotopic dilution of the $^{13}CH_3$-precursor [31] and the incorporation curves must then be modified accordingly.
9. One crucial step to optimize in the protein production process is the bacterial cell lysis. Different lysis methods should be tested. However, when tens or hundreds of mutants are produced, only a few lysis methods can be reasonably used in parallel. Sonication proved problematic unless using High Throughput sonicators (which are expensive) or ultrasonic bath (but the ultrasonic energy is not always equally distributed throughout the plate and the results may not be reproducible from well to well). For these reasons, we chose to

optimize lysis conditions using chemical treatment (all the preliminary tests were performed on the native protein). This allowed us to achieve lysis directly on the culture plates, in parallel. We tested different lysis buffers and selected the BugBuster®, as it showed the best solubility yields for our protein of interest.

10. The NaCl concentration of the lysis buffer has to be settled according to the protein of interest. Decrease it if the protein does not bind to the ion exchange resin.
11. The protein studied is an aminopeptidase from a thermophilic organism, adapted to high temperatures. The protein is thus stable at 85 °C for 15 min, whereas most of other *E. coli* proteins will precipitate. Bacterial contaminants are then removed by centrifugation.
12. Fill your filter plate with any suitable resin (anion or cation exchange resin, Ni-NTA, StrepTactin, etc.). Adapt bead volume to the capacity of the resin. Typically, we prepare wells containing 400 μL of anion exchange resin for a 10-mL cell culture. This can be scaled up or down to suit your needs based on the expected protein yield.
13. In almost 50 % of cases (using conservative mutations (*see* **Note 1**)), the only difference between mutant and reference spectra is a single missing cross-peak in the spectrum of the mutant (Fig. 3a–c). In such instances, the missing peak can be unambiguously assigned to the methyl group of the mutated residue. In the remaining spectra, the disappearance of the signal is accompanied by small changes in the chemical shift of a few additional correlations (Fig. 3d–f). This effect is expected and has previously been observed [18, 32]. Peak movements that do not directly concern the mutated resonance can complicate the process of obtaining a sequence-specific assignment from a single experiment, especially in an overcrowded region of the spectrum. Conservative mutations enable minimization of secondary chemical shift perturbations. In the same way, stereospecific labeling of a single methyl group (Valine pro-*S*) versus labeling of 4 methyl groups (Val/Leu) using α-ketoisovalerate [26] reduces peak overlapping as well as secondary chemical shift up to a factor of 4.

    Nonetheless, secondary chemical shift perturbations reflect modifications in the local electronic environment and can therefore provide complementary information that can be used to confirm the proposed assignment. The key point is that the information provided by secondary chemical shift changes only becomes interpretable when data from a full library of methyl group mutants is considered [18]. Any ambiguous assignment can therefore be readily cross-validated using structurally close, straightforward-assigned resonances. Using an incomplete library of mutants would not permit the same level of confidence in the final assignments.

Acknowledgments

We would like to thank Dr P. Macek, M. Plevin, O. Hamelin, P. Gans, I. Ayala, C. Amero, and A. Favier for stimulating discussions and assistance in sample preparation or analysis. This work used the RobioMol, High-Field NMR, Isotopic Labeling, and Seq3A platforms of the Grenoble Instruct centre (ISBG; UMS 3518 CNRS-CEA-UJF-EMBL) with support from FRISBI (ANR-10-INSB-05-02), and GRAL (ANR-10-LABX-49-01) within the Grenoble Partnership for Structural Biology (PSB). The research leading to these results has received funding from the European Research Council under the European Community's Seventh Framework Programme FP7/2007-2013 Grant Agreement no. 260887.

References

1. Sprangers R, Kay LE (2007) Quantitative dynamics and binding studies of the 20S proteasome by NMR. Nature 445:618–622
2. Gardner K, Kay LE (1997) Production and incorporation of ^{15}N, ^{13}C, ^{2}H (^{1}H-δ1 methyl) isoleucine into proteins for multidimensional NMR studies. J Am Chem Soc 119: 7599–7600
3. Ayala I, Sounier R, Use N et al (2009) An efficient protocol for the complete incorporation of methyl-protonated alanine in perdeuterated protein. J Biomol NMR 43:111–119
4. Tugarinov V, Kanelis V, Kay LE (2006) Isotope labeling strategies for the study of high-molecular-weight proteins by solution NMR spectroscopy. Nat Protoc 1:749–754
5. Ruschak AM, Kay LE (2010) Methyl groups as probes of supra-molecular structure, dynamics and function. J Biomol NMR 46:75–87
6. Plevin MJ, Boisbouvier J (2012) Isotope-labelling of methyl groups for NMR studies of large proteins. In: Recent developments in biomolecular NMR. Royal Society of Chemistry. doi:10.1039/9781849735391
7. Stoffregen MC, Schwer MM, Renschler FA et al (2012) Methionine scanning as an NMR tool for detecting and analyzing biomolecular interaction surfaces. Structure 20:573–581
8. Religa TL, Ruschak AM, Rosenzweig R et al (2011) Site-directed methyl group labeling as an NMR probe of structure and dynamics in supramolecular protein systems: applications to the proteasome and to the ClpP protease. J Am Chem Soc 133:9063–9068
9. Bax A (2011) Triple resonance three-dimensional protein NMR: before it became a black box. J Magn Reson 213:442–445
10. Gelis I, Bonvin AM, Keramisanou D et al (2007) Structural basis for signal-sequence recognition by the translocase motor SecA as determined by NMR. Cell 131:756–769
11. Turano P, Lalli D, Felli IC et al (2010) NMR reveals pathway for ferric mineral precursors to the central cavity of ferritin. Proc Natl Acad Sci USA 107:545–550
12. Marassi FM, Ramamoorthy A, Opella SJ (1997) Complete resolution of the solid-state NMR spectrum of a uniformly ^{15}N-labeled membrane protein in phospholipid bilayers. Proc Natl Acad Sci USA 94:8551–8556
13. Xu Y, Liu M, Simpson PJ et al (2009) Automated assignment in selectively methyl-labelled proteins. J Am Chem Soc 131:9480–9481
14. Venditti V, Fawzi NL, Clore GM (2011) Automated sequence- and stereo-specific assignment of methyl-labeled proteins by para-

magnetic relaxation and methyl-methyl nuclear overhauser enhancement spectroscopy. J Biomol NMR 51:319–328

15. Seven A, Rizo J (2012) Assigning the methyl resonances of the 73 kDa Munc13-1 MUN domain by mutagenesis. 25 th ICMRBS-poster n° P302 TU, Lyon
16. Yang X, Welch JL, Arnold JJ et al (2010) Long-range interaction networks in the function and fidelity of poliovirus RNA-dependent RNA polymerase studied by nuclear magnetic resonance. Biochemistry 49:9361–9371
17. Rosenzweig R, Moradi S, Zarrine-Afsar A et al (2013) Unraveling the mechanism of protein disaggregation through a ClpB-DnaK interaction. Science 339:1080–1083
18. Amero C, Asuncion Dura M, Noirclerc-Savoye M et al (2011) A systematic mutagenesis-driven strategy for site-resolved NMR studies of supramolecular assemblies. J Biomol NMR 50:229–236
19. Goto NK, Gardner KH, Mueller GA et al (1999) A robust and cost-effective method for the production of Val, Leu, Ile (δ1) methyl-protonated ^{15}N-, ^{13}C-, ^{2}H-labeled proteins. J Biomol NMR 13:369–374
20. Fischer M, Kloiber K, Hausler J et al (2007) Synthesis of a ^{13}C-methyl-group-labeled methionine precursor as a useful tool for simplifying protein structural analysis by NMR spectroscopy. Chembiochem 8:610–612
21. Gans P, Hamelin O, Sounier R et al (2010) Stereospecific isotopic labeling of methyl groups for NMR spectroscopic studies of high-molecular-weight proteins. Angew Chem Int Ed Engl 49:1958–1962
22. Ayala I, Hamelin O, Amero C et al (2012) An optimized isotopic labelling strategy of isoleucine-gamma2 methyl groups for solution NMR studies of high molecular weight proteins. Chem Commun (Camb) 48:1434–1436
23. Isaacson RL, Simpson PJ, Liu M et al (2007) A new labeling method for methyl transverse relaxation-optimized spectroscopy NMR spectra of alanine residues. J Am Chem Soc 129:15428–15429
24. Sinha K, Jen-Jacobson L, Rule GS (2011) Specific labeling of threonine methyl groups for NMR studies of protein-nucleic acid complexes. Biochemistry 50:10189–10191
25. Mas G, Crublet E, Hamelin O et al (2013) Specific labeling and assignment Strategies of valine methyl groups for the NMR Studies of high molecular weight proteins (submitted)
26. Tugarinov V, Kay LE (2003) Ile, Leu, and Val methyl assignments of the 723-residue malate synthase G using a new labeling strategy and novel NMR methods. J Am Chem Soc 125:13868–13878
27. Amero C, Schanda P, Dura MA et al (2009) Fast two-dimensional NMR spectroscopy of high molecular weight protein assemblies. J Am Chem Soc 131:3448–3449
28. Delaglio F, Grzesiek S, Vuister GW et al (1995) NMRPipe: a multidimensional spectral processing system based on UNIX pipes. J Biomol NMR 6:277–293
29. Henikoff S, Henikoff JG (1992) Amino acid substitution matrices from protein blocks. Proc Natl Acad Sci USA 89:10915–10919
30. Chan PH, Weissbach S, Okon M et al (2012) Nuclear magnetic resonance spectral assignments of α-1,4-galactosyltransferase LgtC from Neisseria meningitidis: substrate binding and multiple conformational states. Biochemistry 51:8278–8292
31. Plevin MJ, Hamelin O, Boisbouvier J et al (2011) A simple biosynthetic method for stereospecific resonance assignment of prochiral methyl groups in proteins. J Biomol NMR 49:61–67
32. Sprangers R, Gribun A, Hwang PM et al (2005) Quantitative NMR spectroscopy of supramolecular complexes: dynamic side pores in ClpP are important for product release. Proc Natl Acad Sci USA 102:16678–16683

Chapter 18

High-Throughput SAXS for the Characterization of Biomolecules in Solution: A Practical Approach

Kevin N. Dyer, Michal Hammel, Robert P. Rambo, Susan E. Tsutakawa, Ivan Rodic, Scott Classen, John A. Tainer, and Greg L. Hura

Abstract

The recent innovation of collecting X-ray scattering from solutions containing purified macromolecules in high-throughput has yet to be truly exploited by the biological community. Yet, this capability is becoming critical given that the growth of sequence and genomics data is significantly outpacing structural biology results. Given the huge mismatch in information growth rates between sequence and structural methods, their combined high-throughput and high success rate make high-throughput small angle X-ray scattering (HT-SAXS) analyses increasingly valuable. HT-SAXS connects sequence as well as NMR and crystallographic results to biological outcomes by defining the flexible and dynamic complexes controlling cell biology. Commonly falling under the umbrella of bio-SAXS, HT-SAXS data collection pipelines have or are being developed at most synchrotrons. How investigators practically get their biomolecules of interest into these pipelines, balance sample requirements and manage HT-SAXS data output format varies from facility to facility. While these features are unlikely to be standardized across synchrotron beamlines, a detailed description of HT-SAXS issues for one pipeline provides investigators with a practical guide to the general procedures they will encounter. One of the longest running and generally accessible HT-SAXS endstations is the SIBYLS beamline at the Advanced Light Source in Berkeley CA. Here we describe the current state of the SIBYLS HT-SAXS pipeline, what is necessary for investigators to integrate into it, the output format and a summary of results from 2 years of operation. Assessment of accumulated data informs issues of concentration, background, buffers, sample handling, sample shipping, homogeneity requirements, error sources, aggregation, radiation sensitivity, interpretation, and flags for concern. By quantitatively examining success and failures as a function of sample and data characteristics, we define practical concerns, considerations, and concepts for optimally applying HT-SAXS techniques to biological samples.

Key words High-throughput, SAXS, Conformation, Structure, Structural genomics, Macromolecules

1 Introduction

Small angle X-ray scattering (SAXS) has reemerged in its application to the study of biological macromolecules. SAXS from biomolecules was an early application of synchrotron radiation [1] in part because of its simplicity in terms of sample preparation. However

Yu Wai Chen (ed.), *Structural Genomics: General Applications*, Methods in Molecular Biology, vol. 1091, DOI 10.1007/978-1-62703-691-7_18, © Springer Science+Business Media, LLC 2014

with the realization of degree to which biomolecules could be crystallized yielding atomic resolution structures, macromolecular crystallography (MX) quickly became a focus of structural biologists. Relatively speaking, the application of SAXS and the development of analytical tools languished. Over the course of the last 10 years, SAXS has reemerged as a powerful complimentary tool to MX.

Three factors have contributed to the emerging power of SAXS. First, not all macromolecules of interest are amenable to crystallization. Even when a macromolecule has been crystallized and modeled to atomic resolution, biologically relevant alternate conformations can, at best, be inferred. Through a genomic analysis, 35–48 % of human gene products are predicted to have significant flexible regions when isolated [2]. SAXS provides an avenue to capture critical structural information from biomolecules even after an atomic resolution model is available. SAXS results suggest conformational variation is a general functional feature of macromolecules, so biologically relevant structural analyses will require a comprehensive approach that assesses both flexibility, as seen by SAXS, and detail, as determined by X-ray crystallography and NMR [3]. Indeed, SAXS also provides three-dimensional arrangements and oligomeric state for full-length proteins in solution, which is typically the functional assembly state, as seen for DNA break response framework proteins [4, 5], thermophilic superoxide dismutase [6], ATPase motors [7], and abscisic acid receptor [8]. Second, analysis tools have been developed and made accessible for the extraction of structural information. Shapes of macromolecule may be determined to ~15 Å resolution. Higher resolution information may be probed by complimenting SAXS with information from an atomic resolution model. Building upon the promise of early tools [9], the EMBL ATSAS [10] package has been transformative. Others have further contributed to the expanding suite of software available for analysis [11–14]. Additionally, the practical implementation of the Porod-Debye law in SAXS experiments of biopolymers provides a tool for assessing flexibility and for validation of SAXS models [15]. Flexible regions of macromolecules are often involved in interactions, as seen for antibody–protein binding [16, 17], and SAXS provides a means to define solution conformations with flexible regions. As generally appreciated, crystal contacts and constructs with missing regions may cause structural changes in the crystal structure relative to the SAXS solution results [18]. SAXS has recently been used to provide similarity maps of the functional conformational states of macromolecules independent of shape reconstructions [19]. Third, high signal to noise SAXS profiles are routinely collected from small quantities of sample with short exposure times. High-quality SAXS profiles are the result of advances in X-ray detectors and high brilliance synchrotron light with beam dimensions that match sample dimensions. Thus the motivation to move beyond the limits of MX, improved analysis tools and collection capabilities have all contributed to the increase in structural reports utilizing SAXS.

The advent and wide spread availability of high-throughput SAXS is relatively new. Pipelines for high-throughput SAXS have been reported at SSRL [20], SOLEIL [21], PETRA3 [22], and CHESS [23]. Several additional beamlines have developed these capabilities and are yet to be reported. SAXS at SIBYLS has been dedicated to HT-SAXS for the last 3 years with the initial application to structural genomics pipelines [24]. SIBYLS has leveraged tools developed for crystallography such as data control software and optimized features for SAXS [25].

A distinction of the SAXS at SIBYLS is that a significant fraction of samples are collected via mail-in/hand-in. Once an investigator's samples have been delivered to the beamline their samples are placed into a queue and collected by beamline staff. The data output is a SAXS profile which tabulates the q value (X-ray momentum transfer) versus X-ray intensity with an error bar. This three-column format is electronically delivered post collection. One advantage to the mail-in/hand-in approach is an increase in flexibly arranging data collection times. Optimal sample preparation is often challenging and difficult to coordinate for a specific time. A second advantage is that "beamtime" is spent collecting data rather than training; thus increasing throughput. The disadvantage is that the investigators themselves are not there to guarantee every sample. Thus the guiding principle for development of the mail-in/hand-in program has been to enable data collection at as high qualities as if the investigator was present themselves. Over 160 laboratories have since taken advantage of this opportunity. Several results have been included in high profile reports [8, 26–29]. Our goal here is not to review post-processing analysis tools used to determine structural details. We suggest other sources for this purpose [10, 30–32]. We've also recently described more technical aspects of the control system and hardware elsewhere [25, 33]. Here we focus on optimal input and a detailed description of the output to improve coordination between investigators and synchrotron beamlines as required for true high-throughput. HT-SAXS appears rigid given the reduced interaction between the beamline and the investigator. In reality both data collection and data processing are flexible. Investigators are empowered to reprocess data by varying from the automated processing steps. By optimally taking advantage of HT-SAXS, new opportunities continue to be developed for the investigation of biomolecules, such as comprehensive mapping of conformational states without requiring shape reconstructions [19].

2 Materials

HT-SAXS opportunities extend beyond experiments preformed at lower throughput. Optimal samples and procedures depend on the type of experiment being performed. Here we will provide general requirements for low signal samples acknowledging that at high concentrations, requirements may be relaxed.

2.1 Concentration

Concentration is an important parameter that impacts signal, problems from aggregation, and data collection requirements.

For organic macromolecules in an aqueous solvent, a useful rule of thumb for determining the required concentration for high-quality signal is concentration in mg/ml multiplied by molecular weight in kDa must be greater than 100 (mg/ml × kDa > 100).

With HT-SAXS the required concentration can be experimentally evaluated, as the desired signal to noise will vary from facility to facility and by the scattering power of the solvent.

2.2 Isolating the Solute Signal

The proper subtraction of background signal is often critical. Background includes the halo of the primary X-ray beam, scattering from windows in the beam path and scattering from solvent. To focus analysis on a solute (the macromolecule of interest), the SAXS from a solution containing all but the macromolecule of interest (referred to from here forward as the buffer) may be subtracted from the SAXS profile of the solution containing the macromolecule. This subtraction removes all three background components mentioned above.

2.3 Matching Buffers

Everything in solution scatters X-rays so having the appropriate matching buffers is critical.

Adequately matched buffers can be prepared by dialysis, size exclusion chromatography (SEC) or from a spin concentrator. However, these procedures must be carefully attended to, for example, filters in concentrators are typically covered in preservatives which must be washed at least three times before the flow through can be used as a proper buffer. Dialysis requires more time with viscous solvents. Some SEC fractions contain small amounts of column matrix so are not appropriate for use as a buffer.

Pipetting of cofactors into both the buffer and the sample, as a modification, is also possible provided the added volumes are equal to high accuracy (usually requires a minimum of 4 μL).

Added signal from improper buffer subtraction will typically reduce the apparent rate of intensity decay as a function of angle; giving the appearance of an unfolded polymer. Over subtracted signal often results in negative intensities at high values of q.

Because of the importance of proper buffer subtraction and because buffer is typically inexpensive, we recommend preparing larger buffer volumes than required for samples and collecting identical buffers both before and after the sample.

2.4 Sample Format

Robotic sample loading from 96-well plates requires decisions regarding shipping, seal against evaporation, and safe volumes for loading the sample cell. If frozen, the plate should be transported in sub-freezing conditions. If unfrozen, care must be taken so that samples do not slow freeze during transport but remain cool. A kilogram of Blue Ice at 5° packed on both sides of the sample plate in a well-sealed (taped) Styrofoam box is a reasonable option.

HT-SAXS facilities have specific sample formats as precise sample locations in three dimensions are required for robotic loading. The sample format at SIBYLS is a specific, commercially available, full-skirt 96 conical well plate. Samples sent in alternate plate types cause delay as samples must be transferred to the proper plate type.

A safe volume for filling the sample cell above the incident beam path is 24 μL.

Plates must also be covered with an appropriate seal for transport to prevent mixing between wells, evaporation and contamination from the sealing material. Plates are typically covered with a commercially available silicone mat.

Once samples are sealed they are ready for shipment or delivery. Flash freezing of samples is possible but usually unnecessary with 24 h shipping times and a maximum of two additional days between delivery and collection. Flash freezing may be accomplished by placing the plate over a shallow bath of liquid nitrogen. Practice with plates containing water is recommended.

2.5 Homogeneity Requirements for Shape Determination

Shape reconstruction requires homogeneous samples and removal of concentration-dependent signals.

A significant fraction of investigators use SAXS data for shape determination. Strategies for data collection for this purpose have been reported [34]. Important procedures include collecting a concentration series to identify and possibly remove concentration-dependent signals contaminating the signal characterizing macromolecular shape.

SAXS by itself cannot determine heterogeneity so supporting data such as elution profiles from chromatographic purification, native gels or multi-angle light scattering are required for quality assessment of homogeneity. Many problems with SAXS experiments on RNA samples derive from heterogeneity of the folded RNA so separation by sizing chromatography or other means is important [35]. The reporting of a single shape representing an entire population of macromolecules that contribute to the SAXS signal assumes homogeneity.

2.6 Organizing Data Collection

An organized plan for sample and washing steps impacts efficiency.

The SIBYLS HT-SAXS pipeline utilizes formatted spreadsheets, filled out by investigators, for organizing data collection. The spreadsheet describes the order of data collection, the desired naming of output experimental files from each sample, which wells contain buffers and at which points in the data collection washes are necessary.

Washing is not required between every well, if sample collection order is strategically chosen. For example a concentration series collected in the order of lowest to highest does not need washing steps. Washing is a significant bottleneck in data collection so the fewer washes the higher the throughput.

3 Methods

3.1 Instrument Calibration

Significant calibration of the SAXS instrumentation is applied prior to data collection. Investigators should be aware of four important calibration procedures which will affect all data sets.

The incident beam orientation, sample position, and detector orientation must all be accurately defined in order to calculate scattering plots of Intensity versus q. This is typically done through the collection and analysis of a crystalline powder pattern. Inaccuracy in this calibration will result in blurred SAXS curves where sharp peaks are broadened and the small q scattering may have larger variation.

The incident X-ray wavelength is calibrated typically by measuring absorbance from metal filters with fluorescence near an electron orbital edge. Inaccuracy in wavelength leads to shifted and stretched SAXS profiles with peaks occurring at an alternate apparent q value.

The beamstop and other shadows blocking scattering from the beamline to the detector are masked out. Inaccuracy in defining these regions will lead to large drops in intensity at small q near the beamstop. If the mask is too large, valuable low q data may be obscured.

A solute of known molecular weight and concentration is collected to enable plotting data on an absolute scale. This calibration can be valuable for calculating molecular weight when the concentration of the macromolecule is known. However the scattering contrast between buffer and solute must be considered relative to the calibrant. Including a calibrant on the sample plate is an alternative. These calibration files are readily available if desired.

3.2 Sample Handling

Communicating sample handling procedures is important as the assumption is that samples are to be stored in cool conditions and centrifuged prior to data collection.

Once samples have been delivered to the facility they are stored at an appropriate temperature (−80 °C for frozen and 4 °C for unfrozen).

Just prior to data collection they are spun in a centrifuge to condense the sample and sediment large aggregates. Once centrifuged, the sealing mat is replaced with a thinner pierceable seal for better sample delivery by the sample loading needle.

3.3 Sample Temperature Control

Temperature is an important and underutilized parameter.

The plate deck and the sample cell are cooled to 15 °C during data collection using a water chiller. The temperature can be decreased, but the dew point must be considered as condensation on the sample cell windows can negatively affect buffer subtractions.

Helium can be added to the sample cell environment to minimize the surrounding humidity, effectively lowering the dew point. The sample cell can also be heated up to 70 °C using a Peltier; however, the temperature is typically kept at 15 °C.

3.4 Data Collection

Strategic data collection and guarding against interfering bubbles is key for efficiency and data quality.

Three plates may be held on the SAXS instrument at one time. At a rate of 4 h/plate this conveniently enables unsupervised overnight collection.

Procedures are in place to automatically stop data collection and alert the beamline scientists when problems occur. If the X-ray source is shutdown for example, the system stops and sends a text message alert. Sample loading and data collection can be monitored by beamline staff remotely.

A snapshot of every loaded sample is taken so that samples with bubbles can be diagnosed after data collection. Often, sufficient volume remains in the plate to recollect these samples.

Samples are pipetted one at a time from the plate into the sample cell, exposed, then pipetted back into the plate.

Typically, the aspiration rate for sample delivery is set at 4 μl/s but can be decreased for viscous, low volume, or bubble-prone samples.

Samples are exposed with a 10^{11} photon/s, 12 keV monochromatic beam in a series of exposures: 0.5, 1.0, 2.0, and 4.0 s in that order. A range of exposure times are collected to identify radiation damage and overcome the limited dynamic range of the detector.

Images from the sample are named using a prefix designated in the investigator prepared spreadsheet followed by the well location, followed by the exposure number. Results from these images are later merged together by the investigator to maximize quality.

3.5 From Images to SAXS Profiles

Once the images are collected from each sample, data processing begins.

Automated scripts subtract the images of the closest collected buffer before the sample and the closest collected buffer collected after the sample. The two profiles are averaged creating a total of three scattering profiles for each sample exposure.

The subtraction process requires normalization for the number of X-rays during the exposure of the buffer and the sample. X-ray flux is monitored by a diode within the beamstop. Extracting an accurate value for the flux during the exposure to the high accuracy required is not a trivial procedure and is a source of error.

Once a subtracted image is created a mask is applied blocking out unwanted pixels for integration.

Subtracted and masked images are then integrated utilizing geometric and wavelength parameters determined from pre-collection calibration.

3.6 Sources of Error and Error Bars

The calculation of error bars and examination of the buffer subtraction impacts quality of data analysis.

Since SAXS images contain many observations at equivalent q, an error bar may be calculated using the standard deviation and average intensity.

A second error of the subtraction process involves slight but random variations in detector background between sample and buffer. In some cases these can be significant.

Mechanisms are in place to enable investigators to repeat the subtraction and integration process using alternate pairings of sample and buffer.

Raw images are rarely desired, thus investigators typically receive the one dimensional SAXS profile of X-ray intensity as a function of q with error bars.

4 Preliminary Visualization and Interpretation of Results

4.1 Sample Report

A sample report and assessment of scattering profiles provides the basis for appropriate data processing.

Besides receiving scatting data files, investigators also receive an html formatted sample report. The report is viewable utilizing web browser software and enables mouse click based zooming for visualization of individual profiles. A partial example is shown in Fig. 1.

Using this comprehensive view of the data, beamline staff provides guidance on which of the three profiles from each sample to use for further processing.

4.2 Judging Buffer Subtraction

Data redundancy and consistency of buffer subtraction guide further data processing.

If the SAXS profile from the sample analyzed with a buffer collected before the sample agrees to within noise to that analyzed with a buffer collected after then the average is used. If the two do not agree then a judgment is made.

Above we described errors that may occur during data collection and may cause this disagreement between buffer subtraction (improperly matched buffer, incorrect measure of the incident X-ray flux, and detector background oscillations). These errors create obvious features in the data.

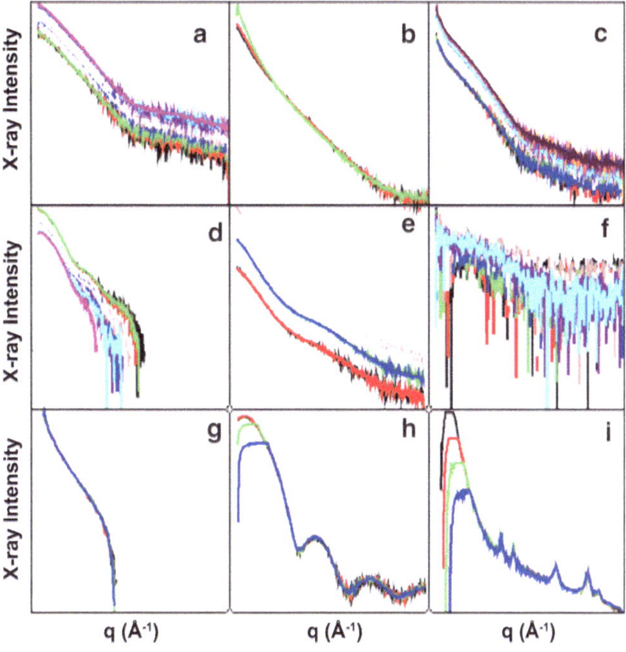

Fig. 1 Exemplary SIBYLS output format of data sets collected from a sample plate. Scattering profiles are grouped by concentration series and graphed on log plots. In the web-enabled version, individual plots can be enlarged for easier viewing. (**a**) A concentration series of a well-behaved sample. (**b**) A sample flagged as radiation sensitive. Aggregation induced through damage has occurred during the highest exposure shown in green. (**c**) The extrapolation of X-ray intensity at $q = 0$ is impossible for the curves shown assuming a particle size smaller than 600 Å. Particles of larger size are considered aggregates at SIBYLS. (**d**) Profiles are over subtracted indicating an error in buffer subtraction (either an inappropriate buffer or instrumental error). (**e**) A slight concentration dependence can be observed as the low q region that increases with concentration (SAXS curves from higher intensity plots). This effect can also be seen in plot. (**f**) The low signal to noise indicates low concentration or insufficient exposure times. (**g**) A sharp drop to negative intensity at low q is characteristic of bubbles or insufficient volume in the sample cell. Images of the sample cell during these exposures may be referenced for further diagnosis. (**h**) The *red* and *black curves* show a smooth downturn in intensity approaching Izero, indicating the presence of inter-particle repulsive forces. The effect is masked by detector saturation in the long exposures (*green and blue curves*). (**i**) Aside from major detector saturation, the curve shows the rare presence of micro-crystals as indicated by sharp peaks of intensity

Significant redundancy often exists in collected data. For example, in concentration series, the q dependent intensity decay rate of high q data is nearly always consistent. Thus outliers can often be identified and eliminated.

When an obvious choice is not possible, the average is taken.

4.3 Red Flags for Further Analysis

Once all scattering profiles are selected and plotted, further comments are added. Comments are based on a visual inspection of the data. These comments are meant to serve as flags of concern rather than a definitive judgment on further processing of data. The following lists typical comments and examples are shown in Fig. 1.

4.3.1 Aggregation or Undefined Guineir Region

The intensity at zero scattering angle ($I(0)$) cannot be extrapolated from aggregated data. Similarly particles of size greater than 600 Å cannot be fully characterized with the available q range at SIBYLS. The scattering angles required for Guineir analysis are smaller than can be measured. Further analysis of data without a Guineir region is limited from a shape restoration perspective as the Guineir region is valued for quality control.

4.3.2 Radiation Sensitivity

X-ray radiation damages samples, but the damage rate cannot be determined a priori. Some samples show no noticeable differences in SAXS for all exposure lengths. Others are damaged by the first exposure. Radiation damage is identified as increase in $I(0)$ with exposure toward features of aggregation. Use of the low exposure data in this q region is thus critical for further analysis.

4.3.3 Detector Saturation

Extremely high concentration samples will scatter with intensities that saturate the detector in some regions of q. Data in these regions cannot be analyzed and must be compensated by utilizing shorter exposures or more dilute concentrations.

4.3.4 Low Concentration

At low concentration the difference between sample and buffer approaches zero. The small q region may have sufficient intensity to identify the radius of gyration R_g. However scattering features quickly blend in to flat, near zero values.

4.3.5 Bubble, Low Volume, or Empty Sample Cell

Bubbles, low volume, and empty sample cells often resemble profiles with over subtracted buffers. Radial streaks near the detector beamstop indicate that the incident X-ray beam is hitting a liquid/air surface. High q is the most clearly affected region.

4.3.6 Bad Buffer Subtraction

See Subheadings 3.5, 3.6, and 4.2 above for identification and causes of this error.

4.3.7 Repulsion

Repulsion is indicated by a gradual dip at low q and is caused by inter-particle interference. This effect most often occurs at high concentration. Unless the additional structure factor is of experimental interest, an extrapolation to zero concentration using a concentration series is often necessary.

4.3.8 Concentration-Dependent Effects

Concentration dependence includes multimerization, aggregation, or inter-particle interference, all of which contribute to characteristic changes in the scattering profiles from different concentrations.

4.3.9 *Micro Crystals*

Sharp peaks along the scattering curve indicate micro-crystal formation in the sample solution (Fig. 1d).

5 Conclusions and Perspectives

By compiling statistics over the course of 2 years (2011 and 2012), below we provide a picture of data collection using the mail-in/hand-in system. SIBYLS collected 267 plates from 106 different labs. Of these labs, 73 % requested subsequent data collection. While most plates were shipped at 4 C, 10 % were shipped frozen. Figure 2 also breaks down the frequency at which each comment was made. The scattering from the samples was sufficient to cause detector saturation in 39 % of samples, usually during the longest exposure. 45 % of samples were sensitive to radiation after 8 s of exposure, while 16 % showed significant radiation damage after only 3 s. 10 % of samples had an undefined Guineir region due to aggregation or molecular dimensions too large for our SAXS configuration. 7 % of samples had poorly matching buffer blanks. Concentration dependence affected 6 % of samples. Another 6 % were below the required concentration. Approximately 1 % of samples were lost by bubbles in the beam path or because of insufficient volume. Repulsion and micro-crystal formation were observed in less than 1 % of samples. Through visual inspection of each scattering curve by the SIBYLS staff, it was estimated that 78 % of all data could be used for further processing after a merging of different exposures and concentrations.

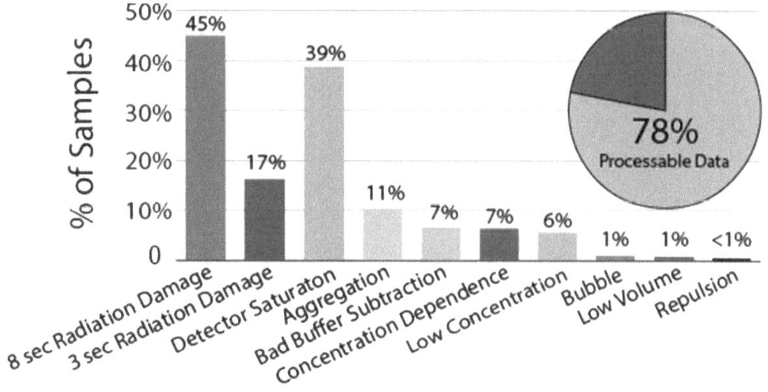

Fig. 2 SIBYLS SAXS sample quality statistics for 2 years of data collection. Each SAXS profile generated through the mail-in/hand-in system is visually inspected by beamline staff and commented upon for sample quality. Though many samples receive comments, when further merged and processed with other exposures and concentrations 78 % are estimated to be suitable for further analysis (*pie chart inset*)

HT-SAXS systems enable wide spread use of SAXS for structural characterization. The introduction of HT-SAXS data collection has been accompanied with criticism for being metric driven rather than science driven. Looking forward, we'd like to connect HT-SAXS efforts with problems being addressed in biology. Biological macromolecules are increasingly appreciated as parts of larger networks. Frequently, even components of these networks are challenging to work with and require specific laboratory expertise. Few single laboratories can successfully purify, characterize, and study many interacting components within a network. HT-SAXS facilities complement efforts to compose more comprehensive pictures of networks by drawing upon samples from many laboratories and enabling facile structural characterization.

SAXS is a solution-based technique so components may be examined individually, in the presence of partners or under a host of chemical conditions. Besides providing access to SAXS, HT-SAXS facilities continue to develop tools to aid in the analysis and integration of information collected; the staff at these facilities thus play a key part of the broader effort of post-genomic science. Further, new opportunities have been enabled with HT-SAXS [19] and by analysis of HT-SAXS data [36]. We anticipate more high impact results in the near future from HT-SAXS as well as from the combination of HT-SAXS with crystallography, NMR, and other biophysical methods.

Acknowledgments

This work and the operation of the SIBYLS beamline has been supported by the Integrated Diffraction Analysis Technologies (IDAT) program, the DOE Office of Biological and Environmental Research plus the National Institutes of Health grant MINOS (Macromolecular Insights on Nucleic Acids Optimized by Scattering) GM105404.

References

1. Koch MHJ (2010) SAXS instrumentation for synchrotron radiation then and now. J Phys Conf Ser 247:012001
2. Fukuchi S, Hosoda K, Homma K et al (2011) Binary classification of protein molecules into intrinsically disordered and ordered segments. BMC Struct Biol 11:29
3. Rambo RP, Tainer JA (2010) Bridging the solution divide: comprehensive structural analyses of dynamic RNA, DNA, and protein assemblies by small-angle X-ray scattering. Curr Opin Struct Biol 20:128–137
4. Hammel M, Yu Y, Fang S et al (2010) XLF regulates filament architecture of the XRCC4. ligase IV complex. Structure 18:1431–1442
5. Hammel M, Rey M, Mani RS et al (2011) XRCC4 protein interactions with XRCC4-like factor (XLF) create an extended grooved scaffold for DNA ligation and double strand break repair. J Biol Chem 286:32638–32650
6. Shin DS, Didonato M, Barondeau DP et al (2009) Superoxide dismutase from the eukaryotic thermophile Alvinella pompejana: structures, stability, mechanism, and insights into

amyotrophic lateral sclerosis. J Mol Biol 385:1534–1555
7. Yamagata A, Tainer JA (2007) Hexameric structures of the archaeal secretion ATPase GspE and implications for a universal secretion mechanism. EMBO J 26:878–890
8. Nishimura N, Hitomi K, Arvai AS et al (2009) Structural mechanism of abscisic acid binding and signaling by dimeric PYR1. Science 326:1373–1379
9. Chacon P, Moran F, Diaz JF et al (1998) Low-resolution structures of proteins in solution retrieved from X-ray scattering with a genetic algorithm. Biophys J 74:2760–2775
10. Petoukhov MV, Franke D, Shkumatov AV et al (2012) New developments in the ATSAS program package for small-angle scattering data analysis. J Appl Crystallogr 45:342–350
11. Pelikan M, Hura GL, Hammel M (2009) Structure and flexibility within proteins as identified through small angle X-ray scattering. Gen Physiol Biophys 28:174–189
12. Bernado P, Mylonas E, Petoukhov MV et al (2007) Structural characterization of flexible proteins using small-angle X-ray scattering. J Am Chem Soc 129:5656–5664
13. Grishaev A, Wu J, Trewhella J et al (2005) Refinement of multidomain protein structures by combination of solution small-angle X-ray scattering and NMR data. J Am Chem Soc 127:16621–16628
14. Schneidman-Duhovny D, Hammel M, Sali A (2010) FoXS: a web server for rapid computation and fitting of SAXS profiles. Nucleic Acids Res 38:W540–W544
15. Rambo RP, Tainer JA (2011) Characterizing flexible and intrinsically unstructured biological macromolecules by SAS using the Porod-Debye law. Biopolymers 95:559–571
16. Getzoff ED, Tainer JA, Lerner RA et al (1988) The chemistry and mechanism of antibody binding to protein antigens. Adv Immunol 43:1–98
17. Tainer JA, Getzoff ED, Alexander H et al (1984) The reactivity of anti-peptide antibodies is a function of the atomic mobility of sites in a protein. Nature 312:127–134
18. Tsutakawa SE, Hura GL, Frankel KA et al (2007) Structural analysis of flexible proteins in solution by small angle X-ray scattering combined with crystallography. J Struct Biol 158:214–223
19. Hura GL, Budworth H, Dyer KN et al (2013) Comprehensive macromolecular conformations mapped by quantitative SAXS analysis. Nat Methods 10:453–454
20. Martel A, Liu P, Weiss TM et al (2012) An integrated high-throughput data acquisition system for biological solution X-ray scattering studies. J Synchrotron Radiat 19:431–434
21. David G, Perez J (2009) Combined sampler robot and high-performance liquid chromatography: a fully automated system for biological small-angle X-ray scattering experiments at the Synchrotron SOLEIL SWING beamline. J Appl Crystallogr 42:892–900
22. Franke D, Kikhney AG, Svergun DI (2012) Automated acquisition and analysis of small angle X-ray scattering data. Nucl Instrum Methods A 689:52–59
23. Nielsen SS, Moller M, Gillilan RE (2012) High-throughput biological small-angle X-ray scattering with a robotically loaded capillary cell. J Appl Crystallogr 45:213–223
24. Hura GL, Menon AL, Hammel M et al (2009) Robust, high-throughput solution structural analyses by small angle X-ray scattering (SAXS). Nat Methods 6:606–U683
25. Classen S, Rodic I, Holton J et al (2010) Software for the high-throughput collection of SAXS data using an enhanced Blu-Ice/DCS control system. J Synchrotron Radiat 17: 774–781
26. Williams RS, Dodson GE, Limbo O et al (2009) Nbs1 flexibly tethers Ctp1 and Mre11-Rad50 to coordinate DNA double-strand break processing and repair. Cell 139:87–99
27. Christie JM, Arvai AS, Baxter KJ et al (2012) Plant UVR8 photoreceptor senses UV-B by tryptophan-mediated disruption of cross-dimer salt bridges. Science 335:1492–1496
28. Chao LH, Stratton MM, Lee IH et al (2011) A mechanism for tunable autoinhibition in the structure of a human Ca2+/calmodulin-dependent kinase II holoenzyme. Cell 146:732–745
29. Dueber EC, Schoeffler AJ, Lingel A et al (2011) Antagonists induce a conformational change in cIAP1 that promotes autoubiquitination. Science 334:376–380
30. Putnam CD, Hammel M, Hura GL et al (2007) X-ray solution scattering (SAXS) combined with crystallography and computation: defining accurate macromolecular structures, conformations and assemblies in solution. Q Rev Biophys 40:191–285
31. Petoukhov MV, Svergun DI (2007) Analysis of X-ray and neutron scattering from biomacromolecular solutions. Curr Opin Struc Biol 17:562–571
32. Rambo RP, Tainer JA (2013) Super-resolution in solution X-ray scattering and its applications

to structural systems biology. Ann Rev Biophys 42:415–441

33. Classen S, Hura GL, Holton JM et al (2013) Implementation and performance of SIBYLS: a dual endstation small-angle X-ray scattering and macromolecular crystallography beamline at the Advanced Light Source. J Appl Crystallogr 46:1–13

34. Jacques DA, Guss JM, Svergun DI et al (2012) Publication guidelines for structural modelling of small-angle scattering data from biomolecules in solution. Acta Crystallogr D 68: 620–626

35. Rambo RP, Tainer JA (2010) Improving small-angle X-ray scattering data for structural analyses of the RNA world. RNA 16: 638–646

36. Rambo RP, Tainer JA (2013) Accurate assessment of mass, models and resolution by small-angle scattering. Nature 496:477–481

Chapter 19

Measuring Spatial Restraints on Native Protein Complexes Using Isotope-Tagged Chemical Cross-Linking and Mass Spectrometry

Franz Herzog

Abstract

Mass spectrometric analyses of proteins affinity-purified from cell lysates are routinely used by cell biologists to characterize the composition and the modifications of protein complexes. Here, we describe a protocol that combines affinity-purification with chemical cross-linking and mass spectrometry (CXMS) in order to detect spatially proximate lysine residues on protein complexes isolated from human tissue culture cells. These cross-links are interpreted as distance restraints that aid in elucidating protein binding interfaces and the topology of protein complexes. In contrast to established high-resolution structural biology techniques, CXMS analysis has the potential to acquire structural information of small amounts of structurally flexible and heterogeneous protein preparations. We recently demonstrated on a network of modular protein phosphatase 2A complexes that restraints obtained by CXMS analysis hold great promise in supporting hybrid structural analysis of endogenous protein complexes by integrating structural data from different sources with computational molecular modeling.

Key words Mass spectrometry, Chemical cross-linking, *N*-Hydroxysuccinimide ester, Isotopic labeling, Protein complex, Subunit topology, Spatial restraint, Hybrid structural analysis

1 Introduction

Chemical labeling of surface-exposed functional groups has been applied for the structural elucidation of proteins and their complexes as a complementary method to established high-resolution techniques for decades [1]. It compensates for the inability of X-ray crystallography and nuclear magnetic resonance spectroscopy to resolve structures of macromolecular protein complexes that in many cases cannot be purified as stable and homogenous particles. Initially, chemical labeling experiments were performed by tagging solvent accessible groups with different chemistries such as hydrogen/deuterium exchange or hydroxyl radical labeling indicating surface exposure of theses residues, also known as footprinting.

Yu Wai Chen (ed.), *Structural Genomics: General Applications*, Methods in Molecular Biology, vol. 1091, DOI 10.1007/978-1-62703-691-7_19, © Springer Science+Business Media, LLC 2014

Probing proteins with bi-functional reagents covalently links amino acids which are juxtaposed in the structure of soluble proteins. The identification of the chemically modified residues takes advantage of recent advances in mass spectrometric instrumentation, especially, electrospray ionization and Fourier transform-based mass analyzers facilitate the sequencing of linked peptides with high mass accuracy, resolution, and scan speed. The structural information gained from chemical cross-linking coupled to mass spectrometry (CXMS) is a distance restraint that is defined by the length of the probe spanning the inter-residue distance [2–5]. The majority of studies applied chemical cross-linkers with functional groups reactive against primary amines of the lysine side chain and the protein N-terminus. The most frequently used reactive groups are *N*-hydroxysuccinimide esters that form stable amid bonds with primary amines at physiological conditions. Despite the high-performance detection of cross-linked peptides and their fragments by mass spectrometry, the low abundance of cross-links and the complexity of cross-link fragment ion spectra hamper identification of the cross-link sites. To overcome these limitations several workflows were developed which use a variety of modified cross-linker molecules, different enrichment strategies and dedicated software programs that exploit the unique features of each approach for the identification of the cross-linked lysine residues [6]. Most notably, recent studies implemented an increasing number of reagents with isotope-coded or collision-induced dissociation (CID) cleavable spacer arms. Different types of CID cleavable cross-linkers generate reporter ions and give rise to the linear peptides modified with the remaining cross-linker masses [7–9] which facilitates the identification of the cross-link sites by conventional MS software tools.

Here, we describe a cross-linking approach using an equimolar mixture of d_0- and d_{12}-labeled disuccinimidyl suberate (d_0-DSS and d_{12}-DSS, [d, deuterium]) which yields cross-links with an isotopic mass difference of 12 Da [10]. Potential cross-links are thus detected as isotopic pairs in the precursor (MS1) scan. The isotopic mass shift also allows discrimination between cross-link and linear peptide fragments in the MS2 scans that are composite spectra of fragment ions of the two cross-linked peptides which is key for the identification of the cross-link sites by the search engine *xQuest* [11, 12].

Initially, CXMS analysis was applied to the structural analysis of multi-subunit protein complexes which were purified to high homogeneity [13–15]. To exploit the potential of MS in detecting modified peptides in less defined, heterogeneous samples, we developed a protocol to probe endogenous complexes affinity-purified from human tissue culture cells. Similar to standard pull-down protocols for the analysis of the composition and the posttranslational modifications of protein complexes, our strategy enables the acquisition of distance restraints on protein complexes isolated from their cellular context. This is fundamental for the understanding of signaling pathways and their regulatory circuits.

Active protein phosphatase 2A (PP2A) holoenzymes are heterotrimers composed of a catalytic subunit, a scaffold subunit and one of a large array of regulatory subunits. Systematic CXMS analysis of the modular PP2A complexes delineated the topology of a PP2A network in human cells by identifying 176 interprotein and 570 intraprotein cross-links [16]. For the structural interpretation of chemical cross-links, we integrated the detected distance restraints with subunit structures and electron microscopy density maps through computational molecular modeling. Our structural predictions localized protein–protein binding interfaces, suggested a mode for targeting the PP2A inhibitor SET to the catalytic center of PP2A by the cell cycle adaptor protein shugoshin and described the topology of a 1 MDa chaperonin complex bound to its protein substrate. This study demonstrated the importance of cross-link derived spatial restraints for the hybrid structural analysis of endogenous protein complexes.

2 Materials

2.1 Expression of Affinity-Tagged Bait Proteins in HEK293 Cells

1. Flp-In™ T-REx™ HEK293 cells (Life Technologies) for the tetracycline inducible expression of N-terminal His_6-HA-StrepII-tagged bait proteins (*see* **Note 1**).
2. Dulbecco's Modified Eagle's Medium (DMEM) with high glucose, L-glutamine, and sodium pyruvate (Life Technologies) supplemented with 10 % (v/v) fetal bovine serum (FBS) Gold (PAA Laboratories). Add 50 ml FBS to 500 ml DMEM in a sterile laminar flow hood. Mix gently and store at 4 °C.
3. 0.05 % Trypsin–EDTA (1×) frozen solution with phenol red (Life Technologies). Thaw over night at 4 °C or quickly in water bath at 37 °C with intermediate agitation. Store at 4 °C or freeze aliquots at −20 °C (*see* **Note 2**).
4. Phosphate-buffered saline (PBS) (1×) (Life Technologies).
5. Hygromycin B as 50 mg/ml solution and Blasticidin S HCl as 50 mg powder (Life Technologies). Dissolve blasticidin in sterile water or PBS to 10 mg/ml. Store aliquots of hygromycin and blasticidin at −20°C and keep at 4 °C after thawing.
6. Tetracycline hydrochloride (Sigma-Aldrich) as 5 g powder. Prepare stock solution by dissolving 2 mg/ml in 95 % (v/v) ethanol. Store stock solution in light-protected tube at −20 °C.
7. 245×245×25 mm tissue culture dishes (Bioassay-dishes, Nunclon™Δ). 145×20 mm cell culture dishes (Greiner Bio-One).
8. 15 and 50 ml Falcon™ polypropylene tubes (Becton Dickinson).

2.2 Purification and Cross-Linking of Protein Complexes

1. Lysis buffer: 20 mM Tris–HCl, pH 7.5, 150 mM KCl, 3 mM $MgCl_2$, 1 mM DTT, 0.1 % (v/v) NP-40, 10 % (v/v) glycerol, 1:500 (v/v) protease inhibitor cocktail (all Sigma-Aldrich), and 1 tablet of phosphatase inhibitors per 10 ml (Roche).
2. Buffer P: 20 mM Tris–HCl, pH 7.5, 150 mM KCl, 3 mM $MgCl_2$, 0.02 % NP-40, 5 % glycerol.
3. Buffer X: 25 mM HEPES, pH 8.0, 150 mM KCl, 5 % glycerol.
4. Avidin as 50 mg powder (IBA BioTagnology GmbH). Prepare 25 mg/ml (1.5 mM) stock solution in buffer P.
5. Biotin as 1 g powder (Thermo Fisher Scientific). Prepare 20 mM stock solution in 50 mM HEPES, pH 8.2.
6. Strep-Tactin® Sepharose® as 50 % suspension (IBA BioTagnology GmbH).
7. Disuccinimidyl suberate isotopically coded (d_0-DSS:d_{12}-DSS = 1:1) as powder in 1 mg aliquots (Creative Molecules, Inc.). Dissolve 1 mg DSS in 107 or 53.5 μl DMF immediately before use generating a 25 or 50 mM stock solution, respectively. Use deionized water or buffer X for dilutions of the DSS stock solution.
8. 4× SDS-PAGE sample buffer: 250 mM Tris–HCl, pH 6.8, 8 % (w/v) sodium dodecyl sulfate, 40 % glycerol, 4 % (v/v) β-mercapto-ethanol, 0.04 % (w/v) bromophenolblue.
9. NH_4HCO_3 as powder (Sigma-Aldrich). Store aliquots of 1 M stock solution at −20 °C.
10. Manual tissue homogenizer, 15 ml (Sartorius).
11. Bio-Spin chromatography columns (Bio-Rad).
12. 15 ml Falcon™ polypropylene tubes (Becton Dickinson). 1.5 ml Eppendorf tubes (Eppendorf AG).
13. Ni-NTA agarose as 50 % suspension (Qiagen).
14. Thermomixer (Eppendorf AG).

2.3 Enrichment and MS Analysis of Cross-Linked Peptides

1. Urea as powder (#U6504, Sigma-Aldrich). Prepare 8 M stock solution immediately before use.
2. Tris(2-carboxyethyl)phosphine hydrochloride (TCEP) as a powder (Sigma-Aldrich). Prepare a 500 mM stock solution in 1 M NH_4HCO_3 (*see* **Note 3**). Dilute stock solution with 1 M NH_4HCO_3 to 50 mM TCEP and freeze aliquots at −20 °C.
3. Iodoacetamide as powder (Sigma-Aldrich). Dissolve iodoacetamide in deionized water to 100 mM immediately before use.
4. Sequencing grade modified trypsin 20 μg in 50 μl solution (Promega AG). Store at −80 °C.
5. Lysyl Endopeptidase (Lys-C) as 20 μg powder (Wako Chemicals). Store at −20 °C. Dissolve in 50 mM NH_4HCO_3 to 0.4 μg/μl and store aliquots at −80 °C.

6. Trifluoroacetic acid (TFA) (Thermo Fisher Scientific). Dilute with deionized water to 1 % (v/v).
7. 1.5 and 2 ml Eppendorf Tubes™ (Eppendorf AG).
8. Formic acid (FA) eluent additive for LC-MS (Sigma-Aldrich).
9. Acetonitrile LC-MS Chromasolv grade (Sigma-Aldrich).
10. Water advanced HPLC grade for gradient analysis (Thermo Fisher Scientific).
11. Buffer A: 3 % (v/v) acetonitrile, 0.2 % (v/v) formic acid.
12. Buffer B: 45 % acetonitrile, 0.2 % formic acid.
13. Buffer C: 30 % acetonitrile, 0.1 % TFA.
14. Mobile phase A: 2 % acetonitrile, 0.2 % formic acid.
15. Mobile phase B: 98 % acetonitrile, 0.2 % formic acid.
16. SepPak C18, 1cc, 50 mg (Waters).
17. Vacmaster 10 and PTFE stopcock/needle set, vacuum manifold (Biotage Sweden AB).
18. MicroWell™ plates, 96-well (Thermo Fisher Scientific).
19. Vacuum concentrator (Labconco).
20. Autosampler vials with caps (Thermo Fisher Scientific).
21. ÄKTAmicro equipped with Autosampler A-905, Fraction Collector Frac-950 and a Superdex Peptide PC 3.2/30 column (GE Healthcare).
22. Acclaim* PepMap* RSLC Column, 75 μm inner diameter, 150 mm length, C18 stationary phase with 2 μm particle size and 100 Å pore size (Thermo Fisher Scientific).
23. EASY-nLC 1000 HPLC coupled to Orbitrap Elite mass spectrometer equipped with Nanospray Flex Ion Source (Thermo Fisher Scientific).

2.4 Generation of xQuest Database

1. *MSConvert* [17] (download from http://proteowizard.sourceforge.net/).
2. *X!Tandem* [18] (open source http://www.thegpm.org/TANDEM/index.html) and Trans Proteomic Pipeline (TPP) using *PeptideProphet* [19] and *ProteinProphet* [20] (open source http://tools.proteomecenter.org/wiki/index.php?title=Software:TPP).
3. Mascot, licensed software (http://www.matrixscience.com/search_form_select.html).

2.5 Identification of Cross-Link Spectra by xQuest and Manual Validation

1. *xQuest* search engine for cross-link spectra [12] (download from http://proteomics.ethz.ch/cgi-bin/xquest2_cgi/index.cgi).

3 Methods

3.1 Expression of Affinity-Tagged Bait Proteins in HEK293 Cells

1. Culture HEK293 cells in 145 × 20 mm dishes in 20 ml DMEM supplemented with 10 % FBS, 100 μg/ml hygromycin B, and 15 μg/ml blasticidin S at 5 % CO_2 saturation and 37 °C. Let cells grow confluent before expanding them into 245 × 245 × 25 mm tissue culture dishes.
2. Rinse cells with 10 ml PBS prewarmed to 37 °C. Detach cells by 2 min incubation with 3 ml trypsin at 37 °C. Gently resuspend cells in 9 ml DMEM supplemented with 10 % FBS.
3. Plate cells on a 245 × 245 × 25 mm tissue culture dish by transferring 6 ml of the cell suspension into 64 ml DMEM supplemented with 10 % FBS. Grow cells to 40–50 % confluency at 5 % CO_2 saturation and 37 °C.
4. Induce protein expression by addition of the 2 mg/ml tetracycline stock to a final concentration of 1 μg/ml. Predilute tetracycline stock solution in DMEM 1:100 and add 3.5 ml to one 245 × 245 × 25 mm dish (*see* **Note 4**) and incubate for 24–28 h at 37 °C.
5. Pour off medium and rinse cells carefully with 20 ml ice-cold PBS. Detach cells by pipetting in 20 ml ice-cold PBS. Collect cells of two confluent dishes in 50 ml Falcon tube on ice. Pellet cells by centrifugation at 300 × *g*. Wash once by resuspending cells in 20 ml ice-cold PBS. Pellet cells again and take off the supernatant. Freeze cells in liquid nitrogen and store at −80 °C or keep them on ice for protein purification.

3.2 Purification and Cross-Linking of Protein Complexes

1. Thaw frozen HEK293 cells on ice and resuspend them in 2 pellet volumes of lysis buffer (*see* **Note 5**). Lyse cells with 2× 15 strokes in the cell homogenizer on ice. Transfer whole cell lysate to 1.5 ml Eppendorf tubes. Pellet cell debris by centrifugation at 16,000 × *g* and 4 °C.
2. Optional: add up to 2 μM avidin to the cleared extract and incubate for 20 min at room temperature with rotation (*see* **Note 6**).
3. Transfer 0.8 ml Strep-Tactin suspension per 1 ml cell pellet into a Bio-Spin column and equilibrate 2× with 1 ml buffer P using gravity flow. Apply cleared lysate 2× on the Strep-Tactin resin and wash 3× with 1 column volume ice-cold buffer P.
4. Prepare a 2 mM biotin solution by diluting the 20 mM stock solution in buffer X. Elute bound proteins with 4 times of 1 bead volume of 2 mM biotin. Pool fractions and take 30 μl aliquot for SDS-PAGE (polyacrylamide gel electrophoresis) analysis.
5. Transfer 60–80 μl Ni-NTA slurry (corresponds to 30–40 μl bead volume) per 1 ml cell pellet (*see* **Note 7**) into a 15 ml

Falcon tube containing 10 ml buffer X. Spin down Ni-NTA agarose at 200×*g* for 1 min. Aspirate supernatant and wash agarose with 12 ml ice-cold buffer X. Remove supernatant and incubate Ni-NTA beads with the biotin eluate for 1 h with end-over-end rotation at 4 °C. Pellet Ni-NTA agarose by centrifugation, take 30 μl supernatant for SDS-PAGE analysis and aspirate supernatant. Wash Ni-NTA bound proteins 2 times with 12 ml ice-cold buffer X (*see* **Note 8**).

6. Resuspend the 40 μl bead volume of protein bound Ni-NTA agarose (for 1 ml cell pellet starting material) in 120 μl buffer X and transfer the 25 % slurry into 1.5 ml Eppendorf tube. Transfer an aliquot of 20 μl slurry for the xQuest database generation into a separate tube.
7. Titrate the cross-linker to protein concentration on beads. Pipette aliquots of 10 μl slurry into ten 1.5 ml Eppendorf tubes and keep them at room temperature. Add nine different dilutions of the DSS stock solution to the beads by resuspending the beads (*see* **Note 9**). Incubate the reactions at 37 °C in a Thermomixer at 1,000 rpm for 30 min. Quench the reactions by adding 5 μl of 4× SDS-PAGE sample buffer.
8. Assess the degree of protein conjugation through chemical cross-linking by analyzing the different reactions with SDS-PAGE and silver staining. Choose the reaction where the protein bands just quantitatively shifted as the optimal DSS concentration (Fig. 1).
9. To cross-link the protein preparation for MS analysis resuspend 120 μl bead volume of protein bound Ni-NTA agarose (for 3 ml cell pellet starting material) in 360 μl buffer X. Add the optimal DSS concentration to the slurry, resuspend the beads, and incubate at 37 °C in a Thermomixer at 1,000 rpm for 30 min. Quench the reaction by adding 1 M NH_4HCO_3 to a final concentration of 100 mM and incubate for another 10 min.

3.3 Enrichment and MS Analysis of Cross-Linked Peptides

1. Spin down Ni-NTA agarose bound to cross-linked proteins and aspirate supernatant. Resuspend beads in 200 μl 8 M urea, add TCEP to a final concentration of 5 mM, and mix at 1,000 rpm for 20 min at 37 °C in a Thermomixer. Add 10 mM iodoacetamide, resuspend beads, and incubate for 40 min at room temperature in the dark.
2. Add Lys-C at an enzyme to protein ratio of 1:50 (w/w) and incubate in a Thermomixer at 37 °C and 1,000 rpm for 2 h. Dilute with 1,400 μl of 50 mM NH_4HCO_3 and transfer the reaction into a 2 ml Eppendorf tube. Add trypsin at an enzyme to protein ratio of 1:50 (w/w) and incubate in the Thermomixer at 37 °C and 1,000 rpm overnight.

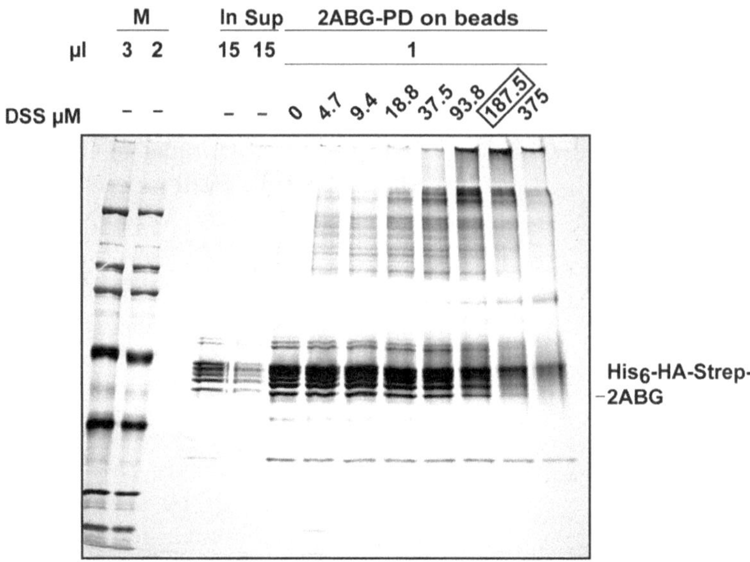

Fig. 1 Titration of DSS to protein concentration. The PP2A regulatory subunit 2ABG was affinity-purified from HEK293 cell lysate using a Strep-Tactin column. The biotin eluate was applied to Ni-NTA beads (In, input; Sup, supernatant) to immobilize 2ABG associated complexes through the N-terminal His_6-tag. Bound proteins were incubated with increasing concentrations of DSS and cross-linked proteins were separated by SDS-PAGE and visualized by silver staining. The DSS concentration applied to the mass spectrometric identification of cross-linked peptides is indicated by the boxed number. Reproduced from ref. [16]

3. Stop digestion by adding TFA to a final concentration of 0.1 % (v/v). Add acetonitrile to a final concentration of 3 % (v/v). Pellet Ni-NTA beads by centrifugation at 16,000 × *g*. Transfer supernatant into fresh 2 ml Eppendorf tube and spin down again at 16,000 × *g*.
4. Assemble SepPak C18 column on the Vacmaster manifold and activate the C18 resin for solid phase extraction by passing through 1 ml acetonitrile at about two drops per second. Wash column with 2 × 1 ml buffer A. Apply acidified protein digest, collect the flowthrough and apply a second time. Wash column with at least 2 × with 1 ml buffer A. Elute bound peptides by passing through 500 µl buffer B. Dry peptides by evaporating the solvent in a vacuum concentrator at 40 °C. Dissolve peptides in 20 µl buffer C and transfer into an autosampler vial.
5. Enrich cross-linked peptides over the non-cross-linked fraction by size exclusion chromatography (*see* **Note 10**) (Fig.2). Inject 15 µl onto a Superdex peptide column connected to an ÄKTAmicro system. Separate peptides with buffer C at a flow rate of 50 µl/min. Collect 100 µl fractions every 2 min in a 96-well plate (*see* **Note 11**). Transfer fractions of cross-linked peptides into 1.5 ml Eppendorf tubes and evaporate solvent in

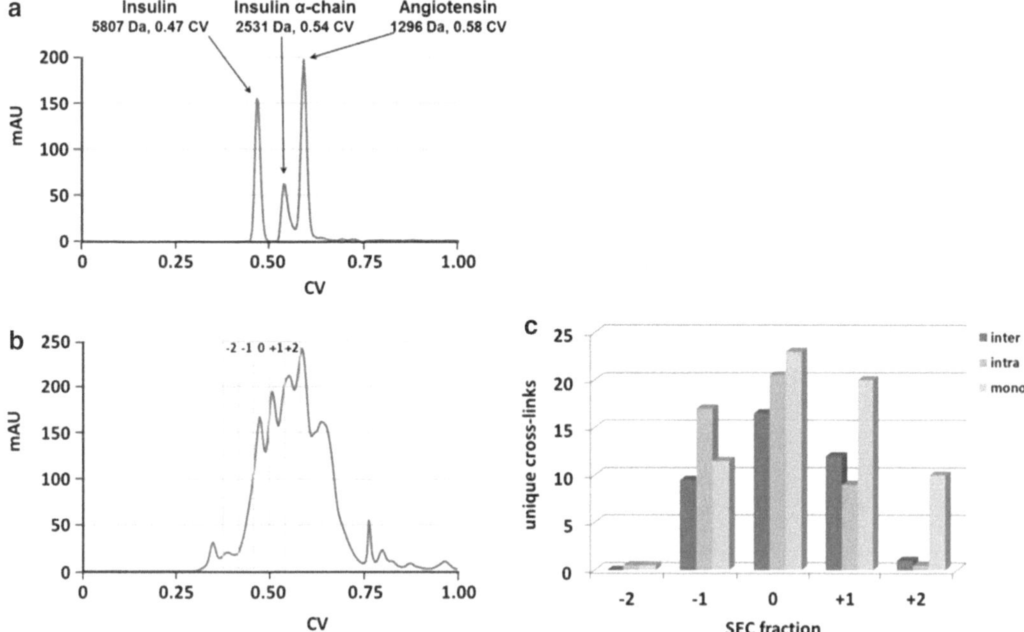

Fig. 2 Enrichment of the cross-linked peptides by size exclusion chromatography. (**a**) The composition of the mobile phase and the flow rate are optimized for the maximum separation of the three standard peptides. The separation of the 3-peptide mixture by size exclusion was monitored by UV absorption at 215 nm (mAU, milli absorption units; CV, column volume). (**b**) The protein complexes copurified with His_6-HA-StrepII-tagged 2ABG were cross-linked with DSS and digested using the described Lys-C/trypsin protocol. The proteolytic digest was separated by peptide size exclusion chromatography (SEC). The indicated fractions −2, −1, 0, +1, +2 were analyzed by MS and the cross-links were identified by the search engine *xQuest*. (**c**) The bar diagram displays the number of unique cross-links (identification with the highest score across all 5 fractions) detected in the individual fractions. About 96 % of the interprotein and intraprotein cross-links eluted within 3 fractions or 12.5 % of the column volume (CV). The increased number of mono-links in comparison to inter- and intraprotein cross-links in fraction +2 (*see* **Note 12**) indicate that mono-links were separated to some degree from inter- and intraprotein cross links by size exclusion [23]

the vacuum concentrator. Reconstitute peptide fractions in 20 μl mobile phase A and transfer into autosampler vials.

6. Analyze fractions enriched for cross-linked peptides with liquid chromatography coupled to tandem mass spectrometry using a EASY-nLC 1000 nano-HPLC system connected to a LTQ Orbitrap Elite mass spectrometer (LIT—Orbitrap, Linear Ion Trap—Orbitrap).
7. Load a volume corresponding to about 1 μg peptide onto the PepMap RSLC column (*see* **Note 13**). Separate peptides at a flow rate of 300 nl/min by running a gradient from 5 to 35 % mobile phase B within 60 min. For the instrument settings for peptide ionization and fragmentation by the LTQ Orbitrap Elite (*see* **Note 14**).

3.4 Generation of xQuest Database

1. Digest the proteins bound to Ni-NTA beads in the 20 μl aliquot taken prior to cross-linking (*see* Subheading 3.2, **step 6**) as described (*see* Subheading 3.3, **steps 1** and **2**).
2. Purify the peptides by solid phase extraction as described (*see* Subheading 3.3, **steps 3** and **4**) and dissolve the dried peptides in 20 μl mobile phase A.
3. Analyze the peptide mixture using the EASY-nLC 1000 nano-HPLC system connected to an LTQ Orbitrap Elite mass spectrometer as described (*see* Subheading 3.3, **steps 6** and 7). Note that MS2 scans of +2 charged precursors have to be enabled (*see* **Note 14**).
4. Convert Thermo Xcalibur .raw files into .mzXML or .mgf files using the MSConvert script. Use the following settings: Output format (mzXML or mgf), Binary encoding precision (32 or 64 bit) and Filters (Peak Picking MS Levels 1–1 or no filters), respectively, uncheck zlib compression and gzip for the generation of both, .mzXML and .mgf files.
5. Search .mzXML or .mgf files with the *X!Tandem/TPP* or *Mascot* software programs and rank the identified proteins according to the quantitation values, spectral counts or emPAI, respectively. Extract the UniProt entries of the 40 most abundant proteins and retrieve a FASTA database from http://www.uniprot.org/ for the xQuest search (*see* **Note 15**).

3.5 Identification of Cross-Link Spectra by xQuest and Manual Validation

1. Convert Thermo Xcalibur .raw files of the MS analysis of the cross-link fractions into .mzXML files using the *MSConvert* script. Use the following settings: Output format (mzXML), Binary encoding precision (32 bit) and Filters (Peak Picking MS Levels 1–1) and uncheck zlib compression and gzip.
2. Download and install *xQuest* by following the detailed instructions at http://proteomics.ethz.ch/cgi-bin/xquest2_cgi/index.cgi.
3. To run the *xQuest* search the following input files have to be generated:
   (a) .mzXML datafile.
   (b) *xQuest* database in .fasta format (*see* **Note 16**).
   (c) For the batch analysis of several MS runs the filenames (without extension and one name per line) have to be listed in a .txt file named "files".
   (d) Retrieve the definition files xmm.def and xquest. def from http://proteomics.ethz.ch/cgi-bin/xquest2_cgi/howtorun.cgi.
   (e) Modify the parameters according to your MS settings and IT infrastructure (*see* **Note 17**).
   (f) To perform the *xQuest* searches follow the detailed instructions at http://proteomics.ethz.ch/cgi-bin/xquest2_cgi/howtorun.cgi.

xQuest/xProphet results viewer

General settings		Filter settings			
Select type of report:	Html Table	Filter by type (top hit):	inter-protein cross-links (all)	Filter by min # seen >:	0 and < 0
Show n ranks per spectrum:	3	Filter hits by max ppm (Range):	From -5 to 5 ppm All hits:	Filter by sequence:	0
Number of hits per page:	100	Filter by unique ids (top hit):		Filter by annotation:	0
Create new index:		Show scores (top hit) >:	22 and < 0	Filter by deltaS (top hit) <:	0
Refresh:	Update Save Selected	Filter by ΔAA (top hit, Range):	From 0 to 0	Filter by FDR <:	0

id 1

Rank	Sequence	Protein1	Protein2
1	GKGAYQDRDKPAQIR-VDNIIKAAPR-a2-b6	sp|P50991|TCPD_HUMAN	sp|P78371|TCPB_HUMAN
3	DVTLEASRESSKPR-VDNIIKAAPRK-a12-b6	sp|Q9Y2T4|2ABG_HUMAN	sp|P78371|TCPB_HUMAN

Spectrum info: MS m/z: 734.9079, Charge: 4, Precursor mass: 2935.6003, I

Matchodds	TIC	wTIC	xcorrx	xcorrb	intsum	deltaS	Id-Score	rel_error [ppm]	type	xl-type	nAA1	nAA2	ΔAA	nseen	view	Sel
8.55	0.72	0.14	0.03	0.12	533	0.80	32.94	0.4	xlink	inter-protein xl	21	522	501	1	view	
7.35	0.52	0.09	0.02	0.05	380	1.00	26.46	-3.0	xlink	inter-protein xl	383	522	139	1	view	

Fig. 3 xQuest results viewer. The viewer offers a platform for filtering the cross-link identifications matched to the acquired spectra. The most effective filtering steps prior to manual validation include selecting the cross-link type, setting the MS1 mass tolerance window (ppm), displaying only the highest-scored identification (unique ids) and setting the minimum threshold score (show scores >) and the maximum deltaS value

4. Upload the results in a web browser by using the results manager. Use the viewer options to filter your results (Fig. 3).
   (a) Filter by type (top hit): select interprotein, intraprotein, or mono-links. Choose whether target or decoy or both cross-link types should be displayed. Decoy hits generated by the target-decoy method are used to calculate the false discovery rate according to ref. 12.
   (b) Filter hits by max ppm (Range): standard values for the MS1 mass tolerance window are −5 to +5 ppm. Check "All hits" to apply the mass error filter to assignments that are ranked as second or third hit for this spectrum.
   (c) Filter by unique ids (top hit): enables displaying the highest-scored hit (identification) for this specific cross-link sequence, in case, this cross-link was identified multiple times in this analysis.
   (d) Show scores (top hit) >: set to preliminary value of 22 and displays all first ranked cross-links of the selected type until a score of 22. The actual minimal score threshold is determined after manual validation (*see* **Note 18**).

(e) The filtered results are exported as .tsv files to process them for manual validation in Excel.

5. All 1st ranked spectra passing the filter step are manually validated.

(a) The score difference between the first and second ranked hit, Δscore, has to be ≥15 %. The Δscore is shown by the column deltaS in the results overview that indicates the score ratio second/first for the first ranked hit or third/second for the second ranked hit. First and second ranked cross-link assignments indicating linkage of the same two peptide sequences that differ in the position of the linked lysine in one peptide are excluded from this rule.

(b) To evaluate the spectrum browse the "view" link in the results overview (Fig. 3). Valid identifications have to exhibit at least four bond cleavages in total or three adjacent ones per peptide and a minimum peptide length of six amino acids. Cross-links containing one peptide of five amino acids and fulfilling the filtering and manual validation criteria are considered as candidate cross-links taking into account that the reduced information content of shorter peptides may result in higher false discovery rates.

4 Notes

1. The generation of these cell lines was described in detail previously [21]. In brief, cDNAs of the bait proteins can be retrieved in Gateway (Life Technologies) compatible entry vectors from existing collections (horfeome v5.1, Open Biosystems, www.openbiosystems.com) or can be easily introduced into the Gateway system by BP clonase recombination (http://invitrogen.com). ORFs in entry vectors were introduced by LR recombination into the destination vector that was constructed by ligating the Gateway recombination cassette and an N-terminal His_6-HA-StrepII-tag into the polylinker of the pcDNA5/FRT/TO vector (Life Technologies). Flp-In™ T-REx™ HEK293 cells containing a single genomic FRT site and expressing the tet repressor protein were used to generate stable isogenic cell lines for the tetracycline inducible expression of tagged bait proteins (http://invitrogen.com).
2. Avoid more than one freeze–thaw cycle of trypsin aliquots.
3. Add 1 M NH_4HCO_3 dropwise to TCEP powder as there is massive CO_2 production due to the acidity of TCEP.
4. Predilution of the tetracycline stock solution prevents high local concentration of the chemical on cells.
5. The starting cell pellet volume depends on the cellular abundance of the over-expressed His_6-HA-StrepII-tagged bait

protein. Approximately, 3 ml cell pellet have to be used for high abundant bait proteins. For low abundant baits the cell pellet volume has to be increased to at least 6–7 ml. One 245 × 245 × 25 mm dish yields about 0.2–0.4 ml cell pellet.

6. Avidin competes off biotin-containing enzymes and thus reduces the levels of contaminating proteins binding to the Strep-Tactin resin.
7. Do not increase the volume of Ni-NTA beads for low abundant bait proteins above the approximately 120 μl bead volume for the 3 ml cell pellet starting material of high abundant proteins.
8. Estimate the protein concentration on beads by SDS-PAGE and silver staining. Run 15 μl of biotin elution prior and subsequent to Ni-NTA incubation and 1–2 μl protein bound Ni-NTA beads together with a BSA dilution series on SDS-PAGE. Estimate the protein concentration by comparing the intensities of silver-stained bands [22].
9. Use the approximation 1 μg protein ≡ 500 pmol lysine ≡ 500 pmol DSS to estimate the equimolar amount of DSS. Use two- to threefold steps to calculate higher and lower concentrations for the titration series (Fig. 1).
10. Optimize the percentage of acetonitrile and the flow rate for the maximum separation of the three standard peptides (insulin, 5,807 Da; insulin chain A (oxidized), 2,531 Da, angiotensin, 1,296 Da; all Sigma-Aldrich) in the range of 3,000–6,000 Da [23] (Fig. 2).
11. Determine the fractions enriched for cross-linked peptides by separating the peptides of cross-linked standard proteins [23] or of your protein complex of interest (Fig. 2). Take the fraction of the insulin peak as fraction 0 as well as −2 and −1 (earlier eluting fractions at higher molecular weight) and +1 and +2 (later eluting fractions at lower molecular weight) and determine the number of non-redundant cross-links in the sample identified in each fraction. The majority of cross-links have to elute in 2–3 fractions to obtain sufficient enrichment of cross-linked over non-cross-linked peptides.
12. In this analysis only mono-links with charge states ≥ +3 were detected. As a significant number of mono-links occur as +2 charged precursors, the actual number of mono-links in fraction +2 might be higher than indicated.
13. Estimate volume of 1 μg peptide by taking into account the total protein amount of the purification and the ratio of the area of the fraction to the total area of the size exclusion chromatogram.
14. Ion source and transmission parameters of the mass spectrometer were set to spray voltage = 2 kV, capillary temperature = 275 °C. The mass spectrometer was operated in data-dependent mode, selecting up to ten precursors from an

MS[1] scan (resolution = 60,000) in the range of m/z 400–2,000 for collision-induced dissociation (CID). Singly (+1) and doubly (+2) charged precursor ions and precursors of unknown charge states were rejected. CID was performed for 10 ms using 35 % normalized collision energy and the activation q of 0.25. Dynamic exclusion was activated with a repeat count of 1, exclusion duration of 30 s, list size of 500 and the mass window of ±10 ppm. Ion target values were 1,000,000 (or maximum 10 ms fill time) for full scans and 10,000 (or maximum 100 ms fill time) for MS/MS scans, respectively [16, 23].

15. Avoid separators like "." and "," and spaces in the protein header line.
16. The *xQuest* search requires the generation of a decoy database (_decoy.fasta) according to the instructions at http://proteomics.ethz.ch/cgi-bin/xquest2_cgi/howtorun.cgi. Copy the root paths of the .fasta and _decoy.fasta databases into the xquest.def file.
17. Definition file settings used in refs. 12, 15, 16: maximum number of missed cleavages (excluding the cross-linking site) = 2, peptide length = 4–40 amino acids, fixed modifications = carbamidomethyl-Cys (mass shift = 57.021460 Da), variable modifications = oxidation-Met (mass shift = 15.99491), mass shift of the light cross-linker = 138.068080 Da, mass shift of mono-links = 156.078644 and 155.096428 Da, MS[1] tolerance = 15 ppm, MS[2] tolerance = 0.2 Da for common ions and 0.3 Da for cross-link ions, search in ion-tag mode.
18. The calculation of the false discovery rate is discussed in detail in refs. 12, 16.

Acknowledgments

I would like to thank Alexander Leitner for comments on the manuscript and Thomas Walzthöni and Ruedi Aebersold for support. The work in the lab of F.H. is funded by the Bavarian Research Center for Molecular Biosystems and by an LMUexcellent Junior grant.

References

1. Fabris D, Yu ET (2010) Elucidating the higher-order structure of biopolymers by structural probing and mass spectrometry: MS3D. J Mass Spectrom 45:841–860
2. Rappsilber J, Siniossoglou S, Hurt EC et al (2000) A generic strategy to analyze the spatial organization of multi-protein complexes by cross-linking and mass spectrometry. Anal Chem 72:267–275
3. Back JW, de Jong L, Muijsers AO et al (2003) Chemical cross-linking and mass spectrometry for protein structural modeling. J Mol Biol 331:303–313
4. Sinz A (2005) Chemical cross-linking and FTICR mass spectrometry for protein structure characterization. Anal Bioanal Chem 381:44–47
5. Leitner A, Walzthoeni T, Kahraman A et al (2010) Probing native protein structures by

chemical cross-linking, mass spectrometry, and bioinformatics. Mol Cell Proteomics 9: 1634–1649
6. Petrotchenko EV, Borchers CH (2010) Crosslinking combined with mass spectrometry for structural proteomics. Mass spectrometry reviews 29:862–876
7. Tang X, Bruce JE (2010) A new cross-linking strategy: protein interaction reporter (PIR) technology for protein–protein interaction studies. Mol Biosyst 6:939–947
8. Kao A, Chiu CL, Vellucci D et al (2011) Development of a novel cross-linking strategy for fast and accurate identification of cross-linked peptides of protein complexes. Mol Cell Proteomics 10:M110.002212
9. Petrotchenko EV, Serpa JJ, Borchers CH (2011) An isotopically coded CID-cleavable biotinylated cross-linker for structural proteomics. Mol Cell Proteomics 10:M110.001420
10. Seebacher J, Mallick P, Zhang N et al (2006) Protein cross-linking analysis using mass spectrometry, isotope-coded cross-linkers, and integrated computational data processing. J Proteome Res 5:2270–2282
11. Rinner O, Seebacher J, Walzthoeni T et al (2008) Identification of cross-linked peptides from large sequence databases. Nat Methods 5:315–318
12. Walzthoeni T, Claassen M, Leitner A et al (2012) False discovery rate estimation for cross-linked peptides identified by mass spectrometry. Nat Methods 9:901–903
13. Jennebach S, Herzog F, Aebersold R et al (2012) Crosslinking-MS analysis reveals RNA polymerase I domain architecture and basis of rRNA cleavage. Nucleic Acids Res 40:5591–5601
14. Lasker K, Förster F, Bohn S et al (2012) Molecular architecture of the 26S proteasome holocomplex determined by an integrative approach. Proc Natl Acad Sci USA 109: 1380–1387
15. Leitner A, Joachimiak LA, Bracher A et al (2012) The molecular architecture of the eukaryotic chaperonin TRiC/CCT. Structure 20:814–825
16. Herzog F, Kahraman A, Boehringer D et al (2012) Structural probing of a protein phosphatase 2A network by chemical cross-linking and mass spectrometry. Science 337: 1348–1352
17. Kessner D, Chambers M, Burke R et al (2008) ProteoWizard: open source software for rapid proteomics tools development. Bioinformatics 24:2534–2536
18. Craig R, Beavis RC (2004) TANDEM: matching proteins with tandem mass spectra. Bioinformatics 20:1466–1467
19. Keller A, Nesvizhskii AI, Kolker E et al (2002) Empirical statistical model to estimate the accuracy of peptide identifications made by MS/MS and database search. Anal Chem 74:5383–5392
20. Nesvizhskii AI, Keller A, Kolker E et al (2003) A statistical model for identifying proteins by tandem mass spectrometry. Anal Chem 75: 4646–4658
21. Glatter T, Wepf A, Aebersold R et al (2009) An integrated workflow for charting the human interaction proteome: insights into the PP2A system. Mol Syst Biol 5:237
22. Herzog F, Peters JM (2005) Large-scale purification of the vertebrate anaphase-promoting complex/cyclosome. Methods Enzymol 398: 175–195
23. Leitner A, Reischl R, Walzthoeni T et al (2012) Expanding the chemical cross-linking toolbox by the use of multiple proteases and enrichment by size exclusion chromatography. Mol Cell Proteomics 11:M111.014126

Part III

Computational Methods and Structural Data Analyses

Chapter 20

Modeling of Proteins and Their Assemblies with the *Integrative Modeling Platform*

Benjamin Webb, Keren Lasker, Javier Velázquez-Muriel, Dina Schneidman-Duhovny, Riccardo Pellarin, Massimiliano Bonomi, Charles Greenberg, Barak Raveh, Elina Tjioe, Daniel Russel, and Andrej Sali

Abstract

To understand the workings of the living cell, we need to characterize protein assemblies that constitute the cell (for example, the ribosome, 26S proteasome, and the nuclear pore complex). A reliable high-resolution structural characterization of these assemblies is frequently beyond the reach of current experimental methods, such as X-ray crystallography, NMR spectroscopy, electron microscopy, footprinting, chemical cross-linking, FRET spectroscopy, small angle X-ray scattering, and proteomics. However, the information garnered from different methods can be combined and used to build models of the assembly structures that are consistent with all of the available datasets, and therefore more accurate, precise, and complete. Here, we describe a protocol for this integration, whereby the information is converted to a set of spatial restraints and a variety of optimization procedures can be used to generate models that satisfy the restraints as well as possible. These generated models can then potentially inform about the precision and accuracy of structure determination, the accuracy of the input datasets, and further data generation. We also demonstrate the *Integrative Modeling Platform* (IMP) software, which provides the necessary computational framework to implement this protocol, and several applications for specific use cases.

Key words Integrative modeling, Protein structure modeling, Proteomics of macromolecular assemblies, X-ray crystallography, Electron microscopy, SAXS

1 Introduction

To understand the function of a macromolecular assembly, we must know the structure of its components and the interactions between them [1–4]. However, direct experimental determination of such a structure is generally rather difficult. While multiple methods do exist for structure determination, each has a drawback. For example, crystals suitable for X-ray crystallography cannot always be produced, especially for large assemblies of multiple

Yu Wai Chen (ed.), *Structural Genomics: General Applications*, Methods in Molecular Biology, vol. 1091, DOI 10.1007/978-1-62703-691-7_20, © Springer Science+Business Media, LLC 2014

components [5]. Cryo-electron microscopy (cryo-EM), on the other hand, can be used to study large assemblies, but it is generally limited to worse than atomic resolution [6–8]. Finally, proteomics techniques, such as yeast two-hybrid [9] and mass spectrometry [10], yield information about the interactions between proteins, but not the positions of these proteins within the assembly or the structures of the proteins themselves.

1.1 Integrative Modeling

One approach to solve the structures of proteins and their assemblies is by integrative modeling, in which information from different methods is considered simultaneously during the modeling procedure. The approach is briefly outlined here for clarity; it has been covered in greater detail previously [11–18]. These individual methods can include experimental techniques, such as X-ray crystallography [5], nuclear magnetic resonance (NMR) spectroscopy [19–21], electron microscopy (EM) [6–8], footprinting [22, 23], chemical cross-linking [24–27], FRET spectroscopy [28], small angle X-ray scattering (SAXS) [29–31], and proteomics [32]. Theoretical sources of information about the assembly can also be incorporated, such as template structures used in comparative modeling [33, 34], scoring functions used in molecular docking [35], as well as other statistical preferences [36, 37] and physics-based energy functions [38–40]. Different methods yield information about different aspects of structure and at different levels of resolution. For example, atomic resolution structures may be available for individual proteins in the assembly; in other cases, only their approximate size, approximate shape, or interactions with other proteins may be known. Thus, integrative modeling techniques generate models at the resolution that is consistent with the input information. An example of a simple integrative approach is building a pseudo-atomic model of a large assembly, such as the 26S proteasome [41–43], by fitting atomic structures of its subunits predicted by comparative protein structure modeling into a density map determined by cryo-EM [44, 45].

The integrative modeling procedure used here [13, 18] is schematically shown in Fig. 1. The first step in the procedure is to collect all experimental, statistical, and physical information that describes the system of interest. A suitable representation for the system is then chosen and the available information is translated to a set of spatial restraints on the components of the system. For example, in the case of characterizing the molecular architecture of the nuclear pore complex (NPC) [13, 14], atomic structures of the protein subunits were not available, but the approximate size and shape of each protein was known, so each protein was represented as a "string" of connected spheres consistent with the protein size and shape. A simple distance between two proteins can be restrained by a harmonic function of the distance, while the fit of a model into

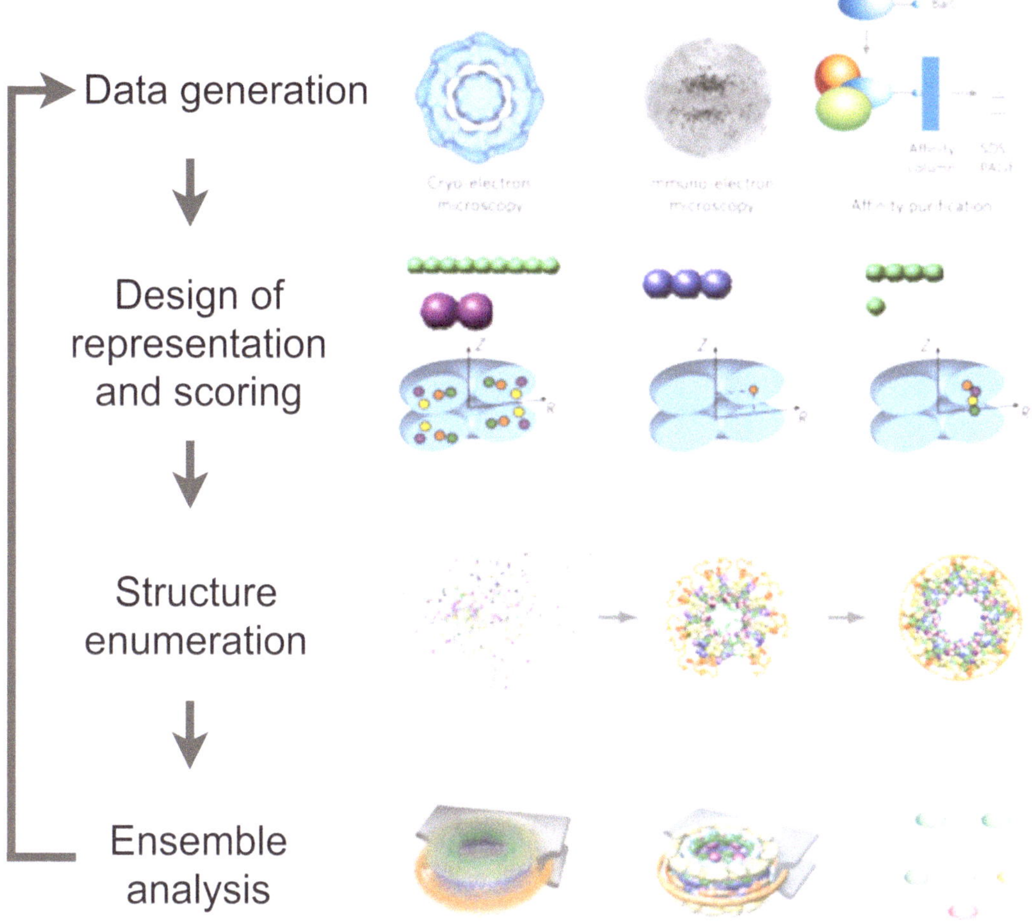

Fig. 1 Integrative modeling protocol. After the datasets to be used are enumerated, a suitable representation is chosen for the system, and the input information is converted into spatial restraints. Models are generated that are optimally consistent with the input information by optimizing a function of these restraints. Analysis of the resulting models informs about the model and data accuracy and may help guide further experiments. The protocol is demonstrated with the construction of a bead model of the NPC [13]

a 3D cryo-EM density map can be restrained by the cross-correlation between the map and the computed density of the model. Next, the spatial restraints are summed into a single scoring function that can be sampled using a variety of optimizers, such as conjugate gradients, molecular dynamics, Monte Carlo, and divide-and-conquer message-passing methods [45]. This sampling generates an ensemble of models that are as consistent with the input information as possible. In the final step, the ensemble is analyzed to determine, for example, whether all of the restraints have been satisfied or certain subsets of data conflict with others. The analysis may generate a consensus model, such as the probability density for the location of each subunit in the assembly.

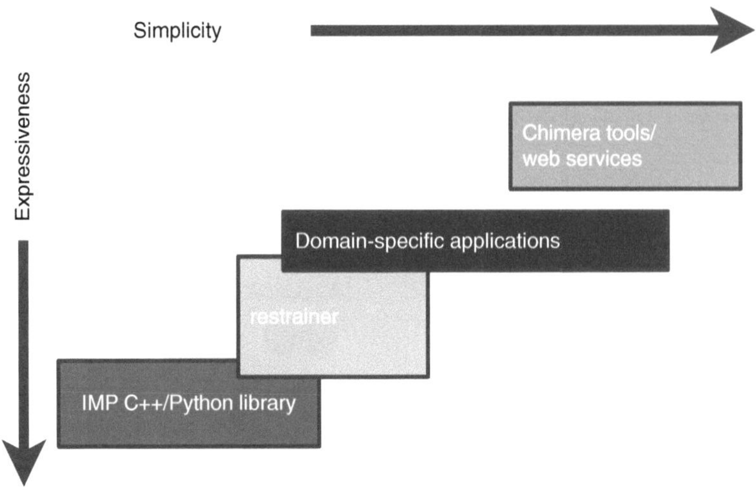

Fig. 2 Overview of the IMP software. Components are displayed by simplicity (or user friendliness) and expressiveness (or power). The core C++/Python library allows protocols to be designed from scratch, at a cost of user friendliness; higher-level modules and applications provide more user-friendly interfaces, at a cost of flexibility

1.2 Integrative Modeling Platform

We have developed the *Integrative Modeling Platform* (IMP) software (http://salilab.org/imp/) [11, 13–16] to implement the integrative modeling procedure described above. Integrative modeling problems vary in size and scope, and thus IMP offers a great deal of flexibility and several abstraction levels as part of a multi-tiered platform (Fig. 2). At the lowest level, IMP provides building blocks and tools to allow method developers to convert data from new experimental methods into spatial restraints, to implement optimization and analysis techniques, and to implement an integrative modeling procedure from scratch; the developer can use the C++ and Python programming languages to achieve these tasks. Higher abstraction levels, designed to be used by IMP users with no programming experience, provide less flexible but more user-friendly applications to handle specific tasks, such as fitting of proteins into a density map of their assembly, or comparing a structure with the corresponding SAXS profile. IMP is freely available as open source software under the terms of the GNU Lesser General Public License (LGPL). Integrative modeling, due to its use of multiple sources of information, is often a highly collaborative venture, and thus benefits from openness of the modeling protocols and the software itself.

2 Materials

To follow the examples in this discussion, both the IMP software itself and a set of suitable input files are needed. The IMP software can be downloaded from http://salilab.org/imp/download.html and is available in binary form for most common machine types and operating systems; alternatively, it can be rebuilt from the source code; either the stable 2.0 release of IMP, or a recent development version, should be used. The example files can be downloaded from http://salilab.org/imp/tutorials/basic_apps.zip. Certain applications also make use of third party software, which must be obtained separately from IMP (download locations for each software package are shown in subsequent sections).

2.1 Typographical Conventions

`Monospaced text` is used below for computer file and folder/directory names, command lines, file contents, and variable and class names.

3 Methods

3.1 The IMP C++/Python Library

The core of IMP is the C++/Python library, which provides all of the necessary components, as a set of classes and modules, to allow method developers to build an integrative modeling protocol from scratch. This core can be used either from C++ (by including the `IMP.h` header file and linking against the IMP libraries) or from Python (by importing the IMP Python module), and provides almost identical functionality in each language, for maximum flexibility. In this text, we will demonstrate the IMP applications that build on top of this core; the core itself has been demonstrated elsewhere [46] and is further described on the IMP website, http://salilab.org/imp/.

3.2 Pairwise Protein–Protein Docking Integrating Data from SAXS and EM

One major computational approach to predicting structures of protein complexes relies on molecular docking of unbound single-component structures. However, even for complexes with two proteins, the docking problem remains challenging despite recent advances [47]. The major bottlenecks include dealing with protein flexibility and the absence of an accurate scoring function [48].

IMP includes an integrative approach to pairwise protein docking, in which additional experimental information about the protein–protein complex is incorporated into the docking procedure to greatly improve the accuracy of predictions. This method succeeds in producing a near-native model among the top ten models in 42–82 % of cases, while state-of-the-art docking methods succeed only in 30–40 % of cases, depending on the benchmark and accuracy criterion [49].

The protocol proceeds as follows (Fig. 3). First, data from one or more of five different experiment types are translated into the corresponding scoring function terms. These data include (1) the pair-distance distribution function of the complex from a SAXS profile, (2) 2D class average images of the complex from negative-stain EM micrographs (EM2D), (3) a 3D density map of the complex from single-particle negative-stain EM micrographs (EM3D), (4) residue type content at the protein interface from NMR spectroscopy (NMR-RTC) [50], and (5) chemical cross-linking detected by mass spectrometry (CXMS). These five experimental

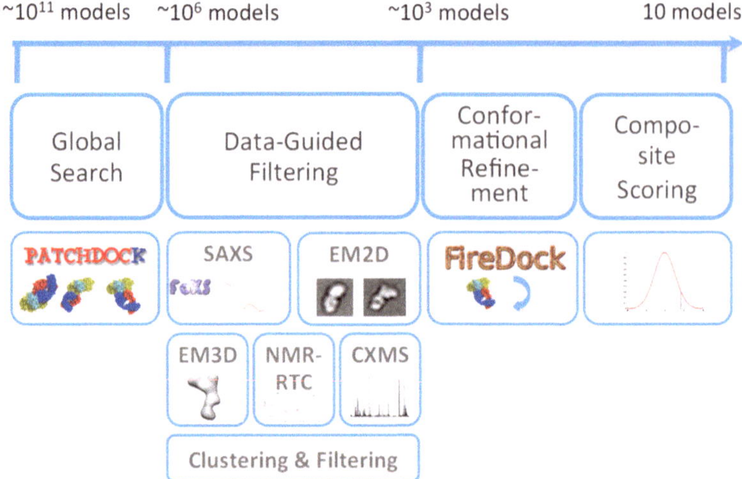

Fig. 3 Schematic representation of the integrative docking method. The number of possible configurations for two docked proteins is on the order of ~10^{11} (three rotational degrees of freedom sampled at 5° intervals and three translational degrees of freedom sampled at 1 Å intervals). As the method proceeds, the number of considered configurations decreases

methods were selected because of their feasibility and efficiency of data collection: a SAXS profile of the complex in solution can be collected in several minutes [30]; a 3D EM density map can be reconstructed from a smaller sample amount than that for SAXS, but the data collection process is significantly longer[6]; 2D class averages can be computed from micrographs more easily and rapidly than performing a full 3D reconstruction; the composition of interface residues from NMR [50] provides information about the interaction interface, unlike the SAXS and EM data; and cross-linking data [51] provide information at intermediate resolution imposing an upper distance bound on inter-molecular pairs of residues. Second, complex models are sampled, relying on efficient global search methods developed for pairwise protein docking, followed by filtering based on fit to the experimental data, conformational refinement, and composite scoring. Third, good-scoring representatives of clusters of models are picked as final models.

Here, we demonstrate the approach by application to the PCSK9 antigen–J16 Fab antibody complex. All input files for this example can be found in the "`idock`" directory of the downloaded zipfile.

1. *Inputs.* The primary inputs are the Protein Data Bank (PDB) [52] structures of the isolated J16 Fab antibody and PCSK9 antigen, `antibody_cut.pdb` and `2p4e.pdb`, respectively; they can be found in the downloaded zipfile. We also collected SAXS, EM2D, and EM3D data on this protein–protein complex, available in the `iq.dat`, `image_*.pgm`, and `complex.mrc` files, respectively. Finally, we added missing residues to

both PDB files, for use in SAXS scoring, yielding `antibody.pdb` and `pcsk9.pdb` (*see* **Note 1**).

2. *Docking*. We can then carry out all steps of the integrative docking by running IMP's `idock.py` application (*see* **Note 2**), giving it the names of our input files:

```
idock.py antibody_cut.pdb 2p4e.pdb --saxs iq.dat --em3d complex.mrc --em2d image_1.pgm --em2d image_2.pgm --em2d image_3.pgm --pixel_size 2.2 --complex_type AA --saxs_receptor_pdb antibody.pdb --saxs_ligand_pdb pcsk9.pdb --precision 2
```

The application makes use of the PatchDock and FireDock programs for docking and refinement, which must be obtained separately from http://bioinfo3d.cs.tau.ac.il/, and the "reduce" program for adding hydrogens to PDB files, available from http://kinemage.biochem.duke.edu/software/reduce.php.

3. *Results*. Once the docking procedure has finished, the primary output file generated is `results_saxs_em3d_em2d.txt`, the first few lines of which look similar to:

```
# | Score | filt| ZScore | SAXS | Zscore | EM2D | Zscore | EM3D | Zscore | Energy | Zscore | Transformation

1 | -5.225 |+| -3.318 | 16.304 | -1.454 | 0.685 | -1.829 | 0.058 | -1.672 | -20.010 | -0.270 | 2.4462 0.7439 2.0137 32.0310 36.5010 74.9757

2 | -4.453 |+| -2.828 | 17.590 | 0.578 | 0.698 | -2.521 | 0.064 | -1.243 | -42.220 | -1.267 | 0.1525 -1.3733 2.1213 -17.2068 -10.3519 13.3553
```

Each line corresponds to one model; the models are ranked by total score, best first. The individual SAXS, EM2D, and EM3D score/z-score pairs are also shown (only docking solutions that were not filtered out by any of three data sources—i.e. they scored well against every source—are included in this file). The last column is a transformation (three rotation angles and a translation vector) that transforms the antibody relative to the antigen (the antigen is not transformed).

3.3 Determining Macromolecular Assembly Structures by Fitting Multiple Structures into an Electron Density Map

Often, we have available high-resolution (atomic) information for the subunits in an assembly, and low-resolution information for the assembly as a whole, such as a cryo-EM electron density map. A high-resolution model of the whole assembly can thus be constructed by simultaneously fitting the subunits into the density map. Fitting of a single protein into a density map is usually done by calculating the electron density map of the protein followed by a search for the protein position in the cryo-EM map that

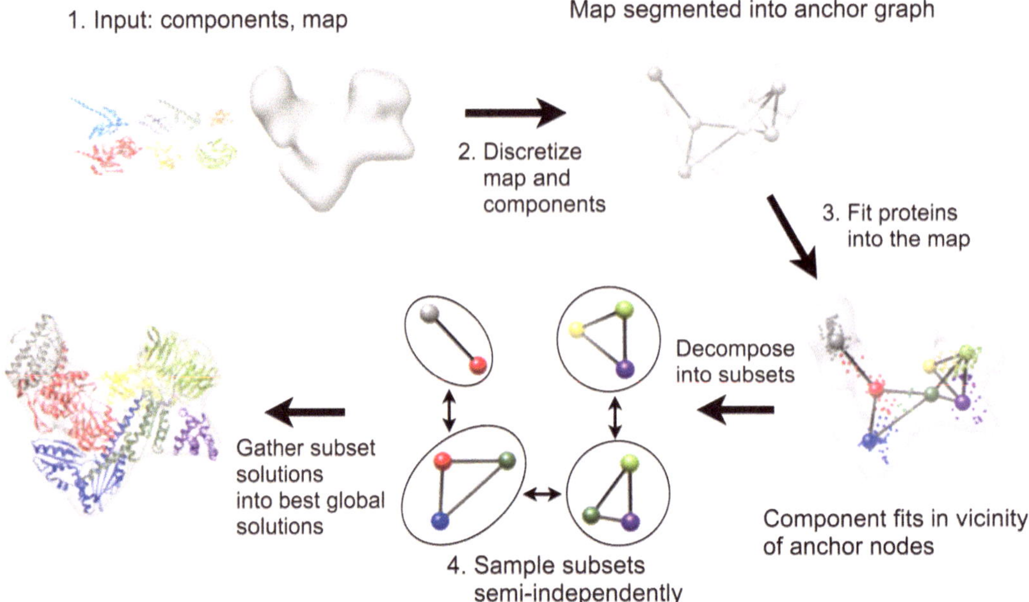

Fig. 4 The MultiFit protocol [45]. Protein subunits are fitted into a density map of the assembly by discretizing both the map and the components, locally fitting each protein, and efficiently combining the local fits into global solutions

maximizes the cross-correlation of the two maps. Simultaneously fitting multiple proteins into a given map is significantly more difficult though, since an incorrect fit of one protein will also prevent other proteins from being placed correctly.

IMP contains a MultiFit [44, 45] application (http://salilab.org/multifit/) that can efficiently solve such multiple fitting problems for density map resolutions as low as 25 Å, relying on a general divide-and-conquer optimizer DOMINO. The application is available both within IMP and as a web interface on the MultiFit website. The fitting protocol is a multi-step procedure that proceeds via discretization of both the map and the proteins, local fitting of the proteins into the map, and an efficient combination of local fits into global solutions (Fig. 4). It is also able to incorporate additional information about interactions between the proteins from proteomics experiments and can take advantage of C_n symmetry to generate structures of such symmetric complexes. Here, we will demonstrate the use of MultiFit in building a model of porcine mitochondrial respiratory complex II (PDB id 3SFD), using crystal structures of its four constituent proteins and a 15 Å density map of the entire assembly. All input files for this procedure can be found in the "`multifit`" subdirectory of the downloaded zipfile. The protocol consists of the following steps:

1. *Setup a subunit list.* We create an input file listing the subunits involved in the complex. The file contains one line per component with the following information: the name that MultiFit will

use for the component, a path to the file containing the atomic coordinates for the component, and a 0/1 fitting flag indicating whether placements of the subunit should be sampled locally (0) or globally (1). The default for the fitting flag is 1 (global search). If the user has prior knowledge or a good hypothesis as to the subunit position, he can provide the proposed subunit placement in the atomic coordinates file and ask for a local search.

In this example, we assume no prior knowledge and provide the following input file (input/3sfd.subunits.txt):

```
3sfdA ../input/3sfd.A.pdb 1
3sfdB ../input/3sfd.B.pdb 1
3sfdC ../input/3sfd.C.pdb 1
3sfdD ../input/3sfd.D.pdb 1
```

2. *Create input files.* We generate two input files to guide the protocol, by running the MultiFit application, multifit.py (*see* **Note 3**). In this case, we use the application's "param" command, by typing on a command line:

```
multifit.py param -i 3sfd.asmb.input --
3sfd.asmb input/3sfd.subunits.txt 30
input/3sfd_15.mrc 15 3. 335 27.0 -6.0 21.0
```

The first file generated by this command, 3sfd.asmb.input, provides information on each of the subunits and their assembly density map, such as names of the files from which the input structures and map will be read, and those to which outputs from later steps will be written. The second file, 3sfd.asmb.alignment.param, specifies scoring and optimization parameters for each step of the MultiFit application. The user is advised to read a detailed description of the different parameters on the MultiFit website (http://salilab.org/multifit/) for better understanding of the algorithm and for troubleshooting difficult modeling cases.

The arguments to the "param" command include the spacing, origin, resolution, and density threshold of the input density map. The spacing and origin are often stored in the map header. To view the map header, run

```
view_density_header.py input/3sfd_15.mrc
```

The resolution is typically not stored in the map header; it is usually provided in the corresponding publication and can also be found in the corresponding EMDB [53] entry. A threshold is often provided by the author in the EMDB entry as "Recommended counter level" under the "Map Information" section. Alternatively, IMP provides a utility to calculate an approximate counter level based on the molecular mass of the complex, which can be run as:

```
estimate_threshold_from_molecular_mass.
py input/3sfd_15.mrc 1092
```

3. *Create the assembly anchor graph*. We determine a reduced representation of the assembly density map using the Gaussian Mixture Model, by running:

```
multifit.py anchors 3sfd.asmb.input 3sfd.asmb.anchors
```

   This command computes a reduced representation of the EM map that best reproduces the configuration of all voxels with density above the density threshold provided in the `3sfd.asmb.input` file as a set of 3D Gaussian functions (*see* **Note 4**). The reduced representation is written out as a PDB file containing fake Cα atoms, where each Cα corresponds to a single anchor point, and also as a Chimera [54] cmm file.

4. *Fit each protein to the map*. We first fit each protein to the map using an FFT search either globally or locally:

```
multifit.py fit_fft -a 30 -n 1000 -v 60 -c 6 3sfd.asmb.input
```

   The output is a set of candidate fits. In each file, a single subunit is rigidly rotated and translated to fit into the density map. Each fit is written out as the transformation (rotation and translation) required to place the original subunit in the density map. The fitting of a subunit into the density map is performed by globally searching for subunit transformations yielding high cross-correlation between the subunit and the map via a fast Fourier transform.

   Second, we create a list of valid fit indexes. By default, this list is simply the top 10 hits from fit_fft, but they could be filtered by other criteria (e.g., proximity to anchor points) if desired. We do this task by running:

```
multifit.py indexes 3sfd 3sfd.asmb.input 10 3sfd.indexes.mapping.input
```

5. *Create a proteomics restraint file*. We create the restraint file used in the next assembly step (*see* **Note 5**). This file instructs MultiFit how to combine the individual subunit fits created above into a global solution of all subunits simultaneously fitted into the map. First, we ask MultiFit to generate a basic proteomics file, indicating between which pairs of proteins a complementarity restraint (i.e., that the surfaces of the proteins should fit and complement each other) should be calculated:

```
multifit.py proteomics 3sfd.asmb.input 3sfd.asmb.proteomics
```

   The user can then add additional information from proteomics experiments to this file. Here, we add seven simulated residue–residue cross-link restraints as indicated in `input/3sfd.xlinks`. We also update the excluded volume (EV) pairs to calculate complementarity restraints between pairs of proteins as indicated by the cross-link restraints (*see* **Note 6**).

After these additions, the final `3sfd.asmb.proteomics` file is as follows:

```
|proteins|
|3sfdA|1|613|nn|nn|
|3sfdB|1|239|nn|nn|
|3sfdC|1|138|nn|nn|
|3sfdD|1|102|nn|nn|
|interactions|
|residue-xlink|
|1|3sfdB|23|3sfdA|456|30|
|1|3sfdB|241|3sfdC|112|30|
|1|3sfdB|205|3sfdD|37|30|
|1|3sfdB|177|3sfdD|99|30|
|1|3sfdC|95|3sfdD|132|30|
|1|3sfdC|9|3sfdD|37|30|
|1|3sfdC|78|3sfdD|128|30|
|ev-pairs|
|3sfdB|3sfdA|
|3sfdB|3sfdC|
|3sfdC|3sfdD|
```

6. *Assemble subunits.* The fits are combined into a set of the best-scoring global configurations by running:

```
multifit.py align 3sfd.asmb.input 3sfd.asmb.proteomics 3sfd.indexes.mapping.input 3sfd.asmb.alignment.param 3sfd.asmb.combinations 3sfd.asmb.combinations.fit.scores
```

   The scoring function used to assess each fit includes the quality-of-fit of each subunit in the map, the protrusion of each subunit out of the map envelope, the shape complementarity between subunits, as indicated in the proteomics file, and distance restraints as defined by proteomics data, also from the proteomics file. The optimization avoids exhaustive enumeration of all possible mappings of subunits to anchor points by means of a branch-and-bound algorithm combined with the DOMINO divide-and-conquer message-passing optimizer using a discrete sampling space [45].

7. *Ensemble analysis.* First, we cluster the top 100 models such that the maximum Cα RMSD between members of a cluster is 5 Å:

```
multifit.py cluster 3sfd.asmb.input 3sfd.asmb.proteomics 3sfd.asmb.mapping.input 3sfd.asmb.alignment.param 3sfd.asmb.combinations -r 5 -m 100
```

   The first cluster consists of 96 models and the second cluster consists of four models. The average Cα RMSD between members of the first cluster is 3.4 Å with a standard deviation of 0.3 Å.

The clustering procedure also generates a new combination file consisting of combinations of the cluster representatives. We can further investigate these cluster representatives by calculating scores of individual restraints:

```
multifit.py score 3sfd.asmb.input 3sfd.asmb.proteomics 3sfd.indexes.mapping.input 3sfd.asmb.alignment.param 3sfd.asmb.combinations.clustered 3sfd.asmb.combinations.clustered.scores
```

Finally, we can generate models (as PDB files) by running:

```
multifit.py models 3sfd.asmb.input 3sfd.asmb.proteomics 3sfd.indexes.mapping.input 3sfd.asmb.combinations.clustered 3sfd.model
```

These models can be visualized in any PDB viewer, such as Chimera [54].

3.4 Assembly of Macromolecular Complexes by Satisfaction of Spatial Restraints from EM Images

Obtaining a high-resolution density map by EM requires a large number of single-particle images and needs an initial low-resolution template density map to perform 3D reconstruction. This procedure is not always possible because in difficult cases the assembly only shows a set of preferred orientations during imaging. However, calculating average 2D images (class averages) from the images of single particles in the same orientation is relatively simple and fast. IMP provides an "EMageFit" application that performs integrative modeling to assemble the subunits of a macromolecular complex using a few class averages, in much the same way MultiFit does for density maps. The class averages can be combined with maximum distance and proximity restraints, such as those from chemical cross-linking and proteomics experiments, respectively. Additionally, an excluded volume restraint prevents the subunits from overlapping. The optimization procedure is a two-step method consisting of building a set of models by Simulated Annealing Monte Carlo (SA-MC) optimization followed by a refinement with DOMINO [55]. An example of the application of EMageFit to the same complex studied above with MultiFit can be found in the "`emagefit`" directory of the downloaded zipfile (*see* **Note 7**).

The inputs for modeling are the three simulated class averages of the complex located in the `em_images` directory, and the subunits to assemble: `3sfdA.pdb`, `3sfdB.pdb`, `3sfdC.pdb`, and `3sfdD.pdb`. The Python file `config_step_1.py` contains all the necessary options and restraint specifications (*see* **Note 8**). Briefly, the class averages are given in the `em2d_restraints` variable; four proteomics restraints are specified in the variable `pair_score_restraints`; and seven cross-links are specified in the variable `xlink_restraints`. The SA-MC optimization is set as a tempering schedule and the best 50 models are selected for refinement with DOMINO. For an extensive description of all the parameters in `config_step_1.py` see `config_example.py`.

Building a model requires four steps: pairwise docking between interacting subunits, SA-MC optimization, SA-MC model gathering, and DOMINO sampling.

1. *Pairwise docking*. The pairwise dockings are calculated with the program HEXDOCK [56], which can be obtained from http://hex.loria.fr/, based on the description of connectivity between subunits given by the cross-linking restraints. To perform this step, run from the command line:

```
emagefit.py --exp config_step_1.py --dock
--log file.log
```

   The docking results will be used during the SA-MC optimization to quickly explore feasible relative positions between pairs of components (although helpful, the dockings are not strictly required and EMageFit can work without them). The command produces multiple files: the PDB files of the initial docking solutions as estimated from the cross-linking restraints (ending in `initial_docking.pdb`); the PDB files with the best solutions from HEXDOCK (ending in `hexdock.pdb`); a set of text files starting with `hex_solutions`, containing all the solutions from HEXDOCK; and four text files starting with `relative_positions`, which contain the relative transformations (in IMP convention) between the subunits participating in each pairwise docking. The latter files are used by the SA-MC optimization. All described files can also be found in the `outputs` directory.

2. *SA-MC optimization*. The parameters controlling the optimization are in the `MonteCarloParams` class in `config_step_2.py`: the profile of temperatures, the number of iterations, number of cycles, and the maximum change in position and orientation tolerated for the random moves. The parameter `non_relative_move_prob` indicates the probability for a component of undergoing a random move instead of a docking-derived relative move. To ignore all docking solutions, or if they are not available, use a value of 1. Other important variables are `dock_transforms`, which specifies the files of relative orientations found previously, and `anchor`, which indicates the components that will not move during the SA-MC optimization. The command for producing one model is as follows:

```
emagefit.py --exp config_step_2.py --o mc_
solution1.db --log file.log --monte_carlo -1
```

   The output is the file `mc_solution1.db`, an SQLite database with the solution. To generate multiple candidate solutions, simply run the script multiple times, changing the name of the output file from `mc_solution1.db` each time (*see* **Note 9**).

3. *Model gathering*. Here we gather all the models produced with SA-MC:

```
emagefit.py --o monte_carlo_solutions.db
--gather {all database files}
```

Here `{all database files}` are the databases to merge and `monte_carlo_solutions.db` is the output database with all the merged results. For this example, we have already included in the zipfile a file `monte_carlo_solutions.db` containing 500 models, so you can skip this step if desired.

4. *DOMINO sampling*. `DominoSamplingPositions` and `DominoParams` in `config_step_3.py` include the relevant parameters: `read` is the file with the SA-MC solutions obtained before, `max_number` is the maximum number of solutions to combine, and `orderby` is the name of the restraint used to sort the SA-MC solutions. The command is as follows:

```
emagefit.py --exp config_step_3.py --o
domino_solutions.db --log file.log
```

This command will produce a database `domino_solutions.db` with all the results. We include the file in the `outputs` directory.

The best solutions can be written out in the PDB format. To write the ten best models according to the value of the em2d restraint, run:

```
emagefit.py --exp config_step_3.py --o
domino_solutions.db --w 10 -orderby em2d
--log file.log
```

The best solution and its fit into the density map of the complex are shown in Fig. 5. Finally, the solutions stored in the database can be clustered with the `emagefit_cluster.py` script. To cluster the first 100 solutions according to the value of the em2d restraint and save the results to `clusters.db`, use:

```
emagefit_cluster.py --exp config_step_3.py
--db domino_solutions.db --o clusters.db --n
100 --orderby em2d --log clusters.log --rmsd 10
```

And to write the elements of the first cluster as PDB files:

```
emagefit.py --exp config_step_3.py --o dom-
ino_solutions.db --wcl clusters.db 1
```

In summary, we have shown with this example how to combine multiple pairwise dockings, EM class averages and distance restraints for assembling the subunits of a macromolecular complex; integrating new restraints into the optimization protocol is also possible.

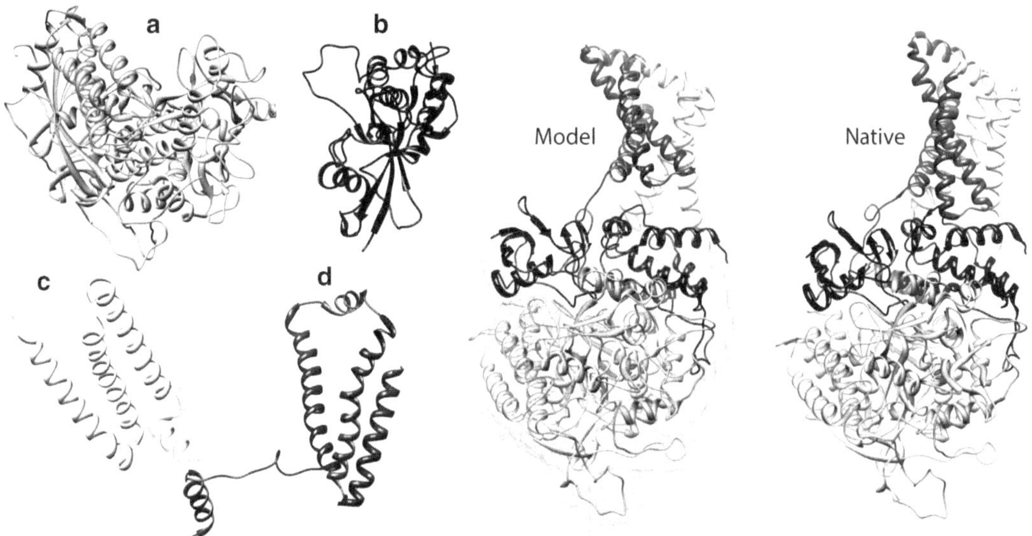

Fig. 5 Results from the application of EMageFit to the macromolecular complex with PDB id 3SFD (porcine mitochondrial respiratory complex II). *Left*: The four subunits of the complex, each labeled with their chain identifier. *Center*: Model for the complex fitted into the simulated density map of the native configuration. *Right*: Native configuration of the complex as stored in the PDB file

4 Summary

The structures of protein assemblies can typically not be fully characterized with any individual computational or experimental method. Integrative modeling aims to solve this problem by combining information from multiple methods to generate structural models. Integrative modeling problems can be tackled by satisfaction of spatial restraints, where information for individual restraints can come from different methods. In this approach, a suitable representation for the system is chosen, the information is converted into a set of spatial restraints, the restraints are simultaneously satisfied as well as possible by optimizing a function that is the sum of all restraints, and the resulting models are analyzed. Further experiments as well as the precision and likely accuracy of both the model and the data can be informed. IMP is an open source and flexible software package that provides all of the components needed to implement an integrative modeling protocol from scratch. It also contains higher-level applications and web services that can tackle specific use cases more conveniently.

5 Notes

1. X-ray crystal structures are often missing coordinates for some of the residues. Since a SAXS profile is typically experimentally determined for the entire structure, including these missing residues, the X-ray structure will not perfectly fit the SAXS profile. To improve the SAXS fit, we added missing loops, N-termini, C-termini, and His tags for both structures using MODELLER.
2. All of the IMP applications demonstrated here are command line tools and must be run by typing at a command line. The user is expected to unzip the downloaded zipfile before running any of the examples, and then to run the command in the directory corresponding to the example ("`idock`" in this case). Each of the command lines shown in this text should be entered as a single line, even though some have been wrapped onto multiple lines.
3. For detailed help on each step of the MultiFit protocol, run "`multifit.py help`" from the command line.
4. The default number of Gaussians is the number of components. However, if the sizes of the subunits differ, it is recommended to use the `-s` option to set the number of residues encapsulated in each Gaussian. For example, if you choose 50 residues per Gaussian, a 170-residue protein should use 3 Gaussians and a 260-residue protein should use 5 Gaussians.
5. A detailed description of the format of the proteomics file can be found on the MultiFit website.
6. The restraints will be used to create DOMINO's junction tree. DOMINO works most efficiently if the size of the intermediate subsets is small. Use the "`multifit.py merge_tree`" command to view the tree defined by the restraints. To reduce the size of the subsets, the user can determine which restraints are used to define the merge tree by setting the first value in the xlink definition. Setting the value to 0 instead of the default 1 specifies that the restraint is evaluated only at the root of the tree and not in an intermediate merging step.
7. An extended version of this manual is available on the IMP website.
8. We have used different configuration files for each step in this example for clarity, but it is possible to use a single one for all steps with all the options.
9. Different models are generated each time the script is run, because Monte Carlo relies on random moves, the specific sequence of which is uniquely determined by a random number seed, and the "`-1`" argument to the "`--monte-carlo`"

option instructs the script to use the current time as the seed. This can be a problem if multiple copies of the script are started at exactly the same time (e.g., on a cluster or a multi-core computer) as they will generate the same model. To avoid this scenario or to generate models that can be exactly reproduced, replace -1 with a specific seed for each model (e.g., 1 for the first model, 2 for the second, and so on).

Acknowledgments

We are grateful to all members of our research group, especially to Frank Alber, Friedrich Förster, and Bret Peterson who contributed to early versions of IMP, and Marc Marti-Renom, Davide Baù, Benjamin Schwarz, and Yannick Spill who currently contribute to IMP. We also acknowledge support from National Institutes of Health (R01 GM54762, U54 RR022220, PN2 EY016525, and R01 GM083960) as well as computing hardware support from Ron Conway, Mike Homer, Hewlett-Packard, NetApp, IBM, and Intel.

References

1. Schmeing TM, Ramakrishnan V (2009) What recent ribosome structures have revealed about the mechanism of translation. Nature 461:1234–1242
2. Sali A, Glaeser R, Earnest T et al (2003) From words to literature in structural proteomics. Nature 422:216–225
3. Mitra K, Frank J (2006) Ribosome dynamics: insights from atomic structure modeling into cryo-electron microscopy maps. Annu Rev Biophys Biomol Struct 35:299–317
4. Robinson C, Sali A, Baumeister W (2007) The molecular sociology of the cell. Nature 450:973–982
5. Blundell T, Johnson L (1976) Protein crystallography. Academic, New York
6. Stahlberg H, Walz T (2008) Molecular electron microscopy: state of the art and current challenges. ACS Chem Biol 3:268–281
7. Chiu W, Baker ML, Jiang W et al (2005) Electron cryomicroscopy of biological machines at subnanometer resolution. Structure 13:363–372
8. Lucic V, Leis A, Baumeister W (2008) Cryo-electron tomography of cells: connecting structure and function. Histochem Cell Biol 130:185–196
9. Parrish JR, Gulyas KD, Finley RL Jr (2006) Yeast two-hybrid contributions to interactome mapping. Curr Opin Biotechnol 17:387–393
10. Gingras AC, Gstaiger M, Raught B et al (2007) Analysis of protein complexes using mass spectrometry. Nat Rev Mol Cell Biol 8:645–654
11. Russel D, Lasker K, Webb B et al (2012) Putting the pieces together: integrative structure determination of macromolecular assemblies. PLoS Biol 10:e1001244
12. Alber F, Kim M, Sali A (2005) Structural characterization of assemblies from overall shape and subcomplex compositions. Structure 13:435–445
13. Alber F, Dokudovskaya S, Veenhoff L et al (2007) Determining the architectures of macromolecular assemblies. Nature 450:683–694
14. Alber F, Dokudovskaya S, Veenhoff L et al (2007) The molecular architecture of the nuclear pore complex. Nature 450:695–701
15. Lasker K, Phillips JL, Russel D et al (2010) Integrative structure modeling of macromolecular assemblies from proteomics data. Mol Cell Proteomics 9:1689–1702
16. Russel D, Lasker K, Phillips J et al (2009) The structural dynamics of macromolecular processes. Curr Opin Cell Biol 21:97–108
17. Alber F, Forster F, Korkin D et al (2008) Integrating diverse data for structure determination of macromolecular assemblies. Annu Rev Biochem 77:443–477
18. Alber F, Chait BT, Rout MP et al (2008) Integrative structure determination of protein

assemblies by satisfaction of spatial restraints. In: Panchenko A, Przytycka T (eds) Protein–protein interactions and networks: identification, characterization and prediction. Springer, London, UK, pp 99–114

19. Bonvin AM, Boelens R, Kaptein R (2005) NMR analysis of protein interactions. Curr Opin Chem Biol 9:501–508
20. Fiaux J, Bertelsen EB, Horwich AL et al (2002) NMR analysis of a 900K GroEL GroES complex. Nature 418:207–211
21. Neudecker P, Lundstrom P, Kay LE (2009) Relaxation dispersion NMR spectroscopy as a tool for detailed studies of protein folding. Biophys J 96:2045–2054
22. Takamoto K, Chance MR (2006) Radiolytic protein footprinting with mass spectrometry to probe the structure of macromolecular complexes. Annu Rev Biophys Biomol Struct 35:251–276
23. Guan JQ, Chance MR (2005) Structural proteomics of macromolecular assemblies using oxidative footprinting and mass spectrometry. Trends Biochem Sci 30:583–592
24. Taverner T, Hernandez H, Sharon M et al (2008) Subunit architecture of intact protein complexes from mass spectrometry and homology modeling. Acc Chem Res 41:617–627
25. Chen ZA, Jawhari A, Fischer L et al (2010) Architecture of the RNA polymerase II-TFIIF complex revealed by cross-linking and mass spectrometry. EMBO J 29:717–726
26. Sinz A (2006) Chemical cross-linking and mass spectrometry to map three-dimensional protein structures and protein-protein interactions. Mass Spectrom Rev 25:663–682
27. Trester-Zedlitz M, Kamada K, Burley SK et al (2003) A modular cross-linking approach for exploring protein interactions. J Am Chem Soc 125:2416–2425
28. Joo C, Balci H, Ishitsuka Y et al (2008) Advances in single-molecule fluorescence methods for molecular biology. Annu Rev Biochem 77:51–76
29. Mertens HD, Svergun DI (2010) Structural characterization of proteins and complexes using small-angle X-ray solution scattering. J Struct Biol 172:128–141
30. Hura GL, Menon AL, Hammel M et al (2009) Robust, high-throughput solution structural analyses by small angle X-ray scattering (SAXS). Nat Methods 6:606–612
31. Schneidman-Duhovny D, Kim SJ, Sali A (2012) Integrative structural modeling with small angle X-ray scattering profiles. BMC Struct Biol 12:17
32. Berggard T, Linse S, James P (2007) Methods for the detection and analysis of protein–protein interactions. Proteomics 7:2833–2842
33. Sali A, Blundell TL (1993) Comparative protein modelling by satisfaction of spatial restraints. J Mol Biol 234:779–815
34. Sali A, Blundell T (1994) Comparative protein modeling by statisfaction of spatial restraints. In: Bohr H, Brunak S (eds) Protein structure by distance analysis. Symposium on distance based approaches to protein structure determination. CTR Biol Sequence Anal. Tech Univ Denmark, Lyngby, Denmark, pp 64–86
35. Vajda S, Kozakov D (2009) Convergence and combination of methods in protein–protein docking. Curr Opin Struct Biol 19:164–170
36. Shen MY, Sali A (2006) Statistical potential for assessment and prediction of protein structures. Protein Sci 15:2507–2524
37. Melo F, Sanchez R, Sali A (2002) Statistical potentials for fold assessment. Protein Sci 11:430–448
38. Brooks BR, Brooks CL 3rd, Mackerell AD Jr et al (2009) CHARMM: the biomolecular simulation program. J Comput Chem 30:1545–1614
39. Case DA, Cheatham TE 3rd, Darden T et al (2005) The Amber biomolecular simulation programs. J Comput Chem 26:1668–1688
40. Christen M, Hunenberger PH, Bakowies D et al (2005) The GROMOS software for biomolecular simulation: GROMOS05. J Comput Chem 26:1719–1751
41. Forster F, Lasker K, Beck F et al (2009) An Atomic Model AAA-ATPase/20S core particle sub-complex of the 26S proteasome. Biochem Biophys Res Commun 388:228–233
42. Nickell S, Beck F, Scheres SHW et al (2009) Insights into the molecular architecture of the 26S proteasome. Proc Natl Acad Sci USA 29:11943–11947
43. Lasker K, Forster F, Bohn S et al (2012) Molecular architecture of the 26S proteasome holocomplex determined by an integrative approach. Proc Natl Acad Sci USA 109: 1380–1387
44. Lasker K, Sali A, Wolfson HJ (2010) Determining macromolecular assembly structures by molecular docking and fitting into an electron density map. Proteins 78: 3205–3211
45. Lasker K, Topf M, Sali A et al (2009) Inferential optimization for simultaneous fitting of multiple components into a cryoEM map of their assembly. J Mol Biol 388:180–194
46. Webb B, Lasker K, Schneidman-Duhovny D et al (2011) Modeling of proteins and their assemblies with the integrative modeling platform. Methods Mol Biol 781:377–397
47. Lensink MF, Wodak SJ (2010) Docking and scoring protein interactions: CAPRI 2009. Proteins 78:3073–3084
48. Ritchie DW (2008) Recent progress and future directions in protein–protein docking. Curr Protein Pept Sci 9:1–15
49. Schneidman-Duhovny D, Rossi A, Avila-Sakar A et al (2012) A method for integrative struc-

ture determination of protein–protein complexes. Bioinformatics 28:3282–3289

50. Reese ML, Dotsch V (2003) Fast mapping of protein–protein interfaces by NMR spectroscopy. J Am Chem Soc 125:14250–14251
51. Rappsilber J (2011) The beginning of a beautiful friendship: cross-linking/mass spectrometry and modelling of proteins and multi-protein complexes. J Struct Biol 173:530–540
52. Berman HM, Battistuz T, Bhat TN et al (2002) The Protein Data Bank. Acta Crystallogr D 58:899–907
53. Lawson CL, Baker ML, Best C et al (2011) EMDataBank.org: unified data resource for CryoEM. Nucleic Acids Res 39: D456–D464
54. Pettersen EF, Goddard TD, Huang CC et al (2004) UCSF Chimera – a visualization system for exploratory research and analysis. J Comput Chem 25:1605–1612
55. Velazquez-Muriel JA, Lasker K, Russel D et al (2012) Assembly of macromolecular complexes by satisfaction of spatial restraints from electron microscopy images. Proc Natl Acad Sci USA 109:18821–18826
56. Ritchie DW, Venkatraman V (2010) Ultra-fast FFT protein docking on graphics processors. Bioinformatics 26:2398–2405

Chapter 21

The Quality and Validation of Structures from Structural Genomics

Marcin J. Domagalski, Heping Zheng, Matthew D. Zimmerman, Zbigniew Dauter, Alexander Wlodawer, and Wladek Minor

Abstract

Quality control of three-dimensional structures of macromolecules is a critical step to ensure the integrity of structural biology data, especially those produced by structural genomics centers. Whereas the Protein Data Bank (PDB) has proven to be a remarkable success overall, the inconsistent quality of structures reveals a lack of universal standards for structure/deposit validation. Here, we review the state-of-the-art methods used in macromolecular structure validation, focusing on validation of structures determined by X-ray crystallography. We describe some general protocols used in the rebuilding and re-refinement of problematic structural models. We also briefly discuss some frontier areas of structure validation, including refinement of protein–ligand complexes, automation of structure redetermination, and the use of NMR structures and computational models to solve X-ray crystal structures by molecular replacement.

Key words Structure quality, Structure validation, Drug discovery, Data mining, Structural genomics

1 Introduction

Structural genomics (SG) programs have greatly expanded our knowledge of the protein structure universe by determining almost 12,000 three-dimensional structures, which constitute approximately 14 % of the protein models that have been deposited to the Protein Data Bank (PDB) [1]. The NIGMS Protein Structure Initiative and NIAID Structural Genomics Centers for Infectious Diseases have alone supported determination of over 7,000 of these structures. However, a vast majority of them were not described in peer-reviewed articles and, taking into account the rate of new structures determined by SG, may never be published. Therefore, the scientific community will be able to access and evaluate them only through the data deposited in the PDB. For that reason the criteria that scientific community applies to model quality of SG structures should be stricter than for those coming from

Yu Wai Chen (ed.), *Structural Genomics: General Applications*, Methods in Molecular Biology, vol. 1091, DOI 10.1007/978-1-62703-691-7_21,

traditional structural biology laboratories. In addition, the overall quality of the deposits, including the completeness and accuracy of the header information, has to be as high as possible, since the remarks in the PDB files provide the only source of information describing the experimental methods that led to structure determination. Indeed, the average quality of 3D models coming from SG projects seems to be higher [2] than that of models coming from traditional structural biology laboratories. There are two reasons: (a) SG projects use very advanced technology and software tools, sometimes developed or enhanced by members of SG consortia; and (b) structural biologists at SG centers may be more experienced in structure determination, model refinement, and validation process than scientists working in traditional laboratories. Analysis of the authorship of PDB deposits shows that 54 % of all first authors served in this capacity for two or fewer deposits. So (perhaps) unlike peer-reviewed publication, PDB deposition seems to be an infrequent event in many biological laboratories. In this text, we discuss the impact of quality of structural models on biomedical research; in particular we address issues that are related to data mining and drug discovery research.

2 Protein Data Bank as a Data Mining Repository

2.1 Data Content of the PDB

The importance and role of the PDB for biomedical research cannot be overestimated. PDB is a unique repository containing atomic structural models of biological macromolecules (protein, DNA, and RNA) obtained by X-ray crystallography, NMR spectroscopy, electron microscopy, and other techniques. As 88 % of all PDB structures were determined by X-ray crystallography, our discussion of structural quality will focus mainly on this subset of the PDB. The PDB deposit for an X-ray diffraction structure usually contains three parts: (a) a header with information about diffraction experiment, structure determination, and refinement protocol; (b) coordinates of the atoms that make up the model of the macromolecules, water sites, and other small molecules in the structure; and (c) a structure factor file that contains diffraction data reduced from X-ray diffraction detector images.

In an ideal world, the first part (the header) would be equivalent to the "Materials and Methods" section in a typical peer-reviewed publication. However, the information in the headers of many PDB files is often contradictory, erroneous, and/or incomplete. The title of a PDB deposit is particularly important, especially if the deposit does not have a published citation, as it may be the only way to clearly identify whether the deposit is relevant to a given area of interest. Many structures have headers of the PDB files containing multiple values of "NULL," indicating that corresponding experimental data parameters are missing. The large

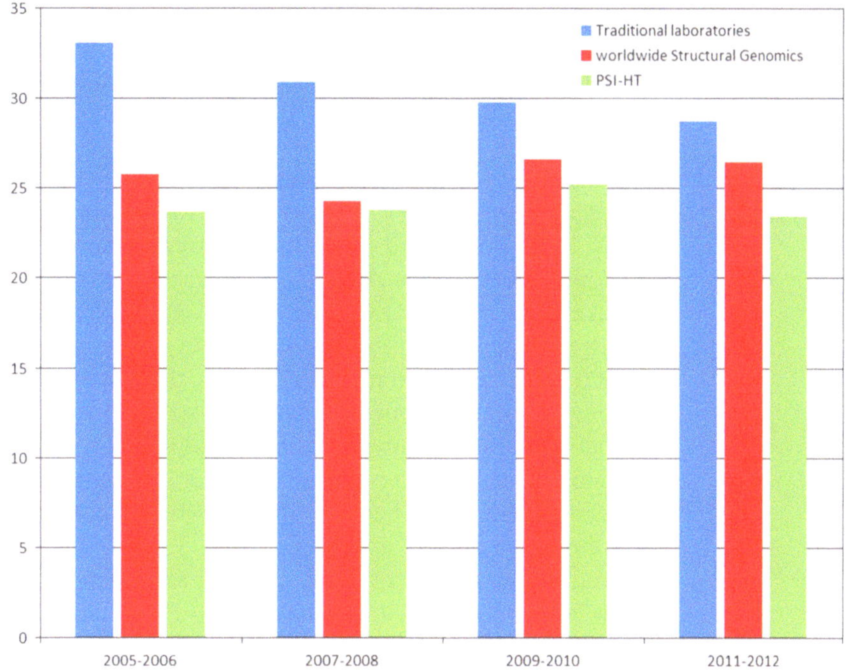

Fig. 1 Average number of missing parameters in the PDB file headers for PSI high-throughput (PSI-HT) centers, structural genomics worldwide (excluding the PSI-HT centers), and traditional structural biology laboratories. A small number of "NULL" values is always present due to generation of PDB file headers—not all parameters are relevant to all kinds of experiments

number of "NULL" data parameters should be alarming as it possibly indicates negligence and/or a lack of knowledge of how the crystallography experiment was performed. The number of "NULL" values for structural genomics centers is lower than average, not only because the depositors are more experienced, but because data needed for completing the header are usually readily available in, and possibly automatically extracted from, an existing database (Fig. 1).

The second part (coordinates of atoms in the macromolecular model) is usually the most reliable, as these coordinates are generated by refinement programs. However, there is no single standard for coordinate quality. For example, there are different methods for dealing with portions of crystallographically derived models corresponding to regions of weak or absent electron density. In some cases, all atoms of an amino acid residue are placed in probable locations regardless of density (with the occupancy parameters of atoms outside the map often reduced or set to 0). In others, atoms may be omitted from the model—perhaps only amino acid main chains are modeled, or only atoms unambiguously identifiable within the map are placed. Both approaches are justified for modeling uncertainty in experimental data, but can lead to very different results when thus derived coordinates are used as input to other programs that calculate, for example, a charge on the surface of the protein. However, it should be noted that many, but not all, programs that read PDB files preprocess coordinates to address some of these ambiguities.

2.2 Inconsistent Quality of Models

Although the protein crystallography community (including structural genomics centers) has had many discussions about model quality, it has not agreed on a single, universal standard that models should meet before deposition. There are many quantitative measures that are clearly correlated with model quality, including resolution, R and R_{free} factors, distribution of deviations from ideal geometry, Ramachandran distribution, Molprobity clashscore, etc., but no single parameter is sufficient to conclusively determine whether a given structure is of high or low quality. "Quality" can also depend on context—the quality of a structural model useful for bioinformatics may be very different from its counterpart for in silico binding studies, for example.

As different depositors have different standards for deposition, mining of PDB data is very challenging. Fortunately, the PDB is unique among biomedical repositories as it contains experimental data as well. Since February 1, 2008 the PDB has required that each deposit based on crystallographic data must include a list of the structure factors used to build the model. When a structure is suspicious, in most cases a PDB user may download the corresponding structure factor file and re-refine the structure until it meets his or her own standards.

It is inevitable that there are differences in model quality standards since, to some degree the structures are based on subjective interpretation of experimental data. However, the X-ray diffraction models and experimental structure factor data in the PDB are generally of high quality, especially when compared to data in other repositories or databases used in biomedical research. In fact, the quality of the models and the ability of PDB users to examine and even re-refine a 3D model makes protein crystallography a "crown jewel" of experimental biomedical research.

Whereas there is some inconsistency in model quality due to a lack of universal deposition standards, much of this inconsistency is also due to the history of the field. For over 40 years more than 17,000 scientists have deposited models derived from experimental X-ray crystallography data of many different resolution limits, determined by various methods, and refined by many different, constantly evolving software packages. The distribution of high-resolution limits for all diffraction-based structures deposited in the PDB is very broad, as shown in Fig. 2. Even if the same software packages are used, quality of structures strongly depends on the design of diffraction experiments, data reduction, structure determination, refinement, and validation, particularly if multiple, weakly diffracting crystals are used. While the handling of diffraction

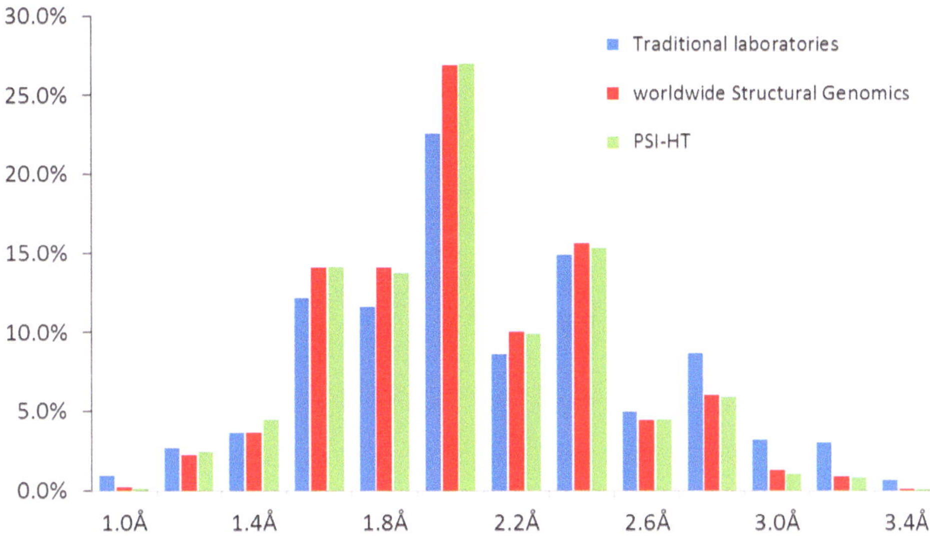

Fig. 2 Normalized distribution of high-resolution limits for X-ray structures solved by PSI high-throughput (PSI-HT) centers, structural genomics worldwide excluding PSI-HT, and traditional structural biology laboratories

experiments clearly depends on the experience and skills of the crystallographers performing them, examination of structures deposited by frequent depositors (i.e., those that are the first authors of more than 100 structures) shows that even different deposits prepared by the same person can vary significantly in quality measures (Fig. 3). Thus one has to acknowledge that the structure quality has to be also affected by factors other than experience, such as the quality of the experimental data. For example, models derived from poor resolution data—an intrinsic property of a crystal over which the crystallographer has little or no control—necessarily contain less information than a model from high-resolution data.

2.3 Selection of the Most Appropriate Deposits in PDB for Data Mining

All nontrivial data mining requires filtering and processing of the data in order to obtain reliable results. Much of this "quality control" work can be done in advance if there is curation, but most biomedical databases are either partially curated or not curated at all. PDB depositions are partially curated, as the authors receive extensive reports about problems in their depositions. The deposition reports produced by the PDB have steadily improved over the years, but there are still some areas for further improvement. For example, the current deposition tool (ADIT) does not yet validate the geometry of small molecules present in macromolecular crystal structures. Moreover, PDB depositors may ignore warnings in the report and ask that the model be deposited "as is." The most common protocol for filtering structures is defining a resolution limit cutoff for exclusion of lower resolution models from further analysis. In principle, this should be an ideal method, at least for the proteinaceous part of a macromolecular model. Unfortunately the

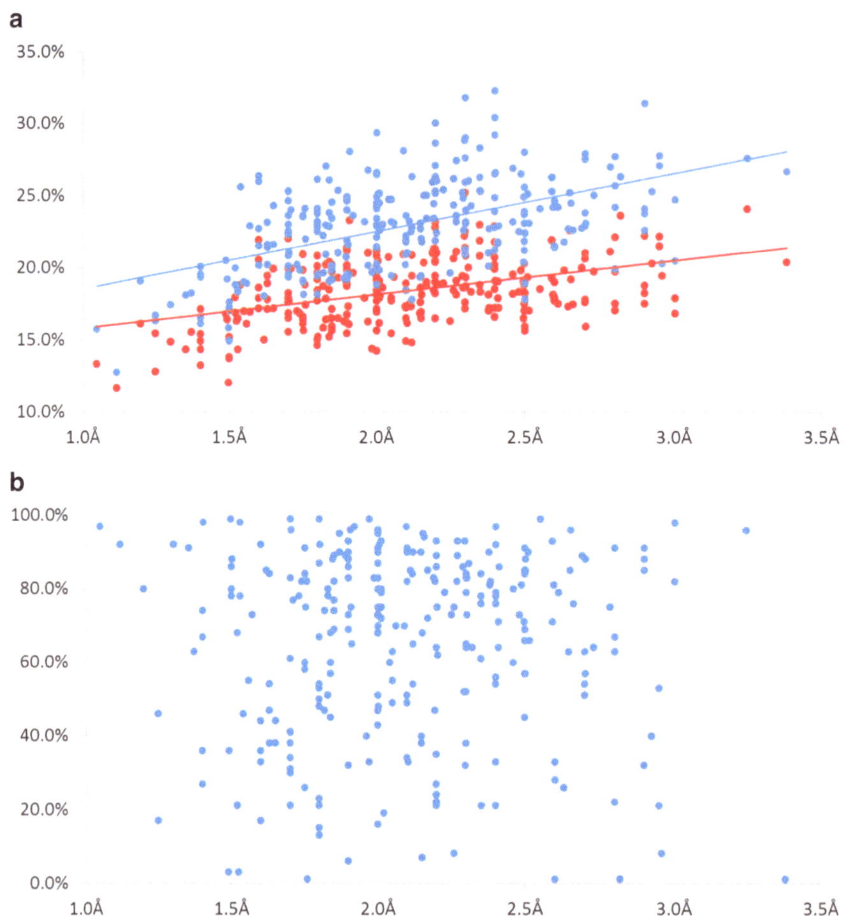

Fig. 3 Selected structure quality metrics of all PDB deposits with the same first author (the author was selected randomly from all such authors with >200 deposits). (**a**) Distribution of R (*red*) and R_{free} (*blue*) as a function of resolution, along with trendlines as determined by linear regression. (**b**) Distribution of Molprobity clashscore percentile (ranking of "raw" clashscore relative to other structures in the PDB of similar resolution)

high-resolution limit reported in a PDB deposit is not always equivalent to the nominal resolution limit of the diffraction data obtained from structure factors. In some cases, it appears that depositors may have chosen a resolution limit higher than is justified by the data. A significant number of PDB deposits include reflections in the highest resolution shell weaker, on the average, than the commonly accepted threshold (mean $I/\sigma(I) \geq 2.0$; *see* Fig. 4) [3]. (It should be noted that the traditional rule of "mean I over sigma ratio greater than 2.0" may not be the ideal way to choose a threshold; Karplus and Diederichs [4] have proposed an alternative statistic that advocates extension of the nominal resolution of a diffraction dataset.) However, analysis of the structure factor data in the PDB shows that the mean $I/\sigma(I)$ in the highest resolution shell of many diffraction datasets is as high as 10, suggesting that usable high-resolution reflections were never collected,

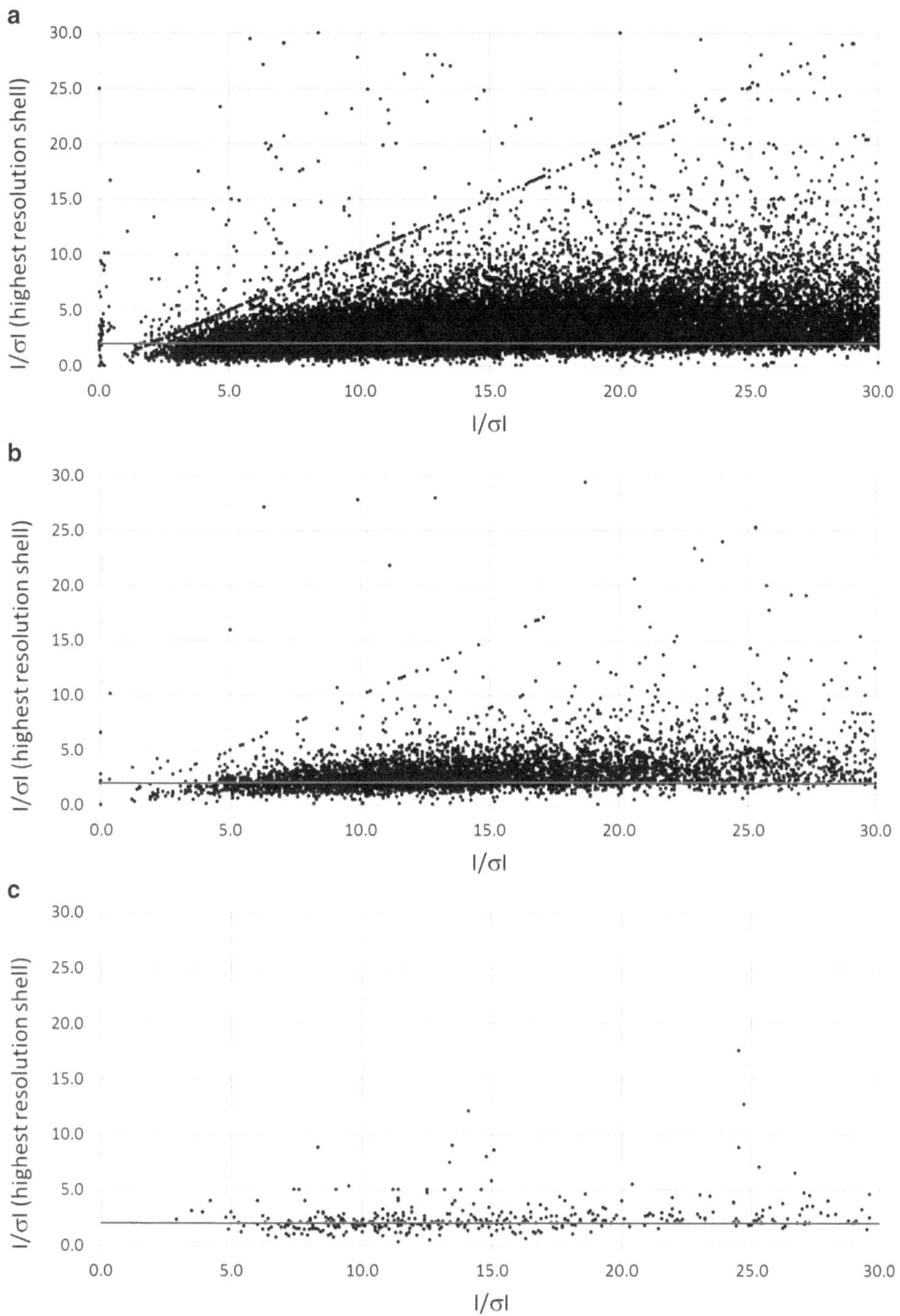

Fig. 4 Distributions of mean $I/\sigma(I)$ for the highest resolution shell vs. mean $I/\sigma(I)$ for all reflections, as determined for different sets of structures in the PDB. (**a**) Distribution for all structures determined by X-ray crystallography. (**b**) Distribution for all X-ray structures solved since April 2011. (**c**) Distribution for all X-ray structures solved since April 2011 by the four high-throughput PSI centers. On all distributions, the conventional threshold of 2.0 of mean $I/\sigma(I)$ for the highest resolution is marked by a *red line*. There are a significant number of structures where the two values are identical, as well as a number where the mean $I/\sigma(I)$ for the highest resolution shell is greater than the mean for all reflections, a physically improbable outcome

despite the tremendous investment of synchrotron beamlines in larger and faster detectors.

Moreover, dependent on the type of data mining analysis needed, validation may focus on either the macromolecular or small molecule portions of structures. In addition to evaluating the agreement between a structure model and the experimental data (R, R_{free}) and the properties of crystal packing (symmetry operations, solvent content), it is a common practice that structure determination, refinement, and validation in macromolecular crystallography are heavily (and necessarily) dependent on prior chemical knowledge of the subject molecule to define the geometry of the corresponding structural features in crystal structures [5]. There has been enormous progress in the development of statistics and tools used to verify the models of macromolecules in crystal structures, which measure agreement with both ideal geometry and experimental electron density. However, validation of small molecule models in macromolecular structures has lagged behind. Recently Rupp et al. demonstrated that a significant number of PDB deposits have ligands with very weak or even no correlation between the small molecule models and the electron density maps [6].

3 Model Quality

3.1 Overall Model Quality in the PDB

The validation tools developed over the years by many software authors [7–12], in addition to the in-house tool developed by the PDB [13], have greatly simplified the process of validation of protein models. Ideal values for bond lengths, bond angles, and dihedral angles within individual amino acid residues and in peptide bonds have been well defined and are incorporated into these programs [5]. Common secondary structural elements in protein structures (helices, strands, coils) can be defined by hydrogen bond patterns [14]. The overall geometrical quality of a protein main chain characterized by a Ramachandran plot [15] is particularly valuable for validation because the dihedral angles of individual peptides are usually not restrained during refinement. Potential steric clashes, which usually indicate problematic regions in a structure model, can be identified by Molprobity [12] and other similar programs. The agreement between diffraction data (structure factors) and a model are described by the R and R_{free} factors. PROSESS provides cross-validation with similar structures in the PDB to identify potential problems [16]. Despite the availability of a large selection of tools for structure validation, there is still no universal way to fully automate the process of model improvement. It is up to the crystallographer to utilize these tools routinely to identify potential problems and improve model quality after structure validation on a case-by-case basis, and it appears that nearly all follow this path. The vast majority of models in the PDB are very good, despite the lack of precise definition of what values of the parameters describing structure quality are acceptable for high-quality structures.

Another often overlooked issue that may affect structure quality is that the structure factors are themselves derived quantities and thus do not represent the "raw" diffraction data used to determine a structure. Structure factors are typically reduced from a set of diffraction images collected in so-called rotation mode, and the way how the individual reflections on the images are indexed, integrated, and scaled together can significantly affect the quality of the structure factor amplitudes produced. Traditionally, the large size of diffraction image files has made it difficult to preserve (let alone distribute) raw diffraction data, but decreases in the cost per terabyte of hard drive storage have made storage and distribution of diffraction images feasible. Four SG centers, namely CSGID, SSGCID, MCSG, and JCSG, have made their diffraction images available for download from the respective servers. Diffraction images for over 2,200 PDB deposits, which comprise nearly 3 % of all X-ray structures in the PDB, are currently accessible. The public availability of original images provides an invaluable resource to determine if structure factors have been optimally reduced. The ability of the scientific community to access and evaluate raw, fundamental data directly from diffraction experiments makes crystallography arguably one of the most reproducible branches of biomedical science, with high transparency and reliability.

3.2 Quality of Macromolecular Structures Complexed with Small Molecules

Small molecules are abundantly represented in the PDB, as 80 % of PDB structures contain one or more residues that do not belong to polymers of amino acids or nucleic acids, or represent ordered water molecules. The presence of ordered small molecules in macromolecular structures usually highlights a specific area of interest or biological relevance. Although small molecules might be unintentionally introduced during sample preparation, the location of a small molecule in a macromolecular structure most often represents a binding site (or active site) that has some topological (concavity) or physiochemical properties suitable for binding. However, validation of small molecule models in protein structures is usually more difficult due to the diversity of small compounds and modes of interaction and conformation, i.e., the chemical sense of the environment. Moreover, even the use of high-resolution diffraction data does not necessarily guarantee high quality of the electron density around the small molecule, especially when there is always a possibility that the ligand may not fully occupy its binding site. For that matter, medium-to-low resolution of diffraction data is certainly insufficient by itself to justify the discovery of novel chemistry.

Small molecule models require specific tools to validate due to their chemical diversity and the fact that ligands are not covalently bound to macromolecule, which can easily result in ambiguity in binding mode [17]. Geometrical parameters derived from the very high-resolution structures in the Cambridge Structure Database (CSD) [18] can be used as restraints in small molecule refinement [19], but in the case of a small molecule–macromolecule complex the procedures implemented to validate atomic resolution small molecule structures (such as the ones in the CSD) no longer apply. There are two main reasons for this. First, the overall resolution is usually significantly lower for a protein–small molecule complex compared to the crystal structure of a small molecule alone. Second, the binding mode of a small molecule needs to be validated, in addition to its conformation. Sometimes the models of small molecules are incomplete due to the degradation or multiple conformations. For that reason, the usage of stricter geometrical restraints is a common technique for the refinement and validation of small molecule binding sites in protein–small molecule complexes [20]. Therefore validation tools that can handle the small molecule portion of the complex are less common and often require substantial manual input to use. Consequently, the quality of small molecule models in PDB varies significantly. Useful tools for their validation include Twilight, which evaluates an agreement between small molecule models and electron density [6], and PURY, which evaluates the geometry [21].

4 Structure Validation

4.1 Validation Tools

Tools to validate structure quality, both overall and within substrate binding sites, are constantly evolving. However, the optimal ways of using these tools vary and are heavily dependent on the user's experience. For example, there is no common standard for a comprehensive set of parameters and threshold values to determine the validity of all structures. In addition, the standard protocols used for validation within most SG consortia are not yet streamlined or well defined. However, two SG centers, CSGID and SSGCID, have agreed that most of their targets should meet a common set of criteria. The structures determined within the HKL-3000 framework [22] may be checked by a standard validation procedure that compares quality parameters with the average values of the parameters as derived from structures deposited in the PDB during the last 2 years. Such a procedure was applied to and tested on more than 2,000 structures. The set of validation parameters, as implemented in HKL-3000 [22], could be easily applied to other software packages to standardize the validation process.

Structure validation is an ongoing, iterative process where model building and refinement are repeated until validation tools and visual inspection no longer reveal any problematic regions that can be further improved. However, no validation tools are perfect, and none can objectively determine when a model cannot be further improved and should be considered "good enough." Therefore differences in knowledge and experience of crystallographers, or sometimes even just differences in opinion, may affect the decision whether or not a model is completed or should be refined further. The involvement of a second person to examine and evaluate the refinement of a structural model is usually considered a more objective approach for structure validation that can partially compensate the limits of experience and/or reduce the potential bias that a crystallographer may have during data interpretation. This approach is working successfully in a number of centers, including JCSG, NYSGRC, and the Structural Genomics Consortium (SGC).

As evaluated by R, R_{free}, Molprobity clash score, and Twilight score for ligands at a given resolution of structures, the average quality of structures determined by SG consortia in the PDB is significantly higher than the average quality of structures determined by other structural biologists over the last 2 years (Fig. 5). This trend is more prominent in overall quality assessment parameters such as the Molprobity clash score but is less prominent in ligand score, indicating that the availability of tools for a particular validation problem varies. Tools for overall validation, such as Molprobity or WHAT IF [23], have been available for a decade, whereas the tools for ligand refinement like Twilight were made available only recently. However, in the current year several SG structures of proteins complexed with small molecule ligands were redeposited, which suggests that SG efforts are promptly taking advantage of the new technologies.

4.2 Validation of Water Structure

As virtually all crystals of biological macromolecules are formed in aqueous solution, ordered water molecules bound to the surfaces of proteins and nucleic acids are commonly observed in X-ray crystal structures. However, at the resolutions of most macromolecular structures, typically only water oxygen atoms are observed. The binding of most waters is relatively weak. For example, NMR relaxation data show that nearly all protein surface water molecules have binding time scales of less than 100 ns, and molecular dynamics calculations predict residence times between 10 and 500 ps [24]. The residence time of even the most buried waters in a small protein BPTI was <20 ms [25]. The positions of most ordered crystallographic waters represent local energy minima into which waters fall reproducibly, appearing as peaks when averaged over all scattering events [26]. Only rarely are crystallographic waters in positions where they can form three or four H-bonds to other ordered atoms in the structure. It has also been noted that the number of crystallographic waters per residue identified in protein structures is inversely proportional to resolution [27, 28].

Although the binding of water molecules in the crystals is weak, their accurate modeling is still important for interpretation

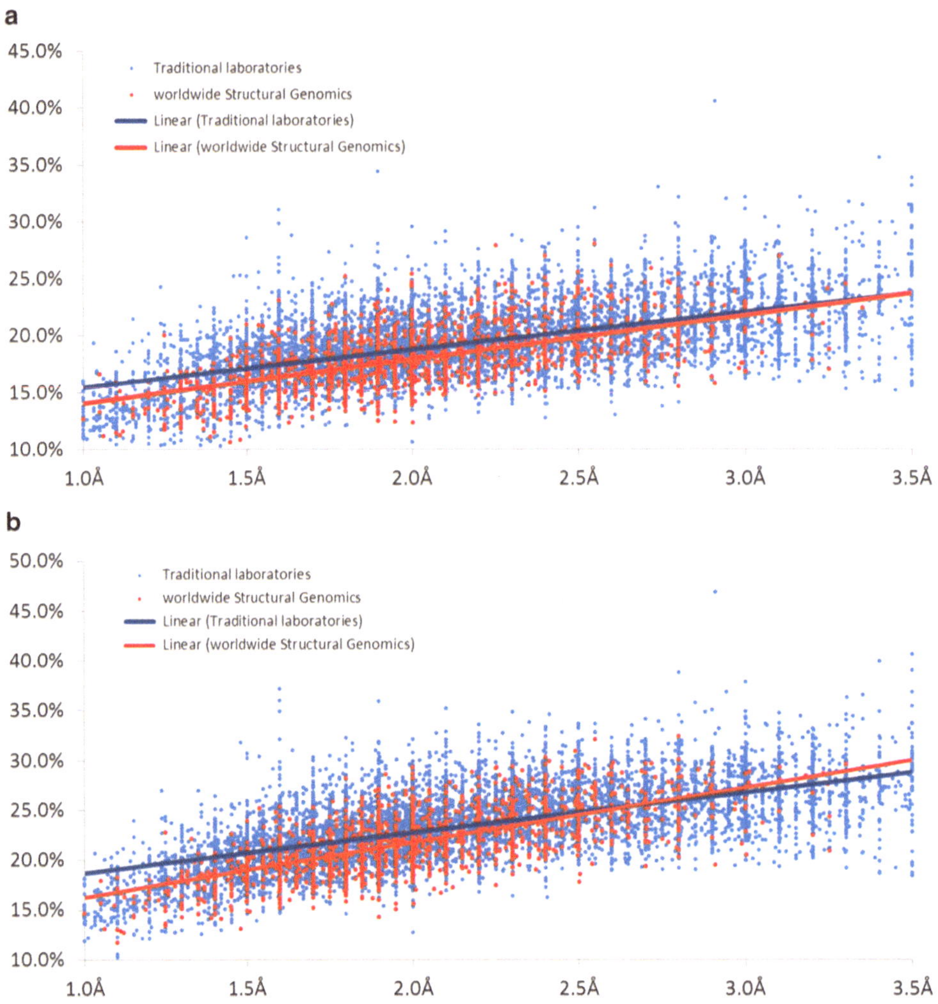

Fig. 5 (**a**) Distribution of *R* factor vs. resolution for all X-ray structures deposited in the PDB since April 2011. Structures solved by SG centers are marked in *red* and structures solved by traditional laboratories are in *blue*. The *lines* represent linear regression trend lines for the two sets of structures in the same color scheme. (**b**) Distribution of R_{free} factors vs. resolution for all X-ray structures deposited in the PDB since April 2011, using the same color scheme as part (**a**)

of results, because incorporation of ordered waters will improve the completeness of the model (and in turn, yield better estimates of the phases of the calculated structure factors). Crystallographically observed waters are not covalently bonded to the macromolecule in a structure, and at most resolution limits only a single peak corresponding to the oxygen atom is observed. Thus some spurious, "ghost" peaks in an electron density map can be mistakenly interpreted as waters, especially in medium-resolution structures. There are a number of tools for validating crystallographic water positions. For example, the interactive "Check Waters" tool in

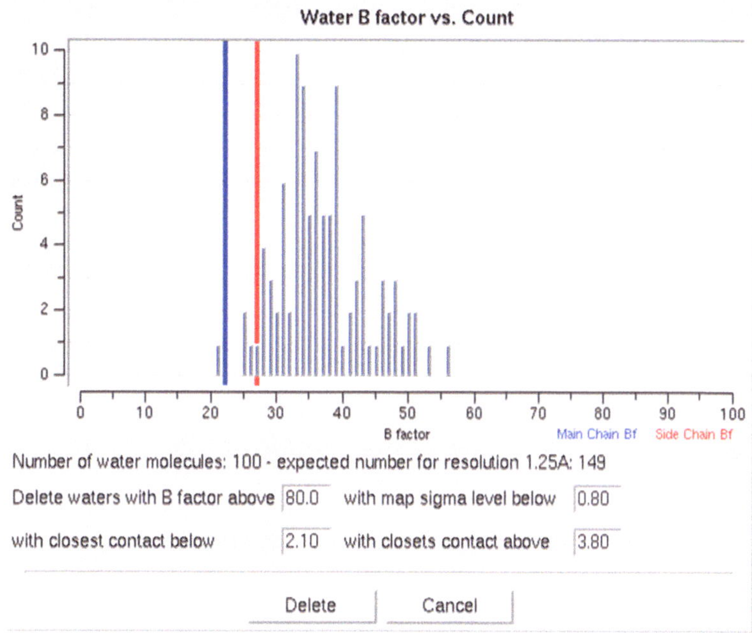

Fig. 6 A screen shot of the "Check waters" tool in HKL-3000

HKL-3000 [22] allows for effective validation of water molecules (Fig. 6). This is accomplished by plotting the distribution of waters as a function of atomic displacement parameters (or B-factors), providing information about the expected number of waters given the number of amino acids and resolution, following the method of Carugo and Bordo [27]. With this tool, all water molecules with B-factors greater than a user-defined threshold can be removed by one click.

5 Rebuilding and Re-refinement of Existing Models

5.1 The Benefits of Re-refinement

As mentioned above, an independent examination of a structure by a second researcher may reduce personal bias in data interpretation. In practice, availability of another expert to examine a structure is always limited, leading to the presence of a significant number of suboptimally refined structures in the PDB. For example, many structures that were determined in the past were refined and validated with tools that were quite primitive compared to the state-of-the-art tools in use today. Since many of these older structures describe important proteins and are frequently utilized as the basis for designing new experiments, it would be beneficial to revisit them using modern validation tools and reinterpret these structures more carefully. This would be especially instructive for structures that are used, for example, as test sets for in silico docking experiments. Although, on average, SG-determined structures have relatively high structural quality, a routine re-refinement process is even more important because the consumer of a structure is likely to be less knowledgeable about X-ray crystallography and may take the structure "as is" in further biomedical research, e.g., as a target for structure-based drug design. Therefore deposition of SG structures of suboptimal quality will have a detrimental effect on subsequent research.

5.2 Automatic Re-refinement: PDB-REDO

One effort to re-refine old crystal structures with new technology on a large scale is the PDB-REDO project [29]. Each structure in the PDB for which structure factor data are available is automatically re-refined by a suite of tools using modern structure refinement and validation procedures, and, even more importantly, all of the different crystal structures processed by the system are handled uniformly, following a standardized refinement protocol. Even though the outcome of the refinement is still somewhat affected by the initial model, to a certain extent the PDB-REDO process removes the bias due to differences in refinement techniques used by different crystallographers. As a result, the quality statistics of the re-refined structures are more comparable. Advances in refinement techniques resulted in significant improvement of the refinement statistics, and in most cases, the values of R and R_{free} were improved by 2–5 %. However, PDB-REDO does not rebuild the original model (i.e., remove or add atoms other than in water molecules), which may be warranted if the electron density map is significantly improved. Whereas automated model building algorithms are becoming available, it has proven very difficult to fully automate this process with consistently reliable results [2]. Therefore, improvement in refinement protocols alone is not a panacea for maximizing the quality of crystal structures. As the majority of serious problems in structures that most affect structure quality require rebuilding of the model, large-scale automated re-refinement projects such as PDB-REDO are still limited. In addition, the PDB does not provide links to the PDB-REDO results. Researchers not familiar with structural biology are far more likely to use data from the PDB, so if PDB-REDO produces a model of higher quality from the same data, the biomedical community may never be aware of it.

In fact, inconsistencies between databases are some of the most significant impediments to effective biomedical data mining and research in general.

5.3 Semiautomatic Ligand Reassignment

As mentioned earlier, the potential presence of a small molecule constitutes a unique feature of the structure of an adjacent macromolecule. Given electron density of reasonable quality and the sequences of the polypeptides or nucleic acids, it is often relatively easy to build or rebuild the macromolecular portions of a structure, and in many cases this process can be automated (albeit with human supervision) [30, 31]. However, correct identification and modeling of ligands is still difficult to automate, as the ligand bound is often unknown a priori and must be identified from an enormous and diverse set of endogenous substances. If the identity of the ligand is known (or limited to a small set), some tools such as RESOLVE [32], ARP/wARP [33], and the "Build ligand" tool in HKL-3000 [22] can search that set and automatically place a ligand in the map and refine it. Careful human examination of the search result is still crucial to verify correct placement. However, a search of a much larger chemical library is necessarily very computationally intensive and the approach described above does not scale. In contrast other programs such as PHENIX (phenix.ligand_identification) [34] or the MCSG-developed LigSearch [35], implement efficient protocols to search for potential physiological or drug-like small molecules in a much larger compound library. The potential ligands identified using a protein structure template may be very informative and may lead to the discovery of physiological ligands when unexplained electron density cannot be interpreted as compounds introduced during the processes of protein production or crystallization.

5.4 Structure Redetermination: Diffraction Images

Sometimes the structure factors that are deposited in the PDB are not sufficient to redetermine the structure. This is especially true when a structure was interpreted in the incorrect space group and the results affect the biomedical context of the structure. In such a case, the access to the original diffraction images is invaluable. Several years ago it was infeasible to store and distribute diffraction data, as building the storage and bandwidth infrastructure required to make diffraction data readily available to the research community was prohibitively expensive at best and impossible at worst. However, as storage media continue to rise in capacity and fall in price (as of this writing a 3 TB hard drive costs $130) and high bandwidth network connections are ubiquitous; the technical and financial barriers become less and less relevant. Application of efficient compression algorithms to diffraction images has further pushed the limits. However, the storage of thousands of datasets or more makes organization of data critical. As led by the four SG centers that make diffraction images available to the public (*see* Subheading 3.1 above), we may hope that, in the future, deposition of images in a public repository will become a requirement for publicly funded X-ray crystallography research. As shown recently, the possibility of reprocessing diffraction data that may not have been processed optimally (for example, by extending resolution limits and improving data quality) will lead to vastly improved models and their better interpretation [36]. Similar reprocessing will be beneficial in many ways, especially for relatively poorly determined structures.

6 NMR Structures

The use of NMR-derived models for solving crystal structures has been postulated and shown in practice over 20 years ago [37], but to date the success rate of such approaches has been somewhat limited. Many NMR models are simply not accurate enough to provide sufficient phasing power for the determination of crystal structures by molecular replacement. This is partly due to the nature of NMR data, which describe quite accurately local structures, but may not contain enough information to unambiguously assign long-range interactions. However, application of computational algorithms such as Rosetta has led to vast improvement in the success of molecular replacement calculations utilizing NMR models [38]. The Rosetta procedure is now part of standard crystallographic software [39]. As an example, it has been shown that its use, together with the involvement of computer games players [40], made it possible to utilize an NMR model for solving a structure by molecular replacement after many years of failure [41, 42].

7 Conclusions and Challenges

The current level of understanding of the biochemical mechanisms affecting living organisms would not be possible without the revolutionary progress of structural biology. The structures deposited in the PDB are only the starting point for many further analyses done by hundreds of thousands of scientists in academia and in industry. Any inaccuracy in a structure, even a small one, has tremendous potential to generate backlash, as the error may proliferate through all analyses that use data from that structure. In other words, a rotten apple can spoil the barrel. In addition, analyses of PDB structures can also be negatively affected by the lack of rigorous standards of data for PDB deposition. The X-ray diffraction structures determined by structural genomics centers worldwide are, on average, of higher quality than structures solved in traditional laboratories. Surprisingly, SG centers, who are unquestioned leaders in high-throughput and high-quality structure determination, have not established a precise definition of the conditions that could be universally used to assess the quality of macromolecular structures. Similarly, SG centers have not set a standard of deposition which could be adopted by the whole structural biology community. It is a serious challenge to establish both deposition standards and quality metrics, but large-scale SG projects are in a good position to propose them, due to the large databases that these efforts have generated. This is the key to success of all large-scale attempts to analyze the vast treasure which is the PDB.

Acknowledgments

The authors would like to thank Maks Chruszcz, Tom Terwilliger, Wayne Anderson, Matt Vetting, and Andrzej Joachimiak for valuable comments on the manuscript. This work was supported by PSI:Biology grants U54 GM094585 and U54 GM094662, as well as grant R01 GM053163, and supported in part by the Intramural Research Program of the NIH, National Cancer Institute, Center for Cancer Research, as well as with federal funds from the National Institute of Allergy and Infectious Diseases, National Institutes of Health, Department of Health and Human Services, under Contract No. HHSN272201200026C.

References

1. Berman HM, Westbrook J, Feng Z et al (2000) The protein data bank. Nucleic Acids Res 28:235–242
2. Chruszcz M, Domagalski M, Osinski T et al (2010) Unmet challenges of structural genomics. Curr Opin Struct Biol 20:587–597
3. Grabowski M, Chruszcz M, Zimmerman MD et al (2009) Benefits of structural genomics for drug discovery research. Infect Disord Drug Targets 9:459–474
4. Karplus PA, Diederichs K (2012) Linking crystallographic model and data quality. Science 336:1030–1033
5. Engh RA, Huber R (1991) Accurate bond and angle parameters for X-ray protein-structure refinement. Acta Crystallogr A 47:392–400
6. Pozharski E, Weichenberger CX, Rupp B (2013) Techniques, tools and best practices for ligand electron-density analysis and results from their application to deposited crystal structures. Acta Crystallogr D 69:150–167
7. Laskowski RA, MacArthur MW, Moss DS et al (1993) PROCHECK—a program to check the stereochemical quality of protein structures. J Appl Crystallogr 26:283–291
8. Hooft RW, Vriend G, Sander C et al (1996) Errors in protein structures. Nature 381:272
9. Oldfield TJ (1992) SQUID: a program for the analysis and display of data from crystallography and molecular dynamics. J Mol Graph 10:247–252
10. Pontius J, Richelle J, Wodak SJ (1996) Deviations from standard atomic volumes as a quality measure for protein crystal structures. J Mol Biol 264:121–136
11. Vaguine AA, Richelle J, Wodak SJ (1999) SFCHECK: a unified set of procedures for evaluating the quality of macromolecular structure-factor data and their agreement with the atomic model. Acta Crystallogr D 55:191–205
12. Chen VB, Arendall WB III, Headd JJ et al (2010) MolProbity: all-atom structure validation for macromolecular crystallography. Acta Crystallogr D 66:12–21
13. Yang HW, Guranovic V, Dutta S et al (2004) Automated and accurate deposition of structures solved by X-ray diffraction to the protein data bank. Acta Crystallogr D 60:1833–1839
14. Colloc'h N, Etchebest C, Thoreau E et al (1993) Comparison of three algorithms for the assignment of secondary structure in proteins: the advantages of a consensus assignment. Protein Eng 6:377–382
15. Ramachandran GN, Ramakrishnan C, Sasisekharan V (1963) Stereochemistry of polypeptide chain configurations. J Mol Biol 7:95–99
16. Berjanskii M, Liang Y, Zhou J et al (2010) PROSESS: a protein structure evaluation suite and server. Nucleic Acids Res 38:W633–W640
17. Malde AK, Mark AE (2011) Challenges in the determination of the binding modes of non-standard ligands in X-ray crystal complexes. J Comput Aided Mol Des 25:1–12
18. Allen FH (2002) The cambridge structural database: a quarter of a million crystal structures and rising. Acta Crystallogr B 58:380–388
19. Lebedev AA, Young P, Isupov MN et al (2012) JLigand: a graphical tool for the CCP4 template-restraint library. Acta Crystallogr D 68:431–440
20. Gront D, Grabowski M, Zimmerman MD et al (2012) Assessing the accuracy of template-based structure prediction metaservers by comparison with structural genomics structures. J Struct Funct Genomics 13:213–225

21. Andrejasic M, Praaenikar J, Turk D (2008) PURY: a database of geometric restraints of hetero compounds for refinement in complexes with macromolecular structures. Acta Crystallogr D 64:1093–1109
22. Minor W, Cymborowski M, Otwinowski Z et al (2006) HKL-3000: the integration of data reduction and structure solution—from diffraction images to an initial model in minutes. Acta Crystallogr D 62:859–866
23. Vriend G (1990) WHAT IF: a molecular modeling and drug design program. J Mol Graph 8:52–56, 29
24. Brunne RM, Liepinsh E, Otting G et al (1993) Hydration of proteins. A comparison of experimental residence times of water molecules solvating the bovine pancreatic trypsin inhibitor with theoretical model calculations. J Mol Biol 231:1040–1048
25. Otting G, Liepinsh E, Wüthrich K (1991) Proton-exchange with internal water molecules in the protein BPTI in aqueous-solution. J Am Chem Soc 113:4363–4364
26. Bryant RG (1996) The dynamics of water–protein interactions. Annu Rev Biophys Biomol Struct 25:29–53
27. Carugo O, Bordo D (1999) How many water molecules can be detected by protein crystallography? Acta Crystallogr D 55: 479–483
28. Wlodawer A, Minor W, Dauter Z et al (2008) Protein crystallography for non-crystallographers, or how to get the best (but not more) from published macromolecular structures. FEBS J 275:1–21
29. Joosten RP, Joosten K, Murshudov GN et al (2012) PDB_REDO: constructive validation, more than just looking for errors. Acta Crystallogr D 68:484–496
30. Terwilliger T (2004) SOLVE and RESOLVE: automated structure solution, density modification, and model building. J Synchrotron Radiat 11:49–52
31. Perrakis A, Morris R, Lamzin VS (1999) Automated protein model building combined with iterative structure refinement. Nat Struct Biol 6:458–463
32. Terwilliger TC (2003) Improving macromolecular atomic models at moderate resolution by automated iterative model building, statistical density modification and refinement. Acta Crystallogr D 59:1174–1182
33. Langer GG, Evrard GX, Carolan CG et al (2012) Fragmentation-tree density representation for crystallographic modelling of bound ligands. J Mol Biol 419:211–222
34. Adams PD, Afonine PV, Bunkoczi G et al (2010) PHENIX: a comprehensive python-based system for macromolecular structure solution. Acta Crystallogr D 66:213–221
35. de Beer TA (2013) LigSearch. http://www.ebi.ac.uk/thornton-srv/databases/LigSearch/index.html. Accessed 23 Apr 2013
36. Perkins A, Gretes MC, Nelson KJ et al (2012) Mapping the active site helix-to-strand conversion of CxxxxC peroxiredoxin Q enzymes. Biochemistry 51:7638–7650
37. Baldwin ET, Weber IT, St Charles R et al (1991) Crystal structure of interleukin 8: symbiosis of NMR and crystallography. Proc Natl Acad Sci USA 88:502–506
38. Ramelot TA, Raman S, Kuzin AP et al (2009) Improving NMR protein structure quality by rosetta refinement: a molecular replacement study. Proteins 75:147–167
39. DiMaio F, Terwilliger TC, Read RJ et al (2011) Improved molecular replacement by density- and energy-guided protein structure optimization. Nature 473:540–543
40. Cooper S, Khatib F, Treuille A et al (2010) Predicting protein structures with a multiplayer online game. Nature 466:756–760
41. Khatib F, DiMaio F, Foldit Contenders Group et al (2011) Crystal structure of a monomeric retroviral protease solved by protein folding game players. Nat Struct Mol Biol 18:1175–1177
42. Gilski M, Kazmierczyk M, Krzywda S et al (2011) High-resolution structure of a retroviral protease folded as a monomer. Acta Crystallogr D 67:907–914

Chapter 22

Navigating the Global Protein–Protein Interaction Landscape Using iRefWeb

Andrei L. Turinsky, Sabry Razick, Brian Turner, Ian M. Donaldson, and Shoshana J. Wodak

Abstract

iRefWeb is a bioinformatics resource that offers access to a large collection of data on protein–protein interactions in over a thousand organisms. This collection is consolidated from 14 major public databases that curate the scientific literature. The collection is enhanced with a range of versatile data filters and search options that categorize various types of protein–protein interactions and protein complexes. Users of iRefWeb are able to retrieve all curated interactions for a given organism or those involving a given protein (or a list of proteins), narrow down their search results based on different supporting evidence, and assess the reliability of these interactions using various criteria. They may also examine all data and annotations related to any publication that described the interaction-detection experiments. iRefWeb is freely available to the research community worldwide at http://wodaklab.org/iRefWeb.

Key words Protein–protein interactions, Interaction networks, Proteomics, Literature curation, IMEx consortium, HUPO PSI-MI standards, Bioinformatics resources, iRefWeb, iRefIndex

1 Introduction

Key cellular processes such as gene transcription, translation, DNA repair, chromatin modification, and signal transduction are carried out by physically interacting proteins often forming multi-protein complexes [1–3]. Characterizing these interactions is therefore a crucial step in unraveling biological function at the level of individual proteins and of the processes in which they participate. Technological advances in the last decade have greatly improved our ability to detect protein interactions (*see* refs. 4, 5 for review). This has led to an explosive growth of protein–protein interaction (PPI) data derived from many small-scale focused studies, as well as from genome-scale interrogations in organisms such as bacteria, yeast, fly, and human [6–11].

Yu Wai Chen (ed.), *Structural Genomics: General Applications*, Methods in Molecular Biology, vol. 1091, DOI 10.1007/978-1-62703-691-7_22,

Curating and archiving PPI data from the scientific literature and making them readily accessible to the scientific community has become a priority, prompting the development of a number of specialized databases [12]. Some of these databases reflect very broad curation efforts, such as BioGRID [13], DIP [14], IntAct [15], or MINT [16]. Other database efforts focus on specific subsets of PPI, such as those of mammalian protein complexes in CORUM [17], PPI related to innate immune response in InnateDB [18], extracellular matrix interactions in MatrixDB [19], or microbial interactions in MPIDB [20]. Typically each PPI database has its own practices for literature curation and data representation, reflecting the needs and priorities of the corresponding database resource [21].

Thanks to the Proteomics Standards Initiative and the efforts of the IMEx consortium, there has been a considerable effort to promote common standards for PPI data representation, common curation policies, and common data access via PSICQUIC Web services [18, 22–24]. Nevertheless, the retrieval of PPI data from the different specialized databases remains a tedious task for most users, not in the least because PPI retrieved from multiple sources often contain redundancies and need to be adequately consolidated. This task is far from trivial: even if two different databases curate the same publication, they may still use different aliases for the proteins involved, different descriptions for the species of origin, and different representations for multi-subunit protein complexes [25–27].

In this chapter we describe how the compendium of archived PPI can be examined and analyzed using the iRefWeb resource [28]. This resource allows researchers to seamlessly access a broad landscape of PPI with all its supporting evidence through one Web interface, regardless of the original source database that curated the data. This landscape is assembled from 14 major public databases using iRefIndex, which implements one of the most rigorous procedures for consolidating redundant PPI records [20]. Users of iRefWeb may retrieve all archived interactions for a given organism, or extract interactions that involve a protein (or a list of proteins) of interest. They may then apply a set of filters to narrow down the search results, taking into account various PPI attributes such as the organisms of origin, the experimental techniques used to detect the interaction, the number of supporting publications, or an estimated confidence score.

We start by briefly describing the nature of the PPI data and their supporting evidence, and follow by instructions on how to explore the consolidated PPI landscape using the versatile range of iRefWeb features.

2 The PPI Data and Their Supporting Evidence

The PPI data archived in the different specialized "source" databases and consolidated in the iRefWeb resource have been curated from published studies that employ a wide range of experimental techniques. Large-scale (high throughput) studies use two main types of techniques: affinity purification combined with mass spectrometry (AP/MS) (for review *see* refs. 2, 29), which detect multiprotein complexes, and techniques that systematically identify binary interactions. Methods of the latter category include the yeast two hybrid (Y2H) [30] and Split-Ubiquitin [31] screens and various protein complementation assays (PCA) [32]. More recently, genome-scale descriptions of human PPIs were also derived from a massive co-fractionation effort [10]. Small-scale hypothesis-driven studies of specific systems tend to use a more diverse set of methods, including, in addition to the methods mentioned above, various co-immunoprecipitation methods, electrophoretic mobility assays, and more rarely, X-ray diffraction, and NMR (nuclear magnetic resonance) spectroscopy.

The PPI data derived from these various studies are described as simple links or "interactions" between proteins (e.g., protein A interacts with B, C, and D; protein C interacts with proteins A, E, F, and G). When a number of proteins are involved, the information is commonly represented as a graph (network) of proteins (nodes) linked to one another whenever an "interaction" has been detected. In PPI data derived from AP/MS or other types of "pull down" experiments, a link between two proteins represents a "co-complex association." It indicates that the proteins belong to the same complex, but does not guarantee that they make direct physical contact. On the other hand, in PPI data derived from Y2H and related techniques, a reported link between two proteins generally indicates that the proteins were found to make a direct physical contact.

It is also important to realize that different experimental techniques suffer from different biases, and that they tend to probe distinct properties of the protein interaction landscape. For instance, some methods are geared more towards identifying stable interactions, whereas others detect more transient ones (*see* ref. 4 for review). It is therefore not uncommon to find very little overlap between genome-scale PPI datasets of the same organism derived using methods such as AP/MS, Y2H, and the PCA techniques [33], and hence to consider these datasets as providing complementary descriptions.

The reliability of the derived interactions (or co-complex associations) is another outstanding issue. Most of the experimental techniques for detecting protein interactions produce noisy data, e.g. they often detect interactions that do not occur in vivo (false positives), and may miss interactions that do occur (false negatives). In large-scale studies, noise is usually dealt with by deriving reliability scores for individual interactions based on statistical analyses of the raw data [34–37]. Results are then reported in the form of a "high confidence" subset of PPIs deemed sufficiently reliable, although authors also tend to provide access to the raw data, which certain source databases may archive. Small-scale studies usually do not have statistical reliability scores associated with the PPI data they report, but tend to provide supporting evidence for their findings based on independent experimental verifications.

It has been assumed until recently that literature curated (LC) protein interactions data are of higher accuracy than those produced by high throughput studies, because they are derived from carefully crafted focused investigations. But recent analyses of literature curated PPI data are suggesting that this is no longer the case, due in part to improvements in high throughput methods, but also to inherent difficulties in extracting relevant information from publications that report results from small-scale studies [26, 27].

Currently, however, most source databases do not provide reliability scores for the PPI data they store, even when such scores have been reported in the original studies. To extract meaningful information from the archived data, users (and some databases) may therefore choose to rely on simple filters, based on the supporting evidence that is usually captured by the source database for each archived PPI record. This evidence includes information on the type of interaction (e.g. a "direct interaction," or an "association"), on the method used to detect it, and on the corresponding literature citation. Most, but not all databases capture this evidence using the PSI-MI controlled vocabulary [38], making it possible to apply filters requiring that an interaction be reported in at least two publications, that it be detected by a given set of methods, or conserved in another organism, as will be discussed below.

iRefWeb computes the MI (MINT-Inspired) score for the interactions it consolidates, by closely following an earlier approach developed by the MINT database team [16], which takes into account some of the above criteria. The development of scoring schemes for literature curated PPI data is a challenging problem, which is currently under study.

Lastly, it is worth mentioning that interactions derived from purely computational procedures [39], or the so-called genetic interactions, which represent phenotype alterations produced by the mutation/deletion of one gene in the background of a mutation/deletion of another gene [40–42], are not considered in this chapter, mainly because only a small subset of the source databases curate them.

3 Materials

In this section we describe how the information necessary to query the PPI landscape in iRefWeb should be represented.

To enable focused iRefWeb searches, prepare a list of identifiers that represent the proteins or genes of interest. As an illustration we shall use the human *neural-restrictive silencer factor* (NRSF), currently known as *RE1-silencing transcription factor* (REST).

It may also be necessary to examine the specific interaction type and experimental detection methods used in the original study. To ensure that the correct terms are used and their meaning is well defined, the user should prepare a list of standard terms that conform to the Molecular Interaction ontology [38]. As an illustration we shall use the interaction detection method "affinity chromatography technology" and the interaction type "physical association."

3.1 Protein Identifiers

1. Access the UniProtKB database at http://www.uniprot.org [43] and search for the protein of interest: human NRSF, or neural-restrictive silencer factor.
2. Note the current protein name: REST, or *RE1-silencing transcription factor*, instead of NRSF (*see* **Note 1**).
3. Record the UniProtKB entry accession Q13127 and name REST_HUMAN, to be used in the iRefWeb search query.

3.2 Gene Identifiers

1. Access NCBI Gene database at http://www.ncbi.nlm.nih.gov/gene [44] and search for the gene of interest: human NRSF, or neural-restrictive silencer factor.
2. Note the current gene name: REST, or *RE1-silencing transcription factor*, instead of NRSF (again, *see* **Note 1**).
3. Record the current gene ID 5978 and gene symbol REST, to be used in the iRefWeb search query.

3.3 Molecular Interaction Ontology: Interaction Detection Method

1. Access the EBI Ontology Lookup Service at http://www.ebi.ac.uk/ontology-lookup/ and select "Molecular Interaction (PSI-MI 2.5) [MI]."
2. Search for an interaction detection method of interest, e.g. "affinity chromatography technology" (*see* **Note 2**).
3. Record the identifier MI:0004 for this MI ontology term, to be used in future iRefWeb explorations.
4. Examine the standard definition of the term "affinity chromatography technology," as well as its various alternative aliases, such as "affinity purification," to determine whether the term was chosen correctly (*see* **Note 3**).

3.4 Molecular Interaction Ontology: Interaction Type

1. Access the EBI Ontology Lookup Service at http://www.ebi.ac.uk/ontology-lookup/ and select "Molecular Interaction (PSI-MI 2.5) [MI]" in the drop-down Search Ontology menu.
2. Search for an interaction type of interest, e.g. "physical association" (*see* **Note 2**).
3. Record the interaction-type code MI:0915 for this MI ontology term, to be used in future iRefWeb explorations.
4. Examine the term definition: "Molecules that are experimentally shown to belong to the same functional or structural complex." Decide if the term was chosen correctly (*see* **Notes 3** and **4**).

4 Methods

This section illustrates how to formulate various commonly used iRefWeb query options, visualize the results and download the data.

4.1 Search for Interactions for a Given Protein of Interest

1. Access the iRefWeb search page at http://wodaklab.org/iRefWeb/search.
2. In the panel *Search Terms*, paste the current gene name into the Left search box: *REST* (*see* **Note 5**). A popup window should appear with the list of matching protein interactors found in iRefWeb.
3. Choose the match that corresponds to the desired organism (*H. sapiens*), protein identifier (REST_HUMAN) and gene ID (5978).
4. Toggle its checkbox "add to search" and click the button "Add your checked selection to your search query terms." The interactor REST_HUMAN should appear below the search box.
5. Click the *Search* button. The Search Results Summary should appear at the bottom of the page, listing the PPI that involve the human REST protein.

4.2 Filter and Download the Interaction Results

1. In the panel Search Filters, click the link *Expand All Filters*. The counts next to each filter option show the number of PPI records that match each filter.
2. In the *Organism* filter panel, select "single organism interaction" to exclude cross-species interactions.
3. In the *Interaction Type* panel, select "physical association," which corresponds to the standard term of the MI ontology as described in the earlier section. This filter operation excludes various weak associations, predicted interactions, genetic interactions (if any) and interactions for which the type was not specified by the source-database curators.

Interacting Proteins		Interaction ID	Distinct Proteins	MI Score	MI Percentile	Supporting PubMeds
● *Homo sapiens* ● *Mus musculus* ● *Other*	● *Drosophila melanogaster* ● *Caenorhabditis elegans* ● *Saccharomyces cerevisiae S288c*	*Click an ID to view the details of an interaction.*	*Min* 2.00 *Max* 2.00 *Mean* ~2.00	*Min* 0.43 *Max* 0.85 *Mean* ~0.51	*Min* 0.00 *Max* 0.00 *Mean* ~0.00	*Min* 1.00 *Max* 3.00 *Mean* ~1.50
● HDAC4_HUMAN	● REST_HUMAN	672415	2	0.43	60	1
● REST_HUMAN	● RCOR1	721583	2	0.85	83	3
● REST_HUMAN	● SIN3B_HUMAN	724329	2	0.59	69	3
● REST_HUMAN	● TF2B_HUMAN	736865	2	0.43	60	1
● REST_HUMAN	● TBP_HUMAN	740210	2	0.43	60	1
● REST_HUMAN	● HDAC1_HUMAN	1017218	2	0.59	69	2
● REST_HUMAN	● SP1_HUMAN	1017356	2	0.43	60	1
● REST_HUMAN	● SIN3A_HUMAN	1080282	2	0.48	62	1
● REST_HUMAN	● Q9HBD4_HUMAN	1084945	2	0.48	62	2
● TCP4_HUMAN	● REST_HUMAN	1096066	2	0.55	67	1
● REST_HUMAN	● CDYL1_HUMAN	1224123	2	0.43	60	1
● HDAC5	● REST_HUMAN	1235639	2	0.43	60	1

Fig. 1 Results of the iRefWeb search for the interaction neighborhood of the human REST protein, after applying a range of filters to exclude cross-organism interactions, low-scoring interactions, and weak-associations

4. Click on the Search button to apply the filters. The Search Results Summary at the bottom of the page should be updated to reveal 18 human PPI pairs.
5. Examine the *MI Score* and the *MI Percentile* columns in the results table. The MI score is a crude reliability score associated with a given PPI. Higher MI scores are assigned to PPI supported by multiple publications, detected by multiple experimental techniques, or conserved across several organisms. Following the original MINT approach [45], direct physical interactions, as well as interactions supported by low-throughput studies, also receive higher scores (*see* the example in Subheading 4.4 below for further details).
6. Click the *Download Interactome* button and select the MINI-TAB option in the popup window. The retained interaction pairs will be downloaded in a simple tab-delimited text file (*see* **Notes 6–8**).
7. After exploring the results the user may choose to retain only the high-scoring interactions from the retrieved list. To enable this option, expand the filter panel *MI Organism Percentile*, and select "50 or more" to exclude low-confidence interactions that fall below the median MI score. Click on the Search button to re-apply the filters, which should retain only the 12 high-confidence human pairs (Fig. 1). Then download this smaller list of PPI by using the *Download Interactome* button again.

4.3 Visualize the Interaction Network

1. Click on any of the REST_HUMAN links in the previous Search Results table to load the detailed interactor page for this protein. The page presents a summary of the PPI and complexes in which REST is involved; a list of its aliases and external ID

with links to the corresponding databases; and several options to explore the REST interaction neighborhood.

2. Click on the link "View a graph of my interaction partners" to load *Netility*, a Web-based visualization tool based on the Cytoscape Web library [46] that is customized to display information from iRefWeb as network graphs. By default this graphics tool shows the protein of interest (a circular node); the corresponding gene (square node); its disease annotations from the DAnCER resource [47], which are labeled using the Medical Subject Headings [48] standard terminology (parallelogram nodes—*see* also **Note 9**); and its homologous genes in human and in several model organisms, such as mouse *M. musculus*, worm *C. elegans*, fly *D. melanogaster*, and yeast *S. cerevisiae* (square nodes with different label colors). It appears that REST is associated with both Neuroblastoma [49, 50] and Down Syndrome [51].
3. To view the protein-interaction network, double-click on the REST_HUMAN protein node. This action should expand the graph to include additional information about REST: its physical interaction partners in human and model organisms (circular nodes), their protein domains (triangular nodes), and their Gene Ontology (GO) [52] annotations (small dark labeled circles). The domain- and GO-related data are retrieved from external databases and seamlessly integrated into the Netility visual engine.
4. To clean up the displayed network, examine the Netility filters on the left side of the graph, and de-select several annotation categories: node types, node organisms, edge labels, domains, and "Chromatin Related" labels (*see* **Note 9**). Also, de-select GO terms from both Biological Process and Molecular Function ontologies to leave only the Cellular Component annotations.
5. In the top left Network Actions panel, select the force-directed layout option and click the button to re-apply the layout.
6. Manipulate the graph by dragging its nodes as needed to create a visually appealing network image (Fig. 2).
7. In the top left Network Actions panel, select the PDF format and click the Export Network button to save an image file (*see* **Note 10**).

4.4 Search for Specific Interaction Partners

The search query of the previous section reveals a presence of several different histone deacetylases in the interaction neighborhood of the human REST protein. In the following sections we shall explore known interactions between REST and various histone deacetylases, which appear in several different contexts. We shall also examine the evidence of REST-HDAC interactions from model organisms such as Mouse and Rat.

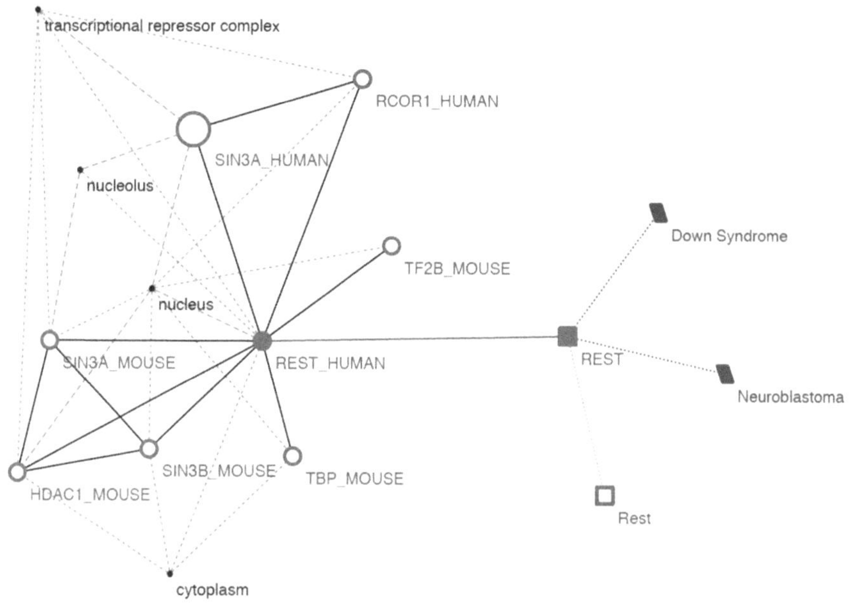

Fig. 2 The interaction neighbors of the human REST protein and the corresponding function and disease annotations. The graph shows the interaction partners (*circular nodes*), genes (*square nodes*), disease annotations from the DAnCER resource (*parallelograms*), and Gene Ontology cellular-component terms (*small dark circles*). It appears that most of the interacting proteins share their cellular localization annotations. Among the human proteins, only those with the MI score of at least 0.6 are displayed. The *red* color of the nodes indicates their involvement in chromatin modification processes. The graph is displayed using the Netility tool in iRefWeb (color figure online)

1. Clear all search results by clicking the button *Clear/New Search.*
2. In the Search Terms panel, input the string *REST* into the Left search box and select REST_HUMAN, REST_MOUSE and REST_RAT.
3. Input the string *HDAC* into the *Full Text* search box (*see* **Note 11**).
4. Click the Search button and examine the results, which include interactions between REST and histone deacetylases 1,2,4 and 5 in the three organisms.
5. Observe that the REST-HDAC4 and REST-HDAC5 interactions have been annotated in both Human and Rat.
6. Examine the MI score for the interaction between REST and HDAC4 in Human, by clicking on the corresponding score value in the Results table. This should load the page *MI Score Calculation* for that interaction.
7. Click on the *View Details* link. The expanded calculation results show that this PPI is scored based on the strength of its own experimental evidence as well as the experimental evidence of its Rat counterpart (*see* **Notes 12** and **13**).

4.5 Compare and Contrast the Supporting Evidence

1. Return to the search results involving REST and HDAC proteins.
2. In the Results table, identify the PPI between Human REST and Mouse HDAC1. This interaction is highly unlikely to occur in vivo. Nevertheless, it may well be a valid experimental result of a study that used orthologous proteins from different organisms interchangeably, as is common in experiments involving human and murine molecular constructs [21].
3. Click on the link *1043810* in the Interaction ID column of the Results table. This should load the interaction page with annotation details, indicating that this PPI was recorded by the BioGRID source database as a result of curating a publication with PubMed ID 10570134.
4. In the PubMed ID column, click on the link corresponding to PMID 10570134. This should load the *PubMed Annotations* page, which summarizes the annotations of the corresponding publication [53] performed by each source database.
5. Examine the significant differences between the annotations reported by the BioGRID and HPRD teams (Fig. 3). The HPRD listed a single interaction between human REST and human SIN3B, whereas BioGRID recorded interactions between human REST and three mouse proteins: Hdac1, Sin3a and Sin3b.
6. Click on the *Abstract* button to reveal the details of the original study, which describes the recruitment of mouse Sin3 and histone deacetylase complex by REST to repress neuron-specific target genes [53]. This suggests that the BioGRID annotations are perhaps better able to represent the information contained in the original paper.

4.6 Examine the Annotation of Protein Complexes

1. Return to the search results involving REST and HDAC proteins.
2. In the Results table, identify the protein complex and click on its Interaction ID *682088* to load the detailed annotation page.
3. Examine the annotations and note that the interaction record represents an association of 18 different proteins, of which 17 are from Mouse (including *Rest* and *Hdac2*) and one from Human. This co-complex association was detected by a *pull down*, an affinity chromatography method in which a protein of interest (bait) is typically bound to a column and then challenged with a solution or cellular extract containing the candidate partner proteins (preys).
4. Click on the Source ID link *EBI-2312516* to access the original IntAct annotation. The IntAct represents this co-complex either as a single list of 18 participating proteins (http://www.ebi.ac.uk/intact/interaction/EBI-2312516), or as a so-called "spoke expansion" into 17 different bait-prey pairs, where the *Nanog* protein is the bait (*see* **Note 14**).

Annotations for PubMed 10570134

Abstract

Click to show/hide the abstract for PubMed 10570134

Summary

	Total	biogrid	hprd
Distinct Interactions Seen	4	3	1
Distinct Proteins Seen	5	4	2
Distinct Organisms Seen	2	2	1

Details

Interaction ID	Accession	Label	Organism	biogrid	hprd
724329	Q13127	REST_HUMAN	H. sapiens		hprd
724329	O75182	SIN3B_HUMAN	H. sapiens		hprd
1043810	Q13127	REST_HUMAN	H. sapiens	biogrid	
1043810	O09106	HDAC1_MOUSE	M. musculus	biogrid	
1082500	Q13127	REST_HUMAN	H. sapiens	biogrid	
1082500	Q62141	SIN3B_MOUSE	M. musculus	biogrid	
1143005	Q13127	REST_HUMAN	H. sapiens	biogrid	
1143005	Q60520	SIN3A_MOUSE	M. musculus	biogrid	

Fig. 3 The iRefWeb *Pubmed Annotations* page for the PubMed ID 10570134 reveals the differences in the annotation of a REST-related protein complex by two different source databases

5. Return to the iRefWeb page of the co-complex and click on the PubMed ID 17093407 to load the *PubMed Annotations* page.
6. Expand the *Abstract* button to reveal that the original publication [54] explored the protein network in which Nanog operates in mouse embryonic stem cells. The authors report using affinity purification of Nanog under native conditions followed by mass spectrometry to identify its interacting partners.
7. Examine how different source databases—in this case IntAct and BioGRID—annotated the original publication. Observe that whereas BioGRID reports only binary interactions (*see* **Notes 15–17**), IntAct reports both binary and multi-protein interactions. However, only a handful of the binary pairs are annotated by both source databases.
8. To sort the annotations by protein, click on the header of the Label column in the Details table. Observe that whereas many proteins appearing in [54] were recorded by both databases, some were only recorded by BioGRID (CDK1_MOUSE, SMRC1_MOUSE), and others only by IntAct (ELYS_HUMAN, SALL1_MOUSE).

9. The curation differences flagged by the iRefWeb page indicate either the ambiguities present in the original publication itself, or possible differences in curation practices or policies of the annotating source databases. The user may therefore proceed to retrieve and examine the original publication in order to clarify these differences.

5 Notes

1. This example demonstrates the problem with using a name or alias instead of a proper molecular identifier: iRefWeb no longer recognizes NRSF in its alias searches because the alias NRSF is now obsolete and has been replaced with REST. Although gene and protein names are widely used by biomedical researchers, it should be considered a good practice to confirm the current names and molecular identifiers of the protein of interest before initiating the iRefWeb search queries.
2. Alternatively you may access the MI ontology webpage and click "Browse," then progressively expand the root term "molecular interaction," the sub-term "interaction detection method" or "interaction type," etc., until a desired ontology term is found in the expanded graph.
3. To further confirm your selection, click on the Browse button to place the term within the MI-ontology hierarchy graph. Then examine the section *Term Hierarchy*, clicking on both *Paths to Root* and *Child relationships* options. This should help to determine whether the desired PPI type or method was chosen correctly, or whether any of its specific varieties should be chosen instead, such as "coimmunoprecipitation" (MI:0019), "pull down" (MI:0096), etc.
4. It is also helpful to examine the ontology terms just above and just below the category of interest. For example, given the ontology term MI:0915 "physical association" (defined as "molecules that are experimentally shown to belong to the same functional or structural complex"), users should be aware of its difference from the parent term MI:0914 "association" ("molecules that are experimentally shown to be associated potentially by sharing just one interactor") as well as from its child term MI:0407 "direct interaction" ("interaction that is proven to involve only its interactors").
5. Instead of the gene or protein name, the user may also provide a UniProt entry name or accession, such as REST_HUMAN or Q13127. A gene ID 5978 will also be recognized.
6. The MINI-TAB is a simple tab-delimited format that provides a minimal description of the retrieved PPI data, namely: the

iRefWeb interaction ID, the standard protein IDs of the two interacting molecules (e.g., Q13127 and P08047) and their iRefWeb interactor IDs. The iRefWeb-specific IDs are maintained across different versions of the iRefWeb, and may therefore be used to form URL links to the iRefWeb interaction or interactor records.

7. Another download option is the MITAB, a rich tab-delimited format that conforms to the current HUPO PSI-MI standards. MITAB columns contain a wealth of information about each PPI or complex, including the primary and alternative aliases for all proteins involved, their organisms, the supporting publication ID, various MI ontology terms that describe the nature of the interaction, the annotating source database, and many more. Beware that the same interaction may appear in several MITAB lines due to multiple annotations (different source databases, supporting publications, detection methods, etc.). The full description of the MITAB format is available at http://irefindex.org.

8. Once the PPI data are downloaded in a tab-delimited format, they may be easily imported into the Cytoscape visualization package [55] or any similar bioinformatics software for further visualization and analysis.

9. Given the importance of chromatin modifications to a variety of cellular processes, genes and proteins related to chromatin modification are shown in red. This option may be turned off by switching off the appropriate Netility filter. The corresponding gene nodes are linked to DAnCER, or the Disease-Annotated Chromatin Epigenetics Resource [47], maintained by our research team. Clicking on the gene node in the graph reveals the link to a DAnCER page of the gene, which contains further functional annotations related to its chromatin-modification function and disease associations.

10. Other image formats are available for network export, such as the Scalable Vector Graphics (SVG), the Portable Network Graphics (PNG), or the eXtensible Graph Markup and Modeling Language (XGMML).

11. The user may also input the string HDAC into the RIGHT search box and select a range of proteins from the matches presented by the popup window.

12. The MI scores are based on the accumulation of appropriately weighted "pieces of evidence," which reflect the presence of multiple supporting publications, multiple experimental detection techniques, or other organisms in which the PPI is conserved. Following the MINT approach, the supporting-evidence terms for indirect PPI are reduced in half, as are the terms derived from high-throughput studies. Thus the MI score

explicitly favors interactions detected by low-throughput experiments (up to 50 PPI). As we have discussed in the introductory section, the latter may or may not be warranted, given the technological advances and statistical techniques used in the analysis of high-throughput PPI data. In the example of Subheading 4.4, the evidence for the human REST-HDAC4 interaction has a weight of 0.5 because the PPI is annotated as a physical association (not a direct interaction) detected using only one experimental technique (affinity chromatography technology). The corresponding rat PPI also has a weight of 0.5 for the same reason. Both human and rat interactions come from a low-throughput study, otherwise their weights would have been further reduced by half.

13. iRefWeb does not compute MI scores to interactions or co-complexes that contain proteins from multiple organisms. Although some of these cross-organism interactions do occur in nature (i.e. between viral and host proteins), many are merely artifacts of the study design, in which proteins from one organism are used as a convenient substitute for homologous proteins from a related organism [21].
14. As the IntAct Web site points out, binary interactions generated by co-complex expansion "will very likely generate some false positive interactions" but allow the data to be presented in a consistent manner.
15. Further exploration of individual PPI pages (by clicking on each interaction ID in the first column of the table) would reveal that BioGRID annotated all binary interactions from reference [54] as "physical associations" (MI:0915) detected by affinity chromatography technology (MI:0004). In contrast, IntAct recorded either "physical associations" or the less stringent "associations" (MI:0914), and provided the more specific descriptions of the chromatography techniques used, such as anti-bait or anti-tag coimmunoprecipitation (MI:0006 and MI:0007 respectively), pull down (MI:0096), tandem affinity purification (MI:0676), and molecular sieving (MI:0071). This difference reflects the curation policies of the two source databases.
16. It is worth noting that BioGRID uses its own Experimental Evidence Codes in its original curations, so that the BioGRID annotations of reference [54] are described as either "Affinity Capture-MS" or "Affinity Capture-Western" on the BioGRID Web site. However, for the purpose of data export and distribution, the BioGRID team translates both of these evidence codes into standard MI ontology terms "physical association" (MI:0915) for the interaction type and "affinity chromatography technology" (MI:0004) for the interaction detection method.

17. Differences in co-complex annotation often make it hard to identify all complexes related to the protein of interest, such as the human REST protein. For example, consider the study that examined the REST/RCOR1/histone deacetylase repressor complex in cells infected by the herpes simplex virus [56]. Although BioGRID recorded an interaction of the REST/RCOR1/HDAC1 complex with a herpesvirus protein ICP0, it represented the co-complex using the spoke expansion, in which the REST corepressor RCOR1 was used as the bait. Hence the search for REST interactions in iRefWeb returned only the REST-RCOR1 binary pair (interaction ID 721583) instead of the full REST/RCOR1/HDAC1 complex. Nevertheless, by examining the iRefWeb evidence for the PubMed ID 15897453, which supports the REST-RCOR1 pair, the users are able to examine the full curation data related to reference [56].

Acknowledgements

This work was supported by the Canadian Institutes of Health Research (MOP#82940), the Ontario Research Fund, and the SickKids Foundation. SJW was Canada Research Chair, Tier 1.

References

1. Alberts B (1998) The cell as a collection of protein machines: preparing the next generation of molecular biologists. Cell 92:291–294
2. Kocher T, Superti-Furga G (2007) Mass spectrometry-based functional proteomics: from molecular machines to protein networks. Nat Methods 4:807–815
3. Chiu W, Baker ML, Almo SC (2006) Structural biology of cellular machines. Trends Cell Biol 16:144–150
4. Shoemaker BA, Panchenko AR (2007) Deciphering protein–protein interactions. Part I. Experimental techniques and databases. PLoS Comput Biol 3:e42
5. Phizicky EM, Fields S (1995) Protein–protein interactions: methods for detection and analysis. Microbiol Rev 59:94–123
6. Krogan NJ, Cagney G, Yu H et al (2006) Global landscape of protein complexes in the yeast Saccharomyces cerevisiae. Nature 440:637–643
7. Gavin AC, Bosche M, Krause R et al (2002) Functional organization of the yeast proteome by systematic analysis of protein complexes. Nature 415:141–147
8. Rual JF, Venkatesan K, Hao T et al (2005) Towards a proteome-scale map of the human protein-protein interaction network. Nature 437:1173–1178
9. Guruharsha KG, Rual JF, Zhai B et al (2011) A protein complex network of Drosophila melanogaster. Cell 147:690–703
10. Havugimana PC, Hart GT, Nepusz T et al (2012) A census of human soluble protein complexes. Cell 150:1068–1081
11. Butland G, Peregrin-Alvarez JM, Li J et al (2005) Interaction network containing conserved and essential protein complexes in Escherichia coli. Nature 433:531–537
12. Bader GD, Cary MP, Sander C (2006) Pathguide: a pathway resource list. Nucleic Acids Res 34:D504–D506
13. Chatr-Aryamontri A, Breitkreutz BJ, Heinicke S et al (2013) The BioGRID interaction database: 2013 update. Nucleic Acids Res 41:D816–D823
14. Salwinski L, Miller CS, Smith AJ et al (2004) The database of interacting proteins: 2004 update. Nucleic Acids Res 32:D449–D451
15. Kerrien S, Aranda B, Breuza L et al (2012) The IntAct molecular interaction database in 2012. Nucleic Acids Res 40:D841–D846
16. Ceol A, Chatr Aryamontri A, Licata L et al (2010) MINT, the molecular interaction

database: 2009 update. Nucleic Acids Res 38:D532–D539
17. Ruepp A, Waegele B, Lechner M et al (2010) CORUM: the comprehensive resource of mammalian protein complexes–2009. Nucleic Acids Res 38:D497–D501
18. Orchard S, Kerrien S, Jones P et al (2007) Submit your interaction data the IMEx way: a step by step guide to trouble-free deposition. Proteomics 7(Suppl 1):28–34
19. Tarcea VG, Weymouth T, Ade A et al (2009) Michigan molecular interactions r2: from interacting proteins to pathways. Nucleic Acids Res 37:D642–D646
20. Razick S, Magklaras G, Donaldson IM (2008) iRefIndex: a consolidated protein interaction database with provenance. BMC Bioinformatics 9:405
21. Turinsky AL, Razick S, Turner B et al (2010) Literature curation of protein interactions: measuring agreement across major public databases. Database 2010:baq026
22. Chaurasia G, Malhotra S, Russ J et al (2009) UniHI 4: new tools for query, analysis and visualization of the human protein-protein interactome. Nucleic Acids Res 37:D657–D660
23. Orchard S, Binz PA, Borchers C et al (2012) Ten years of standardizing proteomic data: a report on the HUPO-PSI Spring Workshop: April 12-14th, 2012, San Diego, USA. Proteomics 12:2767–2772
24. Orchard S, Hermjakob H, Apweiler R (2003) The proteomics standards initiative. Proteomics 3:1374–1376
25. Kamburov A, Wierling C, Lehrach H et al (2009) ConsensusPathDB – a database for integrating human functional interaction networks. Nucleic Acids Res 37:D623–D628
26. Cusick ME, Yu H, Smolyar A et al (2009) Literature-curated protein interaction datasets. Nat Methods 6:39–46
27. Salwinski L, Licata L, Winter A et al (2009) Recurated protein interaction datasets. Nat Methods 6:860–861
28. Turner B, Razick S, Turinsky AL et al (2010) iRefWeb: interactive analysis of consolidated protein interaction data and their supporting evidence. Database 2010:baq023
29. Gingras AC, Gstaiger M, Raught B et al (2007) Analysis of protein complexes using mass spectrometry. Nature reviews. Mol Cell Biol 8:645–654
30. Fields S, Song O (1989) A novel genetic system to detect protein–protein interactions. Nature 340:245–246
31. Stagljar I, Korostensky C, Johnsson N et al (1998) A genetic system based on split-ubiquitin for the analysis of interactions between membrane proteins in vivo. Proc Natl Acad Sci USA 95:5187–5192
32. Morell M, Ventura S, Aviles FX (2009) Protein complementation assays: approaches for the in vivo analysis of protein interactions. FEBS Lett 583:1684–1691
33. Wodak SJ, Vlasblom J, Pu S (2011) High-throughput analyses and curation of protein interactions in yeast. Methods Mol Biol 759: 381–406
34. Bader JS, Chaudhuri A, Rothberg JM et al (2004) Gaining confidence in high-throughput protein interaction networks. Nat Biotechnol 22:78–85
35. Braun P, Tasan M, Dreze M et al (2009) An experimentally derived confidence score for binary protein–protein interactions. Nat Methods 6:91–97
36. Kuhner S, van Noort V, Betts MJ et al (2009) Proteome organization in a genome-reduced bacterium. Science 326:1235–1240
37. Giot L, Bader JS, Brouwer C et al (2003) A protein interaction map of Drosophila melanogaster. Science 302: 1727–1736
38. Kerrien S, Orchard S, Montecchi-Palazzi L et al (2007) Broadening the horizon–level 2.5 of the HUPO-PSI format for molecular interactions. BMC Biol 5:44
39. Brown KR, Jurisica I (2005) Online predicted human interaction database. Bioinformatics 21:2076–2082
40. Collins SR, Miller KM, Maas NL et al (2007) Functional dissection of protein complexes involved in yeast chromosome biology using a genetic interaction map. Nature 446: 806–810
41. Lehner B, Crombie C, Tischler J et al (2006) Systematic mapping of genetic interactions in *Caenorhabditis elegans* identifies common modifiers of diverse signaling pathways. Nat Genet 38:896–903
42. Tong AH, Lesage G, Bader GD et al (2004) Global mapping of the yeast genetic interaction network. Science 303:808–813
43. Boutet E, Lieberherr D, Tognolli M et al (2007) UniProtKB/Swiss-Prot. Methods Mol Biol 406:89–112
44. Sayers EW, Barrett T, Benson DA et al (2012) Database resources of the National Center for Biotechnology Information. Nucleic Acids Res 40:D13–D25
45. Ceol A, Chatr-Aryamontri A, Licata L et al (2008) Linking entries in protein interaction database to structured text: the FEBS Letters experiment. FEBS Lett 582:1171–1177
46. Lopes CT, Franz M, Kazi F et al (2010) Cytoscape Web: an interactive web-based network browser. Bioinformatics 26:2347–2348
47. Turinsky AL, Turner B, Borja RC et al (2011) DAnCER: disease-annotated chromatin epigenetics resource. Nucleic Acids Res 39: D889–D894

48. Lowe HJ, Barnett GO (1994) Understanding and using the medical subject headings (MeSH) vocabulary to perform literature searches. JAMA 271:1103–1108
49. Gillies SG, Haddley K, Vasiliou SA et al (2011) Distinct gene expression profiles directed by the isoforms of the transcription factor neuron-restrictive silencer factor in human SK-N-AS neuroblastoma cells. J Mol Neurosci 44: 77–90
50. Palm K, Metsis M, Timmusk T (1999) Neuron-specific splicing of zinc finger transcription factor REST/NRSF/XBR is frequent in neuroblastomas and conserved in human, mouse and rat. Brain research. Mol Brain Res 72:30–39
51. Canzonetta C, Mulligan C, Deutsch S et al (2008) DYRK1A-dosage imbalance perturbs NRSF/REST levels, deregulating pluripotency and embryonic stem cell fate in Down syndrome. Am J Hum Genet 83:388–400
52. Ashburner M, Ball CA, Blake JA et al (2000) Gene ontology: tool for the unification of biology. The Gene Ontology Consortium. Nat Genet 25:25–29
53. Naruse Y, Aoki T, Kojima T et al (1999) Neural restrictive silencer factor recruits mSin3 and histone deacetylase complex to repress neuron-specific target genes. Proc Natl Acad Sci USA 96:13691–13696
54. Wang J, Rao S, Chu J et al (2006) A protein interaction network for pluripotency of embryonic stem cells. Nature 444:364–368
55. Smoot ME, Ono K, Ruscheinski J et al (2011) Cytoscape 2.8: new features for data integration and network visualization. Bioinformatics 27:431–432
56. Gu H, Liang Y, Mandel G et al (2005) Components of the REST/CoREST/histone deacetylase repressor complex are disrupted, modified, and translocated in HSV-1-infected cells. Proc Natl Acad Sci USA 102:7571–7576

Chapter 23

Mespeus—A Database of Metal Interactions with Proteins

Marjorie M. Harding and Kun-Yi Hsin

Abstract

Modeling and analogy are commonly used to identify the part that a metal may play in the structure or function of a new protein which has been recognized by structural genomics. Mespeus (http://mespeus.bch.ed.ac.uk/MESPEUS_10/) lists metal protein interactions whose geometry has been experimentally determined and allows them to be visualized. This can contribute to the modeling process. The use of Mespeus is described with a series of examples.

Key words Metal, Metal site, Metal coordination group, Metal site geometry, Metalloprotein, Metalloprotein structure, Database

1 Introduction

For a substantial number of proteins (probably about 40 % of known proteins) one or more metal atoms play a key part in the function or structure of the protein, or both. For new proteins, recognized through structural genomics, there is no direct evidence about the part that metals may play in their structure or function; this is usually proposed by modeling and by analogy with known proteins. The Mespeus database [1] of experimentally established geometry of metal protein interactions, with its user-friendly Web interface, allows interactions to be listed and to be visualized; this can make a helpful contribution to the modeling process.

The data in the Mespeus database have been extracted from all metal containing protein crystal structures determined by diffraction methods to a resolution of 2.5 Å or better, which were in the Protein Data Bank (PDB) [2] at January 2010; further updates for Mespeus are intended. The metals Na, Mg, K, Ca, Mn, Fe, Co, Ni, Cu, and Zn are included. Metal–protein interactions are included whenever the distance from metal to a protein donor atom (usually N, O, or S) is within the expected "target" distance [3] plus an allowance of 0.75 Å for experimental errors in their coordinates.

Yu Wai Chen (ed.), *Structural Genomics: General Applications*, Methods in Molecular Biology, vol. 1091, DOI 10.1007/978-1-62703-691-7_23, © Springer Science+Business Media, LLC 2014

Interactions of metals with water molecules or non-protein small molecules within the protein crystal are also included.

Mespeus is thus a useful tool for exploring the variety of metal–protein interactions or for examining these in particular proteins, as shown in the examples below. Mespeus and similar tools have already been used to establish mean distances and likely ranges of distances for many of the commoner types of interaction; the reader may find it useful to look at [4] for these.

For metals not included in Mespeus, or for very recent structures, "Metal Interactions in Protein Structures" (MIPS, provided by the Indian Institute of Science, Bangalore [5]) can be used. This database and interface provides information similar to that in Mespeus. It covers a wider range of metals, it is up to date, and its graphical display is very good. Its search system and the presentation of results are quite different from Mespeus.

2 Materials

With a Web browser go to http://mespeus.bch.ed.ac.uk/MESPEUS_10/. The browser Mozilla Firefox has been used extensively in all the development of Mespeus, but other browsers should be equivalent. The Mespeus interface uses Java; you may be prompted to download some Java material (from www.java.com, provided by Oracle).

3 Methods

On the home page note that there is a link to an introduction, which has explanations of many of the features. There is also a link to Jmol, the package used for graphical display [6], a reference to the original article on Mespeus [1], and advice on how to query the database directly with SQL [7].

It is assumed that the reader is familiar with the PDB conventions of nomenclature. For example Asp, an aspartate residue, coordinates to a metal through either OD1 or OD2, the side chain carboxylate O's; His, histidine, through ND or NE of the imidazole group (Nδ or Nε); etc. Note that when Tyr, tyrosine, coordinates to metal it is through phenolic $-O^-$ but this is named OH. Also, in many structures it happens that the crystal asymmetric unit (the unique part) contains two or more independent protein molecules. Interactions within each of these therefore appear in the lists, usually distinguished as molecule A, B, etc.; normally they have the same geometry, but experimental errors in the atom coordinates lead to small differences in the distances quoted. (*see* **Note 1**, or ref. 4 for more.)

From the home page, click on the title, MESPEUS_10; this will take you to the main query page (Fig. 1).

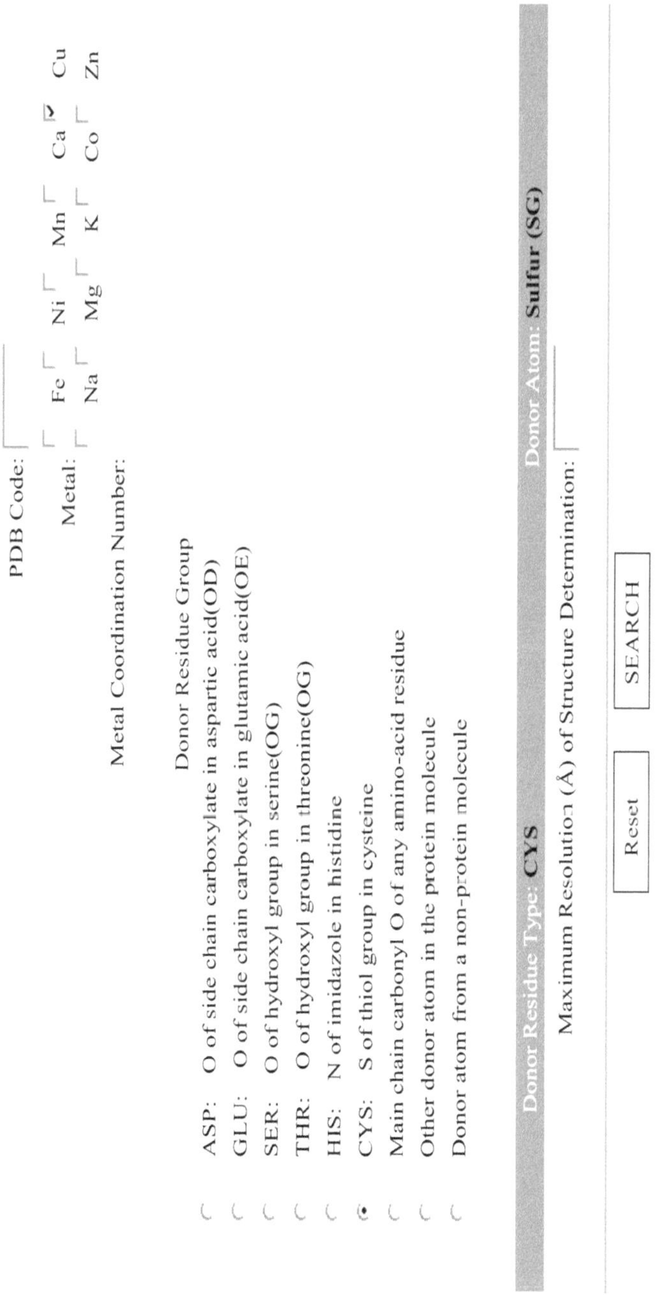

PDB Code:

Metal: Fe Ni Mn Ca Cu Na Mg K Co Zn

Metal Coordination Number:

Donor Residue Group

- ASP: O of side chain carboxylate in aspartic acid(OD)
- GLU: O of side chain carboxylate in glutamic acid(OE)
- SER: O of hydroxyl group in serine(OG)
- THR: O of hydroxyl group in threonine(OG)
- HIS: N of imidazole in histidine
- CYS: S of thiol group in cysteine
- Main chain carbonyl O of any amino-acid residue
- Other donor atom in the protein molecule
- Donor atom from a non-protein molecule

Donor Residue Type: **CYS** Donor Atom: **Sulfur (SG)**

Maximum Resolution (Å) of Structure Determination:

Reset SEARCH

Fig. 1 Main query page of Mespeus, showing Cu selected as metal, and cysteine as donor group

3.1 A Simple Search: For Interactions of Cu with Cysteine Side Chains

On the main query page:

1. Select (by a click) the metal, Cu; its box will be ticked; select the donor group, S of cysteine; it should be marked with a dot; click on search; the results page should appear (Fig. 2), including the number of hits (898 in this case) and one line of details for each. Each hit represents one Cu-S_{Cys} interaction, and the items shown are:
   (a) The Cu–S distance in Å.
   (b) The coordination number of the Cu.
   (c) Information about the coordination shape (follow  in column heading for definitions).
   (d) The metal name with its residue number and chain identifier, as it appears in the PDB.
   (e) The donor atom name, with its residue name, as it appears in the PDB.
   (f) The PDBID, identifying the protein in the PDB.
   (g) The resolution of the structure determination; the lower this number the more precise is the determination likely to be.
   (h) r.m.s.d. (follow in column heading for definition) can also be useful as a quality indicator of the metal site geometry reported.
   (i) Difference (in Å) from the expected or target distance for this interaction.
2. For any one Cu atom in these search results the amino-acid residues and other entities (e.g., water molecules) in its coordination group may be displayed by clicking on the metal name. Figure 3 (left) is an example, for Cu 156A in 1A3Z, the first line of the search results. In this display
   (a) The whole image can be rotated with left mouse button.
   (b) At the bottom of the image, controls allow centering of the metal atom, display of donor names, distances to the metal atom, angles at the metal atom, or reset.
   (c) Jmol facilities can be accessed with the right mouse button; for example zoom 200 or zoom 400 may be helpful before displaying distances or angles.
   (d) Atom names can be shown by hovering over the atom with the cursor.
   (e) Distances can be evaluated; double click on the first atom and click on the second; angles similarly.
   (f) The whole protein molecule may also be displayed, rotated etc. There is an option *Gray* which will show the protein chain trace along with this metal coordination group as in Fig 3 (right) (but it does not show any other metals in the protein).

Search for: CU. Coordination number: =Any. Donor atom: Sulfur. Donor residue: CYS. Maximum resolution: No Specified Å.

Means and Distribution

Download Query Result
Number of Hits: 898

No.	Distance	Coordination Number	Shape	Metal Name	Donor Residue Name	PDB	Reslu.	r.m.s.d.(Å)	Difference	
1	2.251	4	δ tet(11.4°) or δ sqp(40 2°)	CU 156 A	SG CYS A 138	1A3Z	1.90	0.321	0.101	
2	2.162	4	δ tet(14.9°) or δ sqp(30 6°)	CU 130 A	SG CYS A 112	1A4A	1.89	0.143	0.012	
3	2.165	4	δ tet(15.4°) or δ sqp(29.4°)	CU 130 B	SG CYS B 112	-		0.071	0.015	
4	2.247	4	δ tet(14.4°) or δ sqp(31.9°)	CU 130 A	SG CYS A 112	1A4B	1.91	0.122	0.097	
5	2.055	4	δ tet(16.6°) or δ sqp(30.0°)	CU 130 B	SG CYS B 112	-		0.167	-0.095	
6	2.253	4	δ tet(16.9°) or δ sqp(30.5°)	CU 130 A	SG CYS A 112	1A4C	2.45	0.138	0.103	
7	2.057	4	δ tet(17.2°) or δ sqp(31.9°)	CU 130 B	SG CYS B 112	-		0.153	-0.093	
8	2.058	4	δ tet(19.0°) or δ sqp(42.3°)	CU 130 C	SG CYS C 112	-		0.094	-0.092	
9	2.166	4	δ tet(17.0°) or δ sqp(41.9°)	CU 130 D	SG CYS D 112	-		0.186	0.016	
10	2.276	3		CU 901 A	SG CYS A 452	1A65	2.23	0.129	0.126	
11	2.261	3		CU 156 A	SG CYS A 138	1A8Z	2.10	0.097	0.111	
12	2.149	4	δ tet(15.2°) or δ sqp(38.5°)	CU 200 A	SG CYS A 92	1AAN	2.00	0.373	-0.001	

Fig. 2 Part of results page, from search for Cu with S of cysteine

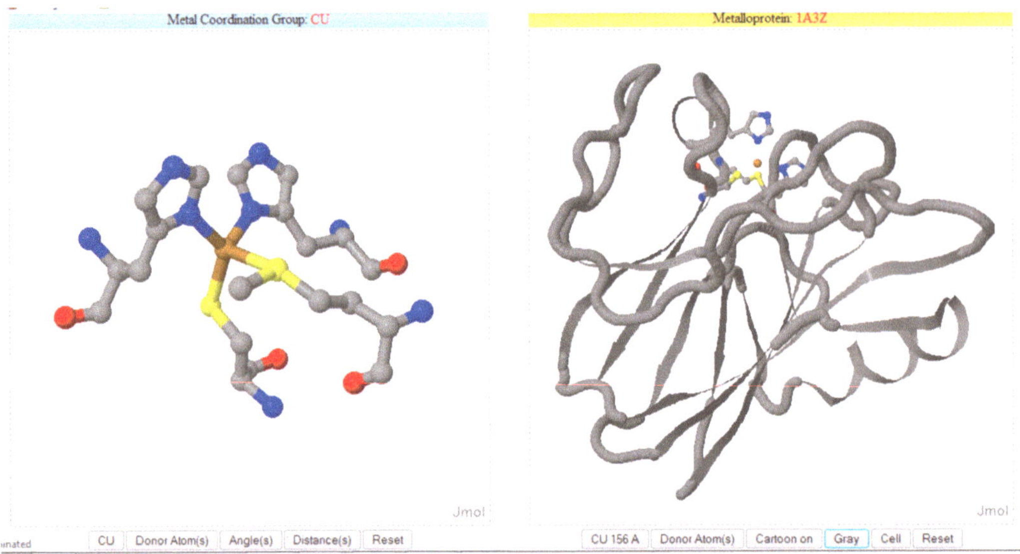

Fig. 3 *Left*: Image of one metal coordination group, that of Cu 156A in 1A3Z; in this coordination group there are ND atoms from two histidine residues, one S of methionine as well as one S of cysteine. *Right*: Image of the whole protein molecule, 1A3Z, including Cu 156A and its coordination group. *See* Subheading 3.1, **step 2**

(g) Below the image, coordination number and shape information are summarized (and below that is the information for this query retrieved by the SQL question to the database).

If metal to donor atom bonds are not displayed in the image; *see* **Note 2**. If the coordination number seems abnormally low (*see* **Note 3**).

(Return to the search results by closing the current tab, or selecting the previous tab.)

3. From the search results page (like Fig. 2) clicking on the PDBID will display the whole protein molecule as in Fig. 4 for the protein structure 1E30 (on line 58). One panel shows the protein molecule, which can be manipulated in the same way as the images above; there are options for different styles for showing the chain. The other panel gives information about the protein and its structure determination, followed by details of each metal coordination group in the protein—full metal and donor atom names as they are given in the PDB, distances, and *B* values (vibration parameters) for metal and donor atoms. (The site occupancies given are the product of metal and donor site occupancy. Most metals have occupancy 1.0; some donor atoms are reported to have occupancy less than 1.0, i.e., in the crystal they are present at the stated position in some molecules but not others.)

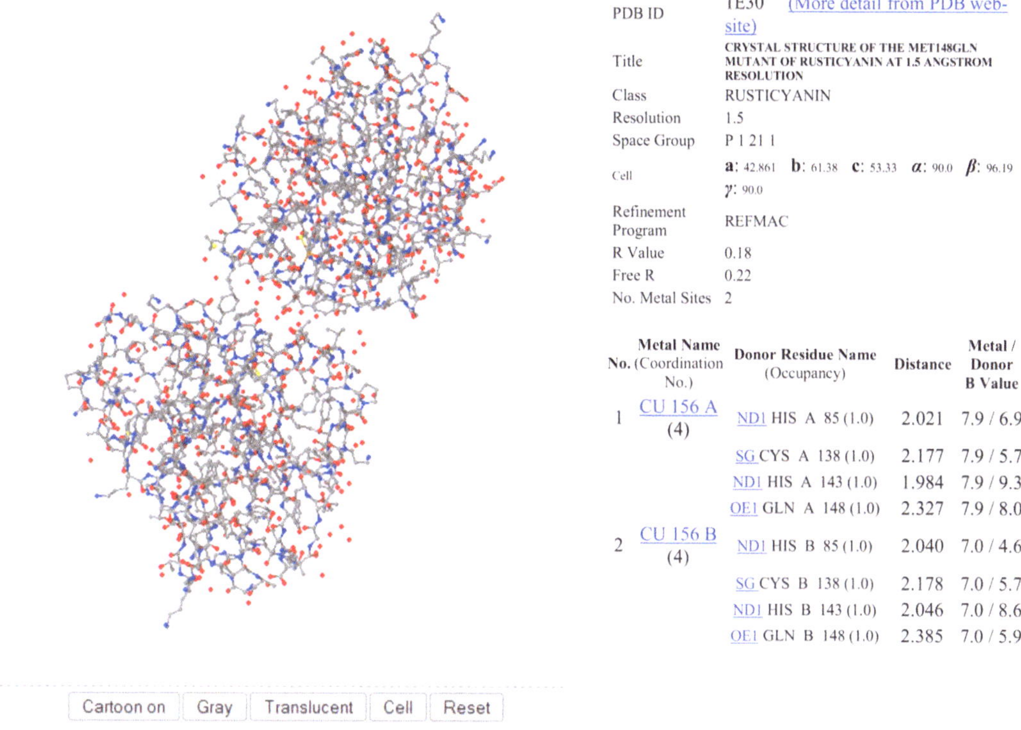

PDB ID	1E30 (More detail from PDB web-site)
Title	CRYSTAL STRUCTURE OF THE MET148GLN MUTANT OF RUSTICYANIN AT 1.5 ANGSTROM RESOLUTION
Class	RUSTICYANIN
Resolution	1.5
Space Group	P 1 21 1
Cell	a: 42.861 b: 61.38 c: 53.33 α: 90.0 β: 96.19 γ: 90.0
Refinement Program	REFMAC
R Value	0.18
Free R	0.22
No. Metal Sites	2

No.	Metal Name (Coordination No.)	Donor Residue Name (Occupancy)	Distance	Metal / Donor B Value
1	CU 156 A (4)	ND1 HIS A 85 (1.0)	2.021	7.9 / 6.9
		SG CYS A 138 (1.0)	2.177	7.9 / 5.7
		ND1 HIS A 143 (1.0)	1.984	7.9 / 9.3
		OE1 GLN A 148 (1.0)	2.327	7.9 / 8.0
2	CU 156 B (4)	ND1 HIS B 85 (1.0)	2.040	7.0 / 4.6
		SG CYS B 138 (1.0)	2.178	7.0 / 5.7
		ND1 HIS B 143 (1.0)	2.046	7.0 / 8.6
		OE1 GLN B 148 (1.0)	2.385	7.0 / 5.9

Fig. 4 Display of a selected protein as described in Subheading 3.7; the protein structure is 1E30

The relationship between metal and protein can often be seen clearly by selecting "translucent" or "grey" for the protein image and then clicking on the metal(s) in the table to the right. Alternatively, any one metal coordination group can be viewed by clicking on one of its donor groups.

3.2 Search with Restricted Resolution Range

This can be useful to select only higher resolution structures, which are normally the more accurate ones. From the main query page, repeat the search for Cu interactions with S of cysteine, but give the maximum resolution (on the second bottom line) as 1.5 Å. There are now only 98 hits.

3.3 Evaluation of Average Distances

Carry out the search for the required kind of distance and the chosen maximum resolution. At the top of the results page click on "Means and Distribution" to display mean, standard deviation, distribution, etc.

If there are inappropriate entries in the list which you wish to remove, tick the delete box on the right, return to the top of the column and press delete, then recalculate Means and Distribution.

3.4 Specifying Coordination Number and Other Properties

On the query page, below the search boxes for different metals, the coordination number can be specified, or a range of coordination numbers can be specified. For some donor groups there are additional questions in the new panel; for example with histidine, coordination by the ND atom of the imidazole group, or the NE atom, or either can be specified. Note that distinction between mono- and bi-dentate carboxylate donors (Asp and Glu) is not always clear or unambiguous [4].

3.5 Other Donor Atoms and Groups

On the query page "Other donor atom in the protein molecule" may be chosen and for this additional options are shown which occur less commonly. If "Donor atom from a non-protein molecule" is chosen the additional options include the water molecule, which is very common, and various others. The database includes every atom except C, H, or P, found within interacting distance of any metal atom; occasionally one will not be a conventional "donor" atom, for example short metal–metal contacts are found and listed. In the search by name for a non-protein donor group, e.g., ADP, it is necessary to give the "het group" name as used in the PDB.

3.6 Search Results Can Be Downloaded for Further Processing

Carry out the search required. On the results page, click on the option "Download query result" near top of page. The resulting file has tab delimiters and can be unzipped and fed straight into Microsoft EXCEL for further manipulation.

3.7 Search for All Metals in One Protein Structure

This can be done by inserting the PDB Code on the top line of the query page. The results are as described in Subheading 3.1, **step 3**.

3.8 More Advanced Queries Using SQL

In the Mespeus database there is some information that is not accessible through the Web interface, but can be accessed with SQL statements; for example a resolution range can be specified, rather than just a maximum resolution, or interactions can be selected according to the protein class (e.g., isomerase), or the number of metal sites in the PDB list of atoms, or the crystallographic space group, etc. Some of the results given in ref. 4 were obtained this way. Details will be provided on request for SQL access.

3.9 Listing and Comparison of Coordination Groups for Many Proteins

The procedure in Subheading 3.1, **step 3**, shows the composition of all the metal coordination groups in one protein. It gives all their constituent donor groups and, if the donor groups are part of the protein molecule, their positions in the protein chain (i.e., residue numbers). This is for one protein structure at a time. It may sometimes be useful to list the information for coordination groups in different proteins all together—for surveys, comparisons, etc.; this can be done at present with a closely associated Web site. (It is hoped that eventually these lists will be updated and incorporated into the Mespeus database.)

(a) In a new browser window go to http://mespeus.bch.ed.ac.uk/tanna/.

(b) Select item 3, "New lists for 10 Metals."

metal	*donors*	*sd1*	*sd2*	*sd3*	*sd4*	*CN*	*othdonors*	*rms*	*resln*	*metid*	*pdb*	*class*
MN	C	-1	-1	-1	-1	5	_NNNN	0.16	1.70	20641	2feu	OXIDOREDUCTASE
MN	C	-1	-1	-1	-1	6	_NNNNW	0.32	1.50	20633	2fe6	OXIDOREDUCTASE
MN	CHE	282	42	-1	-1	3	___	0.20	2.30	18792	2b7o	TRANSFERASE
MN	CHED	170	26	11	-1	5	____W	0.18	2.30	15066	1vs1	TRANSFERASE
MN	CHED	206	34	26	-1	4	____	0.11	2.10	10186	1of6	LYASE
MN	CHED	206	34	26	-1	4	____	0.16	2.00	10197	1ofb	LYASE
MN	DDE	11	197	-1	-1	7	___ONW	0.33	2.00	12146	1r5g	HYDROLASE
MN	DDE	150	45	-1	-1	7	___WW	0.43	2.50	9196	1mwh	VIRUS/VIRALPROTEIN
MN	DDE	3	159	-1	-1	3	___	0.41	2.00	7948	1kz8	HYDROLASE
MN	DDE	3	159	-1	-1	5	___OW	0.11	2.30	3928	1fsa	HYDROLASE
MN	DDEnX	11	187	-99	0	6	_____	0.25	2.30	15371	1wkm	HYDROLASE
MN	DDH	2	108	-1	-1	6	___OO	0.20	1.90	8286	1ll2	TRANSFERASE
MN	DDH	2	108	-1	-1	6	___OO	0.20	2.45	17594	1zdf	TRANSFERASE
MN	DDH	2	108	-1	-1	6	___OO	0.21	2.30	17595	1zdg	TRANSFERASE
MN	DDH	2	139	-1	-1	6	___OO	0.13	2.00	4587	1ga8	TRANSFERASE
MN	DDH	2	139	-1	-1	6	___OO	0.18	2.00	4566	1g9r	TRANSFERASE
MN	DDH	8	7	-1	-1	6	___WWW	0.54	2.30	8365	1lof	LECTIN
MN	DDHD	60	22	52	-1	5	____W	0.07	1.30	30548	1wpn	HYDROLASE
MN	DDHD	61	22	52	-1	6	____OW	0.12	2.20	5870	1i74	HYDROLASE
MN	DDO	16	1	-1	-1	5	___OO	0.13	1.50	13310	1sx5	HYDROLASE/DNA
MN	DDO	44	2	-1	-1	4	___W	0.25	2.10	3629	1ffg	TRANSFERASE/SIGNALINGPROTEI
MN	DDO	44	2	-1	-1	6	___WWW	0.14	2.20	22906	3tmy	CHEMOTAXIS
MN	DDO	44	2	-1	-1	6	___XWW	0.16	2.37	3850	1fqw	SIGNALINGPROTEIN
MN	DDO	52	2	-1	-1	7	___OOW	0.39	2.30	6864	1jlk	SIGNALINGPROTEIN
MN	DDSOEE	2	2	2	2	6	______	0.19	1.80	22348	2pal	CALCIUMBINDINGPROTEIN
MN	DE	1	-1	-1	-1	6	__WWW	0.21	2.40	7598	1kgz	TRANSFERASE

Fig. 5 Example of parts of listing of metal coordination groups for Mn from http://mespeus.bch.ed.ac.uk/tanna/newcngps/mn_cngps_4.htm—*see* Subheading 3.9, and the Web site itself for further explanation

(c) Choose "ALL METAL COORDINATION GROUPS," and then the required metal, e.g., Mn. A table is shown like that in Fig. 5. There are columns for the metal, and for the set of donors which constitute the coordination group given as one letter codes (e.g., HHD for His.His.Asp), and for the relative positions of the donor residues in the amino-acid sequence (sd1, sd2, etc.). Another column gives, in abbreviated form, details of the non-protein donors in each coordination group; for example in the last coordination group listed in Fig. 5, Mn is coordinated by aspartate, glutamate, and three water molecules; in the first Mn coordination group in the list there are four N atoms as non-protein donors—this is likely to be a haem group or similar. Other columns give coordination number, PDB code, etc.; a fuller explanation is given on the Web site, and in the discussion and description of coordination groups in ref. 8.

(d) This table of coordination groups can be sorted or manipulated in many ways in EXCEL. (Copy the contents of the Web page and paste it into an EXCEL file.)

4 Notes

1. There may be errors and uncertainties in the distances given by Mespeus to represent metal–protein interactions. The atom coordinates established in protein structure determination are subject to experimental errors (as well as to other occasional

problems like misidentification of atoms). The resolution of the structure determination is a rough guide to the precision that can be expected. With 1 Å resolution data the standard uncertainty (SU) of an M–O distance can be ~0.07 Å (and confidence limits three times this, ± 0.2 Å); at 2 Å resolution the standard uncertainty rises to 0.2–0.4 Å. (More information is given in ref. 4.)

2. Sometimes metal to donor bonds are not displayed in images, particularly those where the metal is Na, K, Mg, or Ca. The default distances which Jmol [6] uses to decide whether a bond should be drawn are often not completely appropriate for protein structures. In the display Jmol can be used to evaluate the distances (Subheading 3.1, **step 2**e). Distances Na–O < 2.6 Å, Mg–O < 2.3 Å, K–O < 3.0 Å or Ca–O < 2.6 Å should certainly be considered as major interactions (these values are the "target distances" [2] plus an allowance of 0.2 Å for coordinate errors); slightly longer distances would represent weaker interactions. Especially for Na and K these interactions are predominantly electrostatic, rather than the result of covalent bond formation (*see* ref. 4).

3. Occasionally a metal coordination group includes donor atoms from 2, 3, or 4 identical protein monomers, related by exact crystallographic symmetry; in this case since the PDB [2] lists only the atoms from one monomer, Mespeus shows an incomplete coordination group. (Several insulin structures are affected in this way.)

Acknowledgements

Our thanks are due to Professor M. Walkinshaw and Dr. P. Taylor for laboratory space, computer support, and many helpful discussions.

References

1. Hsin K, Sheng Y, Harding MM et al (2008) MESPEUS: a database of the geometry of metal sites in proteins. J Appl Cryst 41:963–968
2. Berman H, Henrick K, Nakamura H et al (2007) The worldwide protein data bank (wwwPDB): ensuring a single, uniform archive of PDB Data. Nucleic Acids Res 35: D301–D303
3. Harding MM (2006) Small revisions to predicted distances around metal sites in proteins. Acta Crystallogr D 62:678–682
4. Harding MM, Nowicki MW, Walkinshaw MD (2010) Metals in protein structures: a review of their principal features. Cryst Rev 16:247–302
5. Hemavathi K, Kalaivani M, Udayakumar A et al (2010) MIPS: metal interactions in protein structures. J Appl Cryst 43:196–199
6. http://jmol.sourceforge.net/ (for Jmol)
7. http://www.mysql.com/ (for SQL)
8. Harding MM (2004) The architecture of metal coordination groups in proteins. Acta Crystallogr D 60:849–859

Chapter 24

High-Quality Macromolecular Graphics on Mobile Devices: A Quick Starter's Guide

Chin-Pang Benny Yiu and Yu Wai Chen

Abstract

With the rise of tablets, truly portable molecular graphics are now available for wide use by scientists to share structural information in real time at a reasonable cost. We have surveyed the existing software available on Apple iPads and on Android tablets in order to make a recommendation to potential users, primarily based on the product features. Among 12 apps, *iMolview* (available on both platforms) stands out to be our choice, with *PyMOL* app (iOS) a close alternative and *RCSB PDB Mobile viewer/NDKmol* (both platforms) offering some uniquely useful functions. Finally, we include a tutorial on how to get started using *iMolview* to do some simple visualization in 10 min.

Key words Protein structure, RCSB PDB, Protein data bank, Macromolecular graphics, Tablets, Mobile devices, iMolview, PyMOL, iPad, iPhone, iOS, Android

1 Introduction

Molecular graphics is the language of structural biologists. In the past few years, the world witnessed the rise of the thin and lightweight handheld tablets. These are portable computers in every sense, without keyboard or mouse, thanks to a touch-sensitive screen. The Apple iPad has a 250 mm (9.7 in.) screen of very high sensitivity and resolution (1,024 × 768 for first and second generations, and iPad mini; 2,048 × 1,536 for third and fourth generations). Since their inception, iPads have been well received by consumers which encouraged software development on the iOS (the operating system on Apple mobile devices) platform. On the other hand, many rivals to iPads have been developed; these devices mostly adopt the Google Android operating system which is based on Linux. Together, these mobile devices have completely revolutionized how users interact with computers, in more intuitive ways using finger gestures.

Yu Wai Chen (ed.), *Structural Genomics: General Applications*, Methods in Molecular Biology, vol. 1091, DOI 10.1007/978-1-62703-691-7_24, © Springer Science+Business Media, LLC 2014

In this article, we shall compare the currently available molecular graphics products on the iPads and the Android tablets. Among these, we shall recommend the best all-round graphics software. Next we shall discuss how to set up and perform some very basic visualization tasks. We aim to get people who are not familiar with molecular graphics to start using it on their mobile devices.

2 Graphics Software

2.1 Hardware Used for Testing

The iOS apps (application software on mobile devices) were tested on an iPad 2 (16 GB, iOS 6.1) which is the most popular device at the time of writing (April, 2013). Most of the software written for the iOS should be able to run on an iPhone. Unfortunately, attempts to loan newer devices from Apple Europe Ltd. for testing were not successful. For the Android platform, we used an inexpensive ($120) tablet (8 GB, Android 4.0 Ice Cream Sandwich).

2.2 Comparison of Graphics Software

We identified 12 mobile graphics apps as listed in Table 1 (*see* **Note 6**). As we shall reveal, the best apps do not cost more than one US dollar. We find it important to consider whether apps are under active development because this is related to support, and

Table 1
Basic information of mobile apps for macromolecular graphics

App	iOS	And.	Price ($)	Developer	Version (updated)
iMolview	●	●	Free (lite)	Molsoft	1.0/1.2 (5/2013)
	●		0.99 (full)		1.8.1 (10/2013)
PyMOL	●		Free	Schrödinger	1.1.1 (11/2012)
RCSB PDB Mobile	●	○	Free	RCSB PDB	3.15 (9/2013)
Ball & Stick	●		Free	MolySym	1.5.2 (9/2013)
CueMol viewer	●		Free	Ryuichiro Ishitani	2.1.0.250 (11/2012)
Molecules	●		Free	Sunset Lake	2.1 (4/2012)
iPharosDreams	●		4.99	EQUISnZAROO	1.0.2 (10/2011)
FinMol	●		4.99	Dubesoft	1.00 (2/2012)
CMol	●		Free	Helen Ginn	1.3.3 (4/2011)
NDKmol/ESmol		●	Free	biochem_fan	0.92/0.74 (11/2012)
Jmol MV		●	Free	Bob Hanson	1.1 (12/2011)
Atomdroid		●	Free	CCB Goettingen	1.5.0 (6/2012)

The respective versions reviewed are the latest at the time of proof (October 2013; *see* **Note 7**). The price indicated is that advertised on developer's Web sites (*see* **Note 6**) and may not match the local App Store price. "And." stands for Android. Open circle: beta version available but not tested here thus the release dates of the latest version are included. For the iOS platform, one can assume that apps that have not been updated since the introduction of iOS 6, a major operating system upgrade in September 2012, are considered not under active development. We picked six apps for detailed comparison. Since there are free or low-cost apps that perform very well, we did not find it worthwhile to pursue the costlier ones. The other apps are not reviewed because they are lacking in some important ways (*see* **Note 1**).

We performed a comparison of the most essential functions offered by the six molecular graphics apps (Table 2). Note that this is a features comparison and computing performance was not vigorously tested. The viewer in *RCSB PDB Mobile* adopted the same software core of *NDKmol* so these are reviewed together. All apps offer the basic control operations (rotate, translate, zoom, and clip), except for *Jmol MV* where zoom is not available. We used the crystal and NMR structures of the p53 tetramerization domain (PDB ID 1AIE and 2J0Z, *see* **Note 2**), a small protein of 31 residues (monomer) or 124 residues (tetramer) for testing.

From Table 2, *iMolview* and *PyMOL* compare similarly and both offer the full set of features to satisfy most structural biologists' needs. We found *iMolview* easier to use and it offers an extremely useful sequence view which enables quick access to any residue in the structure. *PyMOL* on desktop computers is one the most popular molecular graphics software and its app excels in producing ray-traced photorealistic scenes. *Jmol* is a reputable Java program that is widely used for interactive molecular graphics embedded in Web pages. However, at the moment, its app on the Android platform (*Jmol MV*, for *M*olecular *V*isualization) is still under development, with many useful features only accessible via the command console (Table 2). Until these are incorporated into a menu-driven interface, its use is restricted to *Jmol* expert users. *NDKmol* and its counterpart *RCSB PDB Mobile* viewer are unique in being able to display the biological assembly of a crystal structure. This is best illustrated with PDB ID 1AIE (*see* **Note 2**). While all other apps show only the 31-residue monomer in the crystallographic asymmetric unit, *RCSB PDB Mobile* viewer/*NDKmol* displays this as a tetramer (Fig. 1), correctly taking crystallographic symmetry and oligomerization information into account. *Ball* & *Stick* offers an interesting feature which allows view sharing with mobile devices in proximity and in real time. We tested this and it works very well. *CMol*, unfortunately, is not under active development and can only open/import PDB files (e.g., as an e-mail attachment), but cannot download from the RCSB Protein Data Bank.

The quality of the graphic images produced by various apps is ranked in descending order, as follows: *PyMOL* (ray tracing, Fig. 1a), *iMolview*, *Ball* & *Stick*, *RCSB PDB Mobile* viewer/*NDKmol* (Fig. 1b), *Jmol MV*, *CMol*.

Table 2
A comparison of main features of six molecular graphics apps

Features		iMolview[a]	PyMOL	RCSB PDB Mobile/NDKmol[b]	Ball & Stick	Jmol MV	CMol
Structural object styles	Ball-and-stick	●	●	●	●	●	○
	Space filling	●	●	●	●	●	○
	Ribbon/cartoon	●	●	●	●	●	●
	Wire/stick	●	●	●	●	●	●
	Surface	●	●				
	B-factor putty		●	●			
Custom color	Background	●	●			○	●
	Graphical object	●	●	●		○	●
Label		●				○	
Selection	To act on a subset	●	●			○	●
Sequence view		✪					
Biological assembly				✪			
Measure	Distance, angle	●	●			○	
View and render	Center on atom	●	●			○	
	Stereo	●	●			○	
	Ray trace		✪				
	Fog	●	○	◗			●
	Rock/spin	●	●			○	●
Load/import	PDB	●	●	●	●	●	○
	Local import	●	●	●	●	●	●
Save/export	Dropbox	●	●				
	e-mail	●			●		
Wi-Fi share					✪		

Filled circle: feature available, *filled circle with a star*: unique feature, *half-filled circle*: only present in *NDKmol*, *Open-circle*: feature that is problematic or, in *Jmol MV*, needs to be invoked manually from the command console (i.e., no menu interface). Note that each of these software has additional advanced features (e.g., transparency, molecules alignment, scripting) that are not included here, please refer to the respective developer's Web page

[a] *iMolview* full version (iOS)

[b] *NDKmol* and the viewer in the *RCSB PDB Mobile* app share the same core

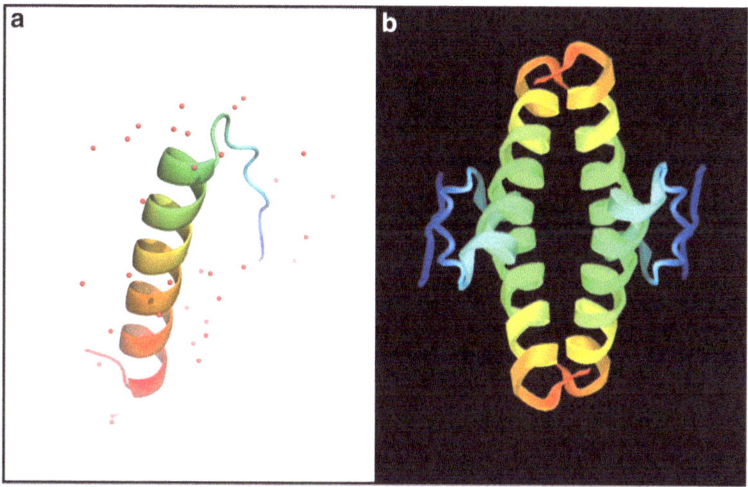

Fig. 1 *PyMOL* app and *RCSB PDB Mobile* viewer compared. The crystal structure of p53 tetramerization domain (PDB ID 1AIE) rendered in (**a**) *PyMOL* app (ray traced) and (**b**) *RCSB PDB Mobile*. By default, *RCSB PDB Mobile* shows the biologically functional tetrameric arrangement, whereas all other graphics software shows the monomer in the crystallographic asymmetric unit

If one needs a simple tool to import a PDB file (*see* **Note 2**) and get an overall view of the protein fold with cartoon or ribbon style, then any of these apps can serve the purpose. The deciding factor of a good structural biologist's tool is whether it allows the user to select a subset of atoms for rendering. For this, only *iMolview*, *PyMOL*, and *CMol* are suitable (*Jmol MV* can do that too but in a hidden way).

3 Methods (Using *iMolview*)

iMolview can be used with or without Internet. An active connection is required to download structures from the PDB. After the PDB file has been imported, structure viewing, analysis and rendering can be performed offline (without Internet).

3.1 Installation

iMolview is available ($0.99) in the Apple App Store for iPhones and iPads. There is also a "Lite" (free) version which is available in both Apple App Store and in Google Play for Android devices. Some features described in the following sections (e.g., molecular surface) are only available in the full version.

3.2 Importing a PDB Entry

1. Make sure that there is an active Internet connection (Wi-Fi or 3G).
2. Tap the top search bar; and enter some search criteria into it. For this tutorial, type "p53 tetramerization." As the text is typed in, a dropdown menu appears listing all the entries that

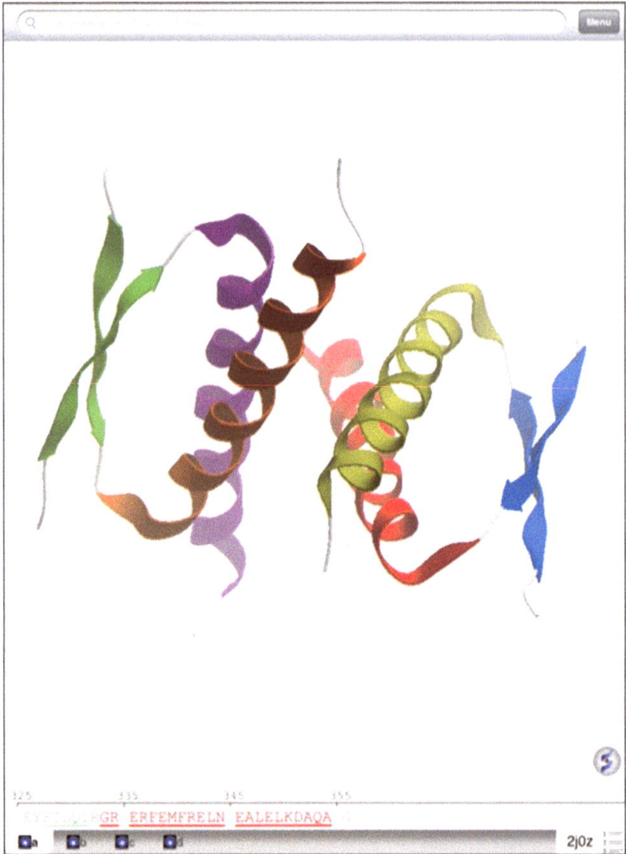

Fig. 2 Default *iMolview* display. The default display in *iMolview* of the p53 tetramerization domain (PDB ID 2J0Z) as Richardson secondary-structure cartoon, in a white background. At the *bottom* of the display screen, the protein sequence is shown, with residue numbers and color-coded by secondary structure that matches the coloring scheme of the cartoon, and chain tabs (a, b, c, d) at the very bottom

satisfy the search text string. Tap on the third entry starting with "2J0Z," which is the PDB ID (*see* **Note 2**). This is the solution NMR structure of the tetramerization domain of the p53 tumor suppressor.

3. A representation of the structure appears on the screen (Fig. 2). The default style is the Richardson protein cartoon (*see* **Note 3**), colored by secondary structure. At the bottom of the screen, the protein sequence is shown, with residue numbers, and color-coded to match the coloring scheme of the cartoon. Multiple protein chains in the crystal structure (e.g., a, b, c, d here) are listed as tabs at the very bottom. One can tab on these chain identifier tabs to quickly show/hide a chain.

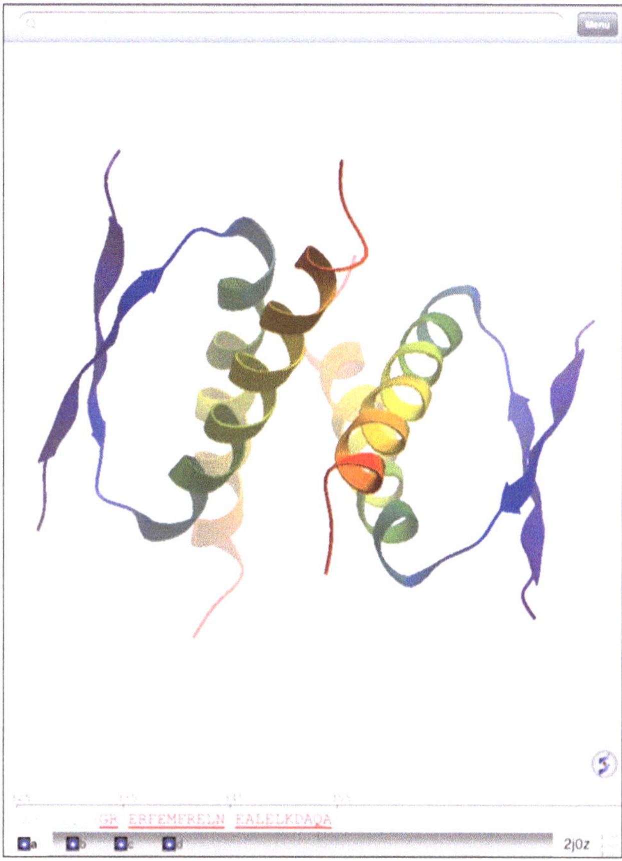

Fig. 3 Rainbow color display in *iMolview*

3.3 Viewing with Different Styles

Set the background to white by tapping the *Menu* button (top right), then tap "Color background, and pick white from the palette.

3.3.1 Rainbow Coloring (Blue to Red) from N- to C-Termini

Tap *Menu* button, tap "Tools >"; on the next menu, tap the first item "Assign Secondary Structure" (Fig. 3).

3.3.2 Transparent Items

Tap *Menu* button, tap the grey "Back" button, then tap "Settings >"; on the next menu, slide the "Transparent Ribbon" *ON* (default is OFF; Fig. 4).

3.3.3 Molecular Surface

1. At the bottom of the screen, tap and hold the "d" chain of the structure. This selects the whole D chain and the selected atoms are represented by small green crosses.
2. Tap *Menu* button, make sure that you are at the top level of the main menu (you will see "Display" as the first item in this menu). If you just follow Subheading 3.3.2 above, you will find yourself at an inner menu level, then you need to tap the *Back* button at the top of the menu to return to the top level.
3. To the right of the "Display", tap the fifth icon (*see* **Note 4**) for surface (Fig. 5).

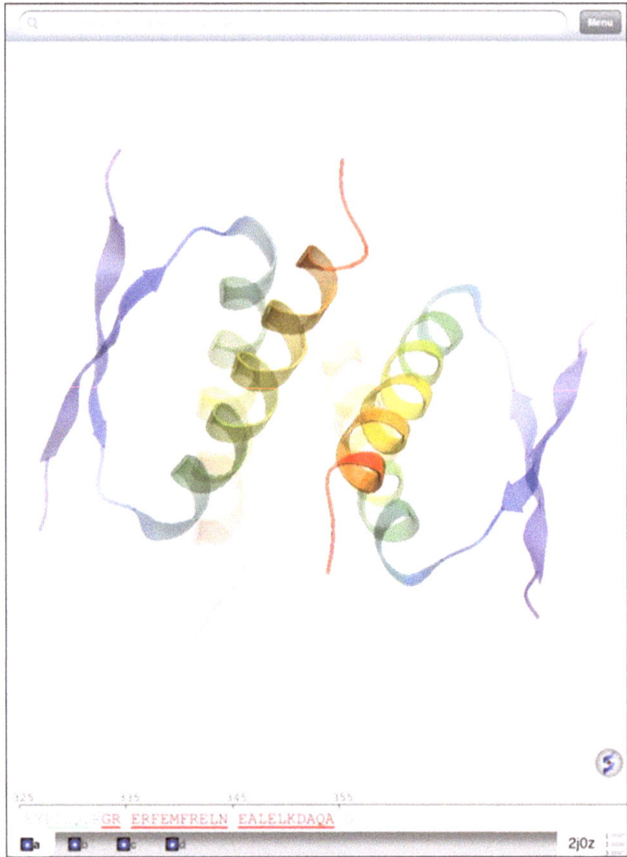

Fig. 4 Transparent ribbon display in *iMolview*

3.4 Exporting an Image

On the iPad, this is very easy. Just press on/off button and the main button together, a screenshot will be saved to the iPad's photos storage. The image can then be shared with other mobile devices (*see* **Note 5**).

4 Conclusion

It is very exciting to see portable molecular graphics developed into the present state. Finally, scientists can carry molecular models around and show these to their colleagues. The models can be examined in real time, using natural hand and finger manipulations. Among the software available, the low-cost *iMolview* tops the list because of its user friendliness and it offers the most complete set of functions for visual communication. *PyMOL* for iOS is not far behind and the developer's Web site announced that a version for Android is under development. *RCSB PDB Mobile* for Android is currently in alpha testing. *Jmol MV* lags behind but it

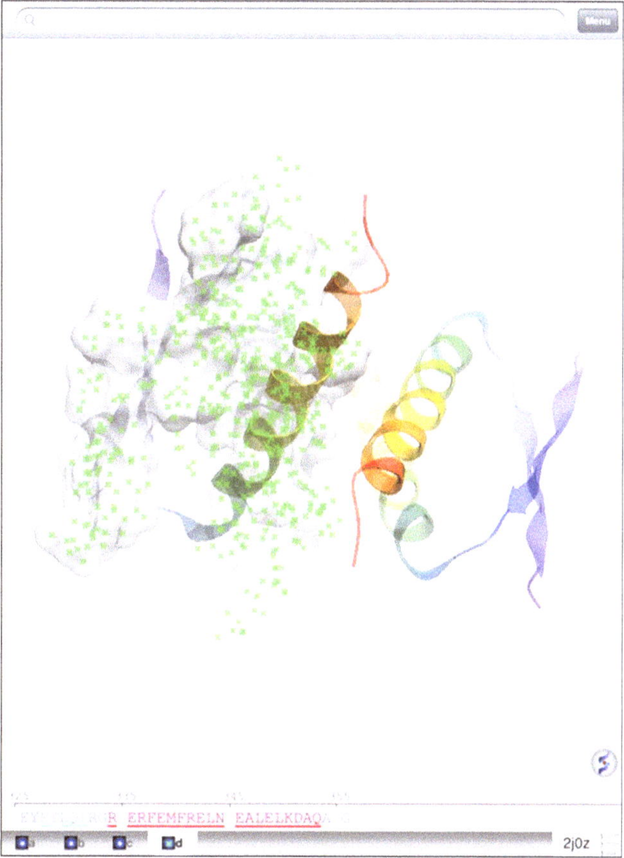

Fig. 5 Composite display in *iMolview*. Molecular surface display of the selected D chain of 2JOZ. The D chain is accessed by the "d" tab at the *bottom*. The molecule has been zoomed in for full display in this mode using two fingers

still shows great potential. We expect to see these products, for free or at very low costs, come to maturity in the near future. We hope that with the primer we have demonstrated how easy it is to use *iMolview* to create a molecular scene of mixed styles, and it can help some colleagues to start using their tablets in visualizing and communicating structures.

5 Notes

1. *CueMol* viewer app (iOS) is only a viewer for scenes created on the desktop versions of *CueMol* and does not offer standalone graphics capability. Several apps are mainly designed for small molecules and lack the full graphical representation styles for protein structures—these include *Molecules* (iOS) and *Atomdriod* (Android). *ESmol* is the little brother of *NDKmol* but for supporting old devices.

2. PDB ID is a unique 4-alphanumeric character combination that is assigned to each deposited structure in the Protein Data Bank (www.rcsb.org). This ID is usually found in the manuscript that describes that particular structure.
3. Richardson cartoon style is a representation of the overall backbone structure of the protein, with secondary structural elements α-helices shown as coiled ribbons, β-strands shown as flat arrows, and coils/loops shown as thin tubes.
4. The complete "user manual" of *iMolview* is accessed by *Menu*, then "Help" inside the app.
5. The users can transfer PDB files or images between mobile devices using Bluetooth-based Apps such as *iShareFiles* (free), without Internet or *Bump* (http://bu.mp; free) with Wi-Fi or 3G. *Bump* is cross-platform and can be used for transferring files between iPhones/iPads, Android devices, and computers. Obviously, the files can also be sent via e-mail.
6. Developers' Web sites:

   iMolview: www.molsoft.com/iMolview.html

   PyMOL app: pymol.org/mobile

   RCSB PDB Mobile: www.rcsb.org/pdb/static.do?p = mobile/RCSBapp.html

   Ball & *Stick*: www.molysym.com

   CueMol: www.cuemol.org/en

   Molecules: www.sunsetlakesoftware.com/molecules

   iPharosDreams: www.pharosdreams.com/mobile3d

   FinMol: www.dubesoft.com

   CMol: cmol.org.uk

   NDKmol/ESmol: webglmol.sourceforge.jp/android-en.html

   Jmol MV: jmol.sourceforge.net

   Atomdroid: www.uni-goettingen.de/en/123989.html
7. This article was written in the first quarter of 2013 but all information has been checked and updated at proofs stage (October, 2013).

Acknowledgement and Declaration

Helen Ginn supplied a trial license of *CMol* (now free) for our review. In addition, we thank the developers of *iMolview* (Andrew Orry), *PyMOL* (Jason Vertrees), *RCSB PDB Mobile* (Greg Quinn)/*NDKmol* (Takanori Nakane), *Ball* & *Stick* (Jon Cody Haines), and *CMol* (Helen Ginn) for responding to our request to check the accuracy of Table 2. The authors share no commercial interests on the software or hardware described in this article.

INDEX

A

B

C

Yu Wai Chen (ed.), *Structural Genomics: General Applications*, Methods in Molecular Biology, vol. 1091, DOI 10.1007/978-1-62703-691-7,

Q

R

S

MIX
Papier aus verantwortungsvollen Quellen
Paper from responsible sources
FSC® C105338

If you have any concerns about our products,
you can contact us on
ProductSafety@springernature.com

In case Publisher is established outside the EU,
the EU authorized representative is:
Springer Nature Customer Service Center GmbH
Europaplatz 3, 69115 Heidelberg, Germany

Printed by Libri Plureos GmbH
in Hamburg, Germany